Studies of Cloud, Convection and Precipitation Processes Using Satellite Observations

Lectures in Climate Change

Print ISSN: 2059-1071
Online ISSN: 2591-7455

Climate change is a portending issue not only for our time, but indeed for generations to come. Informing current and future leaders regarding this issue is a vital task, so that they harbor the knowledge to devise and implement the actions that are required to forestall its consequences. Those actions must be directed towards curtailing the emissions of greenhouse gases to the greatest extent possible, and towards adapting to unavoidable changes of climate.

The goal of this series is to provide resources that can be used by scholars and students worldwide concerning topics related to climate change. Released in both print and online versions, the lectures in this series are targeted at university-level educators and their students across the globe to aid in teaching these particular topics. The series would provide the materials necessary for online learning by individuals seeking to gain a deeper understanding of relevant topics. The lecture presentations with their accompanying written introductions and notes are peer reviewed.

Published:

Vol. 3 *Studies of Cloud, Convection and Precipitation Processes using Satellite Observations*
edited by Zhengzhao Johnny Luo, George Tselioudis and William B Rossow

Vol. 2 *Our Warming Planet: Climate Change Impacts and Adaptation*
edited by Cynthia Rosenzweig, Martin Parry and Manishka De Mel

Vol. 1 *Our Warming Planet: Topics in Climate Dynamics*
edited by Cynthia Rosenzweig, David Rind, Andrew Lacis and Danielle Manley

Every volume in this series comes with its own electronic slides and accompanying notes. Follow the instructions provided in each volume to download the material.

With Electronic Slides and Accompanying Notes

Lectures in Climate Change Vol. 3

Studies of Cloud, Convection and Precipitation Processes Using Satellite Observations

Editors

Zhengzhao Johnny Luo
The City College of City University of New York, USA

George Tselioudis
NASA Goddard Institute for Space Studies, USA
Columbia University, USA

William B Rossow
Franklin, NY, USA

World Scientific

NEW JERSEY • LONDON • SINGAPORE • BEIJING • SHANGHAI • HONG KONG • TAIPEI • CHENNAI • TOKYO

Published by

World Scientific Publishing Co. Pte. Ltd.
5 Toh Tuck Link, Singapore 596224
USA office: 27 Warren Street, Suite 401-402, Hackensack, NJ 07601
UK office: 57 Shelton Street, Covent Garden, London WC2H 9HE

Library of Congress Cataloging-in-Publication Data
Names: Luo, Zhengzhao Johnny, editor. | Tselioudis, George, editor. | Rossow, William Brigance, 1947– editor.
Title: Studies of cloud, convection and precipitation processes using satellite observations / editors Zhengzhao Johnny Luo (The City College of City University of New York, USA), George Tselioudis (NASA Goddard Institute for Space Studies, USA, Columbia University, USA), William B. Rossow (Franklin, NY, USA).
Description: New Jersey : World Scientific, [2023] | Series: Lectures in climate change, 2591-7455 ; vol. 3 | Includes bibliographical references.
Identifiers: LCCN 2022017126 | ISBN 9789811256905 (hardcover) | ISBN 9789811257940 (paperback) | ISBN 9789811256912 (ebook) | ISBN 9789811256929 (ebook)
Subjects: LCSH: Satellite meteorology. | Atmospheric circulation. | Climatology--Observations.
Classification: LCC QC879.5 .S78 2023 | DDC 551.51/7--dc23/eng20220718
LC record available at https://lccn.loc.gov/2022017126

British Library Cataloguing-in-Publication Data
A catalogue record for this book is available from the British Library.

For any available supplementary material, please visit
https://www.worldscienti ic.com/worldscibooks/10.1142/12862#t=suppl

Desk Editors: Balamurugan Rajendran/Amanda Yun

Typeset by Stallion Press
Email: enquiries@stallionpress.com

Instructions for Downloadable Electronic Slides

Included with this publication are downloadable electronic slides of each lecture for students and teachers around the world to be better able to understand and present climate change. These supplementary materials may be retrieved by using the following access Token:

https://www.worldscientific.com/r/12862-SUPP

The Token may be either clicked on (for eBook only), or manually copied/typed into the address bar of your internet browser.

If you purchased the eBook directly from the World Scientific website, your access token would have been activated under your username.

Users will be directed to the sign in page, and prompted to either sign in or register an account with World Scientific.

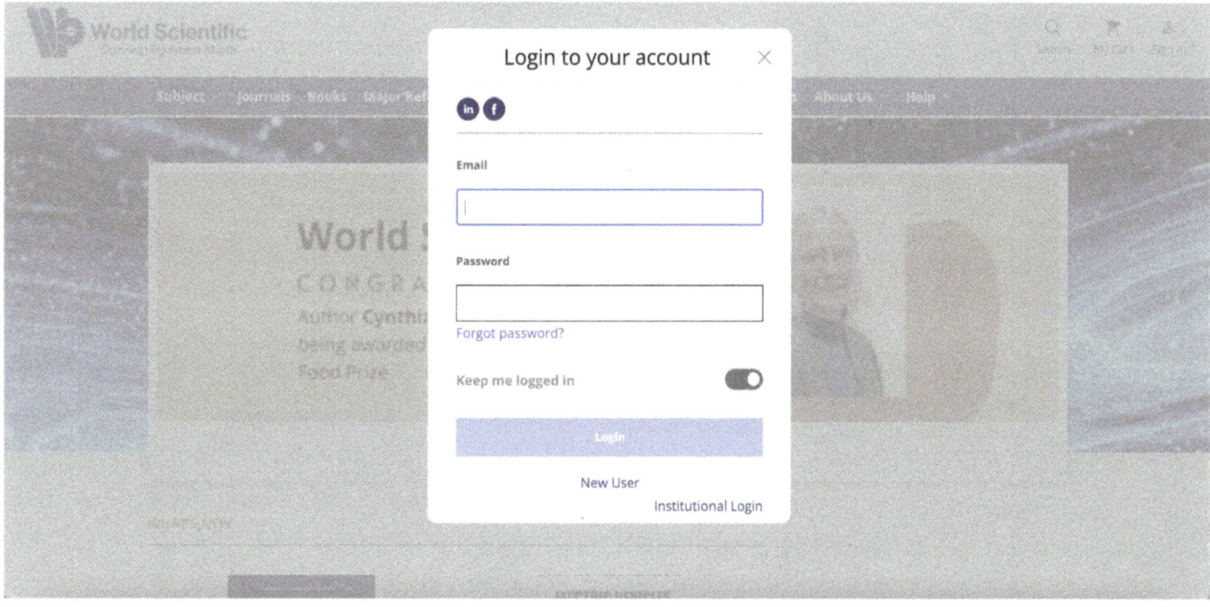

Upon successful login, you will be redirected to the book's page; click on the 'Supplementary' tab (see screenshot below) to locate the .zip file containing all 25 lecture slide sets. Click on the 'Electronic Slides' hyperlink to commence downloading.

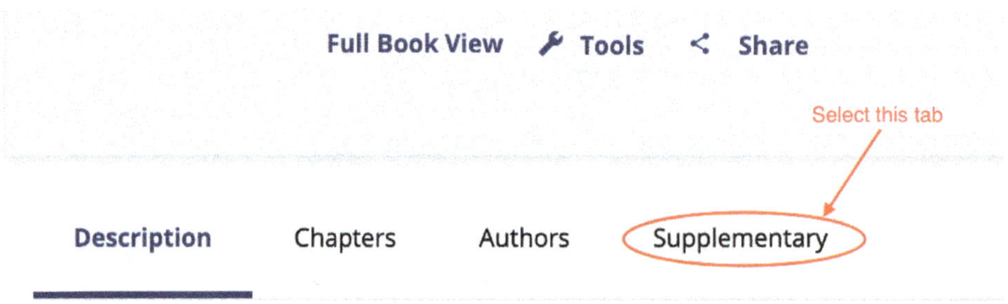

Full Book View Tools Share

Select this tab

Description Chapters Authors Supplementary

For subsequent access to the supplementary materials, simply login with the same username.

For any enquiries on access issues, please email sales@wspc.com.sg.

Contents

Introduction

William B. Rossow

The ceaseless circulations of the Earth's atmosphere are comprised of motions on all space-time scales ranging from three-dimensional dissipating turbulence (meters, minutes) up to global, multi-decadal scales, where the larger and most energetic scales of the general circulations are recognizable in the form of prevailing winds and currents, distinct streams and waves, and storm systems and eddies. The atmospheric and oceanic circulations are driven by and coupled by imbalances in heating-cooling by solar and thermal radiation and by water phase changes (as well as direct sensible heat exchanges at the surface), but they also transport energy and water, forming a complex system of feedbacks that is the central component of the Earth's climate system.

The presence of heating-cooling imbalances over the whole range of scales means that circulation-induced processes are transient, not static — the time derivative is of significant magnitude and cannot be neglected in an analysis of processes. The most important circulation-induced processes are cloud and precipitation formation because they significantly perturb the radiation and latent heating-cooling of the atmosphere and ocean. These perturbations are, in turn, major feedbacks on the atmospheric and oceanic circulations that produce the clouds and precipitation.

• **Global HX Picture**

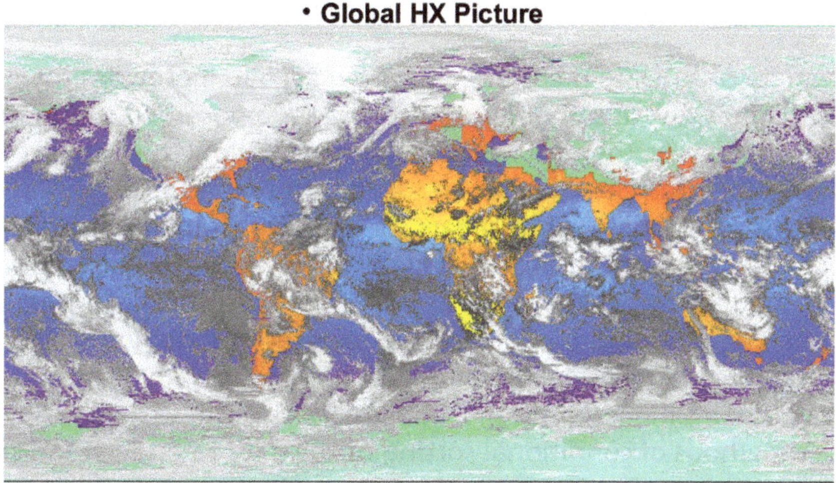

Cloud processes are controlled by the atmospheric motions, so the properties of clouds and precipitation and the distribution of water vapor vary over the whole range of scales of the atmospheric circulation. Before satellite observations became available, these processes were studied with surface- and aircraft-based measurements, but these tools have continued to be the focus of effort, even after satellite observations became available. These kinds of observations provide very limited sampling of the scales of cloud process variability. Surface-based observations are made at a point, mostly limited to land areas, and the observed time variations occur mostly because of the motion of the atmosphere past that point. Aircraft measurements cover a larger range of spatial scales up to hundreds of kilometers, but they provide only instantaneous measurements at many locations because of their rapid motion, not the time variations at a single location. In other words, both these types of measurements do not follow the time evolution of the cloud processes. Moreover, the sum total of many decades of such field experiments still does not provide many samples of each type of atmospheric situation.

What has not been done very often is to study the cloud and water processes at the larger scales of the global circulation of the atmosphere and ocean and, in particular, to investigate the behavior of clouds, water vapor, and precipitation across all the scales of variation at the same time. Only the satellite observing system spans the whole range of the circulation space and time scales. Since the early period nearly 60 years ago, when satellite measurements were simple images at two wavelengths (visible and infrared), the types of satellite measurements have been increasing in comprehensiveness and sophistication. The constellation of satellites in recent times has included imaging at numerous ultraviolet, visible, near-infrared, thermal infrared and microwave wavelengths, infrared and microwave atmospheric temperature-humidity sounding systems, profiling lidars, and cloud and precipitation radars. Satellites also provide information about the ocean surface (temperature, salinity, winds) and its deep circulation. Since the satellite observing system has contained most of these capabilities for at least the last 20 years and generally with global coverage, the number of atmospheric situations sampled is more complete, with many examples of each to provide robust statistics.

Although the variety of satellite measurements now available provides many key pieces of information, the limitations of sampling and coverage imposed by the nature of satellite orbits make it difficult to use them all in combination to follow the time evolution of cloud and precipitation processes. Specifically, the technology (and expense) of larger instruments, like sounders or profiling lidar/radars, necessitates a lower altitude orbit (usually polar) that limits the time sampling to twice daily but usually provides global coverage. In contrast, the imaging instruments can be flown in geosynchronous orbits, which provide very high space and time resolutions (kilometers, minutes) but do not provide global coverage. However, the international collection of both types of satellite together covers the globe and provides high time resolution as well. Analyses of these data to investigate cloud processes require methods of conditional sorting of the data to understand cloud evolution under different conditions. For instance, the higher time resolution imaging data from geostationary satellites can indicate the age of each particular atmospheric event (cloud or storm system) at the time that the less frequent polar orbiting satellites observe it so that a composite lifecycle of the processes can be compiled. Another type of analysis uses either cloud tracking or wind analyses to follow the evolution of individual clouds or air masses, again collecting other lower time resolution measurements into statistical measures of the time derivatives of cloud processes. A different way to sort the observations is by identifying different distinct atmospheric states, different weather conditions, and compiling observations associated with each of these conditions. There is much work that still can be done.

This book summarizes some of the latest research using combinations of satellite observations and conditional sorting and compositing analyses to examine these cloud and water processes to understand and model the interactions of the atmospheric and oceanic circulations. Some model-related results are also included to illustrate the process detail that needs to be verified by comparison with the observational analyses. Exploiting the full range of satellite observations to examine cloud and other atmospheric processes over the whole range of the general circulation variations is now possible and can deepen our understanding of the weather and climate system.

SECTION 1

Clouds, Convection and Precipitation

Lecture 1

Convection as a Regulator of the Earth's Climate

Anthony D. Del Genio

NASA Goddard Institute for Space Studies, New York, USA (Emeritus)

Anthony Del Genio is a Senior Research Scientist Emeritus at the NASA Goddard Institute for Space Studies. He develops the cumulus and stratiform cloud parameterizations for the GISS climate models. His research focuses on climate feedbacks, the hydrologic cycle, interactions with the general circulation, and comparative dynamics of planetary atmospheres.

Bill began his career as a planetary scientist, creating one of the first general circulation models (GCMs) to represent the strong greenhouse effect and circulation of the thick CO_2 atmosphere of Venus. Bill's seminal paper on the time scales of cloud microphysical source and sink processes and how they can be used to understand clouds on a variety of Solar System planets (Rossow, 1978) introduced that community to the physics in clouds and it is still used today by scientists trying to characterize the atmospheres of planets outside our Solar System.

Bill then shifted his focus to climate science, a field that was just beginning to struggle with the idea that clouds play an important role in climate change. Bill entered the field fairly early in the history of satellite remote sensing, a time when satellite data consisted almost exclusively of images or global maps. As the leader of the International Satellite Cloud Climatology Project (ISCCP), Bill revolutionized the field of remote sensing by retrieving systematic, quantitative information about clouds from operational weather satellite datasets that were not designed with that purpose in mind (Rossow and Schiffer, 1991). This was the first comprehensive estimate of the fraction of the Earth covered by clouds, using different approaches that helped quantify uncertainties and gave users data options to make their own decisions on what to include or exclude. In retrospect, one of the more important contributions of ISCCP was its retrievals of cloud optical thickness and top pressure. In the ISCCP dataset, these are available in compact form in mesoscale joint-histograms of frequencies of occurrence of different ranges of optical thickness and top pressure. These histograms are of considerable utility in objectively identifying the cloud morphological types long used by surface weather observers (Hahn *et al.*, 2001).

An objective analysis of the ISCCP histograms defines a set of six tropical "weather states" (WSs) (Rossow *et al.*, 2005) and 11 global WSs (Tselioudis *et al.*, 2013). Three of these states identify convective clouds in their various realizations — shallow fair-weather cumulus that is most prevalent over the subtropical oceans but also in the equatorial regions when they are unusually dry; isolated convective cells with tops from mid-levels to the deep troposphere that occur in warm humid regions of the tropics when the atmosphere is beginning to moisten from previously dry conditions; and organized heavily precipitating mesoscale convective systems that occur under the wettest atmospheric conditions. These clouds are essential to many aspects of the Earth's climate and projected future climate change. We describe some of these below.

Clouds are integral to the Earth's global energy budget at the top of the atmosphere, at the surface, and within the atmosphere (Slide 2). At the top of the atmosphere, approximately 2/3 of the Earth's ~0.3 planetary albedo is due to the reflection of sunlight by clouds, and outgoing thermal (heat) radiation to space is reduced by ~10% due to the greenhouse effect of clouds. A significant fraction of both these effects is due to convective clouds. Convective clouds are central to the Earth's global water cycle as well (Slide 3). Water evaporated from the oceans is carried upward into the free troposphere by shallow cumulus clouds. This allows the general circulation to transport it to other parts of the globe, including continental regions where some of it is delivered to the land surface as precipitation, often by convective storms. In heavily precipitating parts of the world, deep convective storms deliver more precipitation to the ocean than is evaporated by the ocean, freshening (i.e., reducing the salinity of) the ocean and affecting its density, helping to drive its circulation. These regional variations in ocean salinity driven by convection are clearly detectable in satellite data.

Convection helps link the global energy and water cycles to each other (Slide 4) on climatic time scales (months or longer). At the Earth's surface, radiative heating due to absorbed sunlight and net downwelling thermal radiation from the atmosphere is balanced by turbulent fluxes of water and energy from the surface to the atmosphere (i.e., evaporation and sensible heat fluxes) that cool the surface. These fluxes create convective instability, either locally in the places where the fluxes occur or remotely in places to which the resulting water vapor is transported, which produces convective clouds and latent heat release in the atmosphere. The latent heating of the atmosphere by convection must be balanced by the net radiative cooling of the atmosphere. This cooling is produced largely by the thermal radiative flux emitted to space by water vapor carried to the upper troposphere by deep convective clouds and is modulated by the reduction in emission to space caused by the clouds themselves. At the top of the atmosphere, this thermal cooling to space must be balanced by the absorption of sunlight in order for the Earth's climate to be in equilibrium. The absorption of sunlight is limited by bright convective clouds that help determine the Earth's planetary albedo.

Differences among the world's climate GCMs in the sensitivity of climate to external perturbations (e.g., increases in greenhouse gases or changes in the sunlight incident on the planet) can be due to disagreements in their predictions of changes in water vapor, sea ice, snow, and the temperature structure of the atmosphere, or due to different predictions of changes in clouds. Diagnosis of these differences (Cess *et al.*, 1990) shows that the models are largely in agreement in the contributions to the sensitivity from changes in clear-sky regions. Most of the differences between more and less sensitive models occur in cloudy regions (Slide 5), and convective clouds play a large role in this disagreement.

Shallow cumulus clouds (Slides 6 and 7) are common over the subtropical oceans, where the atmospheric boundary layer is convective, but the subsiding branch of the Hadley cell maintains an inversion and a dry free troposphere that prevent convection from rising to a great height. Shallow cumulus updrafts carry air that has been moistened by ocean evaporation up into the inversion layer. Trade winds then carry the moistened air toward the Equator, setting the stage for it to be condensed later in deep convective updrafts in the Intertropical Convergence Zone (ITCZ). Global climate models predict that in a warmer climate, the area covered by shallow cumulus clouds will decrease, causing less sunlight to be reflected. This is positive feedback that amplifies the warming by greenhouse gases. Since these clouds occupy a large area of the open ocean in the subtropics and mid-latitudes, this is an important contributor to the shortwave component of model-predicted cloud feedback (Slide 8). Unfortunately, climate models disagree on the extent of the predicted decrease, explaining a significant amount of the spread in estimates of global climate sensitivity (Zelinka *et al.*, 2012). Models also underestimate stratocumulus clouds and their role in cloud feedback, which may be equally or even more important, though this depends on how sea surface temperature gradients change with warming.

In warmer regions of the tropics, the moistening of the lower free troposphere by shallow cumulus eventually allows convection to penetrate somewhat deeper without losing its buoyancy by the mixing of dry air into the cloud. The resulting clouds with mid-level tops are called congestus (Slide 9). Congestus continues the moistening of the troposphere that started by shallow convection but at higher levels. They also precipitate and bring cold air back to the surface, where it spreads in cold pools. These processes set the stage for the eventual breakout, persistence, and organization of deep convection. The progression from shallow to congestus to deep convective clouds as the tropical atmosphere moistens is clear in ISCCP and CloudSat-CALIPSO satellite data when combined with AMSR-E satellite water vapor data (Del Genio *et al.*, 2012).

Deep convective clouds cover only a small area and have little radiative effect. But in the upper troposphere, where they inject water vapor and ice into the atmosphere, they sometimes form extensive anvil clouds, especially when convective cells cluster into mesoscale systems. The anvils of these systems form the ITCZ that is easily seen as a band of clouds encircling the equatorial region in satellite images (Slide 10). The ITCZ represents the rising branch of the Earth's Hadley cell. The most intense rain events occur when convection organizes into immense systems with strong winds rotating around a central low-pressure area, called hurricanes or typhoons (Slide 13). Hurricanes are born over warm tropical oceans, far from land. Satellites are integral to our ability to detect hurricanes' formation and intensification and track their movement toward land. A major question is whether hurricanes will become more intense in a warmer climate. Convection also determines the Earth's surface water balance (Slide 12), with precipitation exceeding evaporation in the equatorial ITCZ region and the opposite occurring in the subtropical shallow cumulus regions. The resulting density differences between fresher and saltier water partially drive ocean circulation (Slide 14).

Deep convection plays several roles in the Earth's climate, but one such mechanism that has been confirmed to be important to climate change is the cloud height feedback (Slide 11). Convection deepens with warming in all global climate models and in cloud-resolving models as well. But cloud top temperature remains nearly constant (called the *fixed anvil temperature* [FAT] hypothesis) due to a balance between convective heating and H_2O vapor radiative cooling (Hartmann and Larson, 2002). The result is positive feedback on surface temperature because convection communicates the warming to the surface: a moist adiabat extrapolated to the surface from a higher altitude, but the same temperature warms the surface. Satellite data of many kinds, including the ISCCP, confirm that during El Niño events, convective anvil cloud tops adjust in altitude in a manner consistent with a modified version of the FAT hypothesis, lending credence to climate model predictions of a positive cloud height feedback (Zelinka and Hartmann, 2011).

Convection is central to the land biosphere as well (Slide 15). The MODIS instrument, which senses year-to-year variability in vegetation, surface temperature, surface water, and cloudiness, documents how climate variables regulate the health of the biosphere. Over most of the tropics, either the delivery of water to the Earth's surface by precipitating convective systems or the regulation of sunlight incident on the surface by convective anvil clouds is the driver of variations in vegetation productivity (Seddon *et al.*, 2016). Innovative uses of satellite data, such as the examples shown in these slides are either the direct result of Bill Rossow's vision or the distributed impact of that vision to the larger Earth science community.

References

Cess RD, Potter GL, Blanchet JP, *et al.* (1990) Intercomparison and interpretation of climate feedback processes in 19 atmospheric general circulation models. *Journal of Geophysical Research* **95**:16601–15.

Del Genio AD, Chen Y-H, Kim D, Yao M-S. (2012) The MJO transition from shallow to deep convection in CloudSat/CALIPSO data and GISS GCM simulations. *Journal of Climate* **25**:3755–70.

Hahn CJ, Rossow WB, Warren SG. (2001) ISCCP cloud properties associated with standard cloud types identified in individual surface observations. *Journal of Climate* **14**:11–28.

Hartmann DL, Larson K. (2002) An important constraint on tropical cloud-climate feedback. *Geophysical Research Letters* **29**:1951. doi:10.1029/2002GL015835

Kuang Z, Hartmann DL. (2007) Testing the fixed anvil temperature hypothesis in a cloud-resolving model. *Journal of Climate* **20**:2051–57.

Rossow WB. (1978) Cloud microphysics: Analysis of the clouds of Earth, Venus, Mars, and Jupiter. *Icarus* **36**:1–50.

Rossow WB, Schiffer RA. (1991) ISCCP cloud data products. *Bulletin of the American Meteorological Society* **71**:2–20.

Rossow WB, Tselioudis G, Polak A, Jakob C. (2005) Tropical climate described as a distribution of weather states indicated by distinct mesoscale cloud property mixtures. *Geophysical Research Letters* **32**. doi: 10.1029/2005GL024584

Seddon AWR, Macias-Fauria M, Long PR, Benz D, Willis KJ. (2016) Sensitivity of global terrestrial ecosystems to climate variability. *Nature* **531**:229–32.

Stephens GL, Li J, Wild M, Clayson CA, Loeb N, Kato S, L'Ecuyer T, Stackhouse Jr. PW, Lebsock M, Andrews T. (2012) An update on Earth's energy balance in light of the latest global observations. *Nature Geosci.* **5**:691–6.

Tselioudis G, Rossow W, Zhang Y-C, Konsta D. (2013) Global weather states and their properties from passive and active satellite cloud retrievals. *Journal of Climate* **26**:7734–46.

Zelinka MD, Hartmann DL. (2011) The observed sensitivity of high clouds to mean surface temperature anomalies in the tropics. *Journal of Geophysical Research* **116**:D23103. doi:10.1029/2011JD016459

Zelinka MD, Klein SA, Hartmann DL. (2012) Computing and partitioning cloud feedbacks using cloud property histograms. Part II: Attribution to changes in cloud amount, altitude, and optical depth. *Journal of Climate* **25**:3736–54.

Slide 1

Slide 2

Clouds are integral to Earth's energy cycle

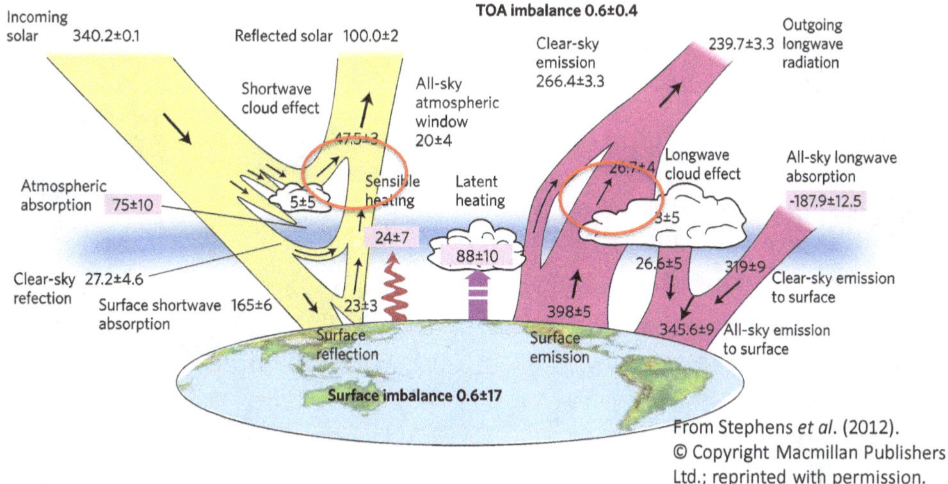

- Approximately 2/3 of Earth's ~0.3 planetary albedo is due to reflection of sunlight by clouds
- Thermal infrared radiation to space is reduced by ~10% due to the greenhouse effect of clouds
- Many of the clouds responsible for this are associated with convection

Slide 3

Convective clouds are central to the water cycle as well

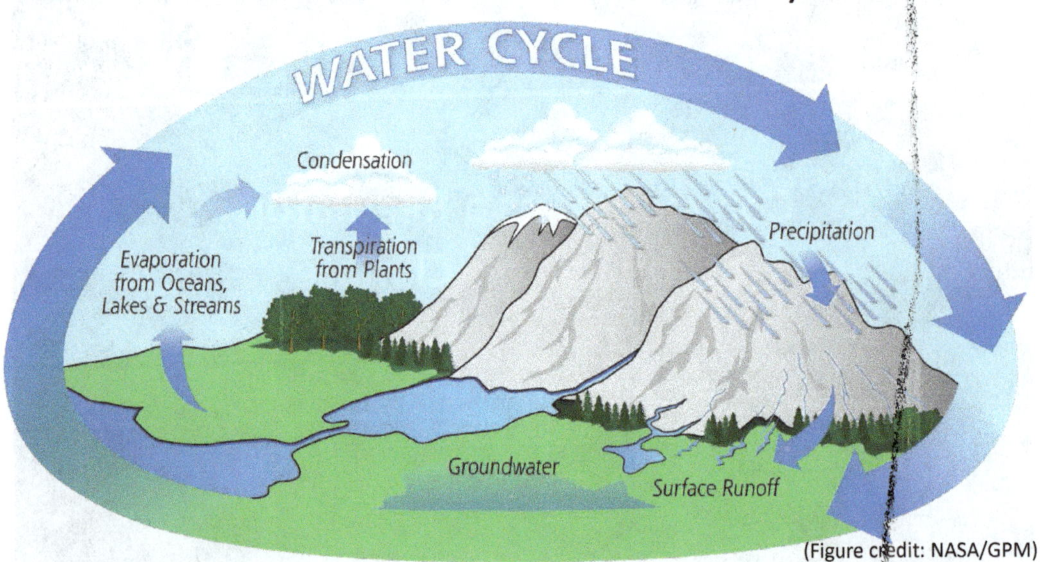

(Figure credit: NASA/GPM)

- The agent by which water vapor is evaporated from the ocean is delivered to the free troposphere.
- And by which some of that water is delivered to the land surface as precipitation.
- The regulator of surface ocean salinity and the thermohaline circulation.

Slide 4

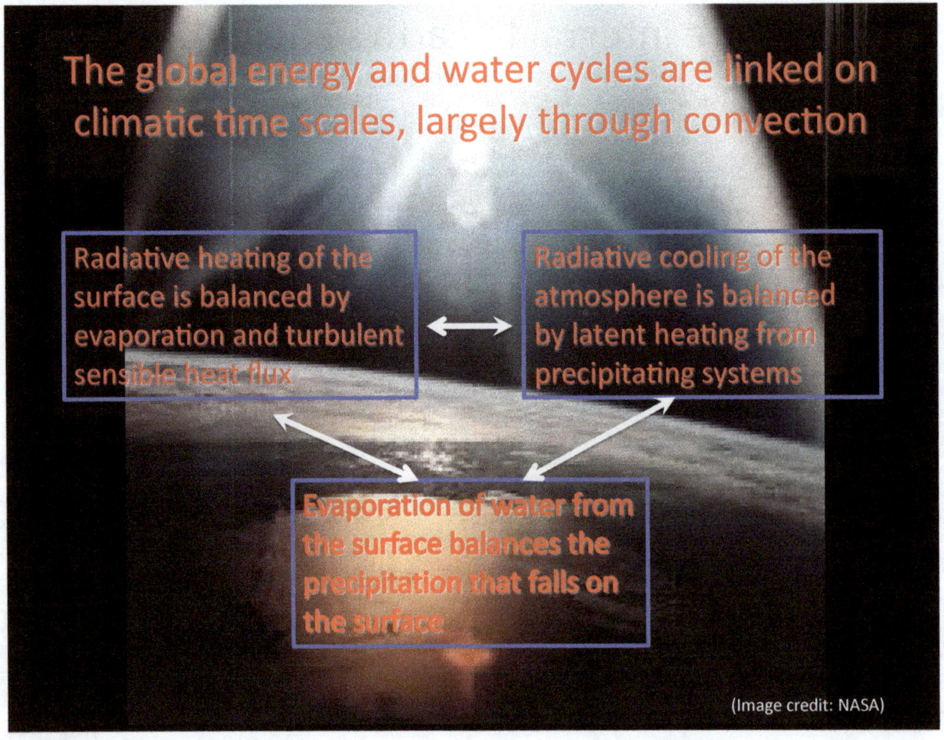

The global energy and water cycles are linked on climatic time scales, largely through convection
Radiative heating of the surface is balanced by evaporation and turbulent sensible heat flux
Radiative cooling of the atmosphere is balanced by latent heating from precipitating systems
Evaporation of water from the surface balances the precipitation that falls on the surface
(Image credit: NASA)

Slide 5

Clouds are the biggest uncertainty in projections of future climate change, partly as a result of convection

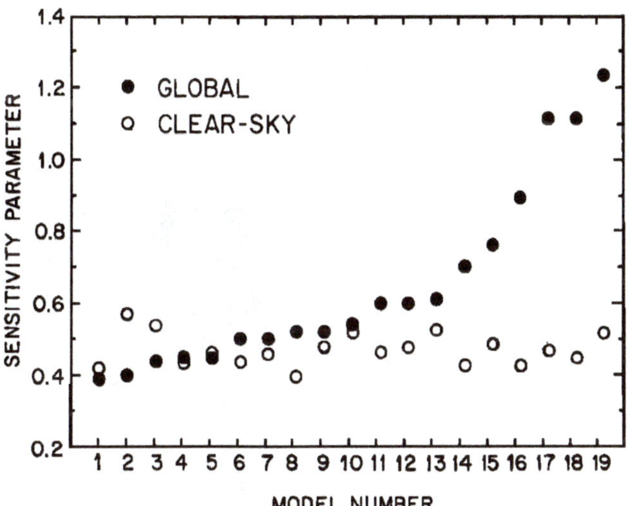

Fig. 1. Clear-sky and global sensitivity parameters (K m^2 W^{-1}) for the 19 GCMs. The model numbers correspond to the ordering in Table 9.

Cess *et al.* © 1990 American Geophysical Union

Slide 6

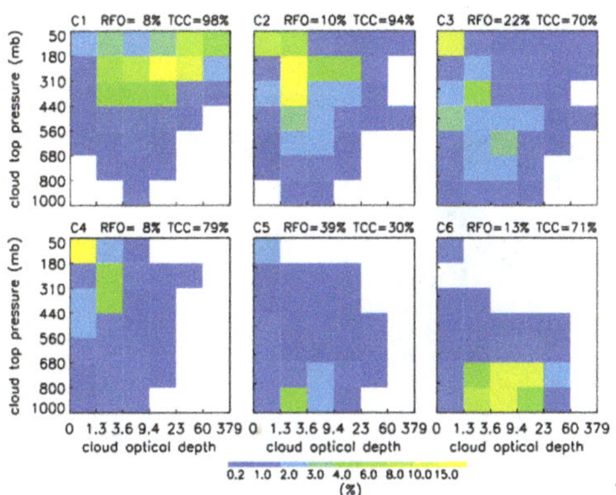

Tropical convective cloud classes can be identified objectively in ISCCP satellite data by their optical thickness (which determines how much sunlight they reflect) and their cloud top pressure (which determines their greenhouse effect).

Three of the six tropical classes are convective clouds:

C1: Organized deep convective cloud systems that occur primarily over warm equatorial ocean and land

C3: Disorganized convective cells, including mid-level congestus, which occur in the same places in different conditions

C5: Shallow cumulus, most frequent over the subtropical oceans

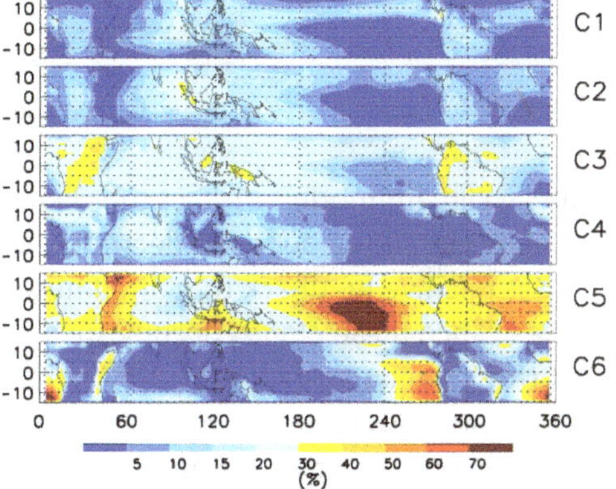

(Figures credits: NASA)

Slide 7

(Image credit: NASA)

Shallow cumulus are common over the subtropical oceans, where the boundary layer of the atmosphere is convective but the subsiding branch of the Hadley cell maintains an inversion and a dry free troposphere that prevents convection from rising to great height.

Shallow cumulus updrafts carry air that has been moistened by ocean evaporation upward into the inversion layer. The trade winds then carry the moistened air toward the equator, setting the stage for it to be condensed later in the deep convective updrafts of the ITCZ.

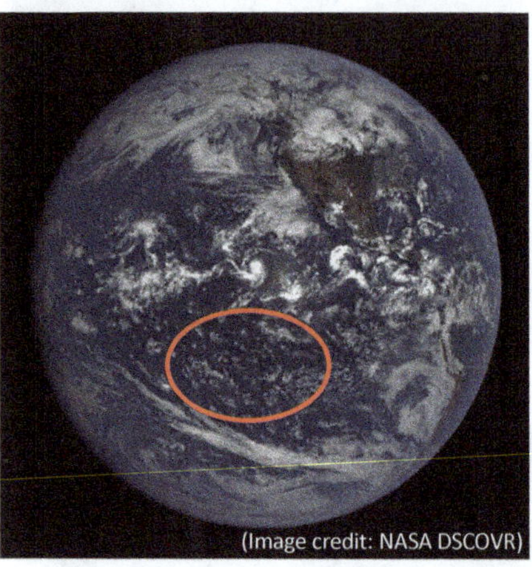

(Image credit: NASA DSCOVR)

Slide 8

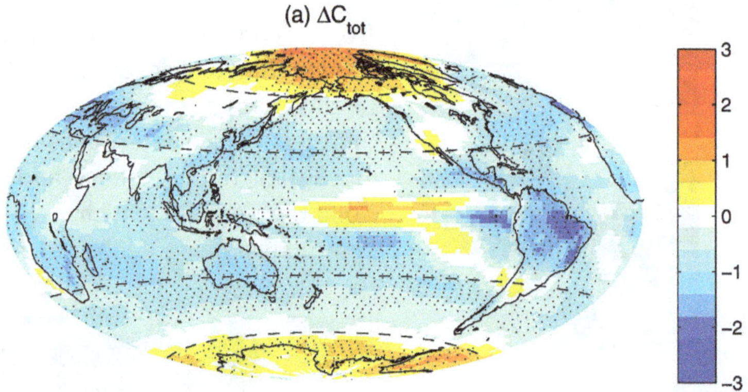

Zelinka *et al.* © 2012 American Meteorological Society. Used with permission.

Global climate models predict that in a warmer climate, the area covered by shallow cumulus clouds will decrease, causing less sunlight to be reflected. This is positive feedback that amplifies the warming due to greenhouse gases. Since these clouds occupy such a large area of the open ocean in the subtropics and mid-latitudes, this is the most important contributor to the shortwave component of cloud feedback. Unfortunately, climate models disagree on the extent of the predicted decrease, explaining a significant amount of the spread in estimates of global climate sensitivity.

Slide 9

(Image credit: NASA)

In warmer regions of the tropics, the moistening of the lower free troposphere by shallow cumulus eventually allows convection to penetrate somewhat deeper without losing its buoyancy by the mixing of dry air into the cloud — the resulting clouds with mid-level tops are called congestus.

Congestus continue the moistening of the troposphere at higher levels. They also precipitate and bring cold air back to the surface, where it spreads in cold pools. These processes set the stage for the eventual breakout and organization of deep convection. The progression from shallow to congestus to deep as air moistens is clear in CloudSat-CALIPSO and AMSR-E satellite data.

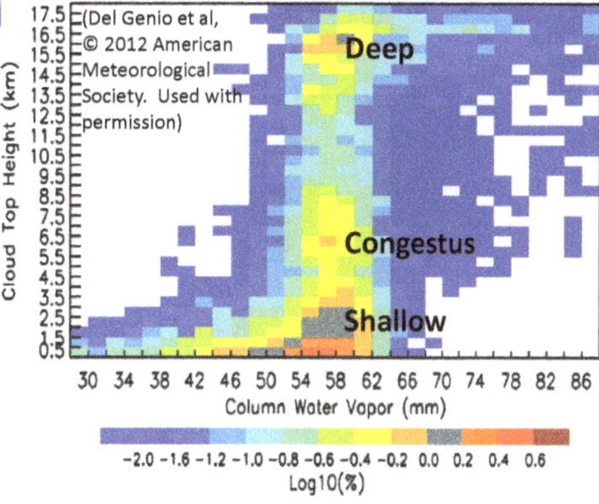
(Del Genio et al, © 2012 American Meteorological Society. Used with permission)
Deep
Congestus
Shallow
Cloud Top Height (km)
17.5 16.5 15.5 14.5 13.5 12.5 11.5 10.5 9.5 8.5 7.5 6.5 5.5 4.5 3.5 2.5 1.5 0.5
30 34 38 42 46 50 54 58 62 66 70 74 78 82 86
Column Water Vapor (mm)
-2.0 -1.6 -1.2 -1.0 -0.8 -0.6 -0.4 -0.2 0.0 0.2 0.4 0.6
Log10(%)

Slide 10

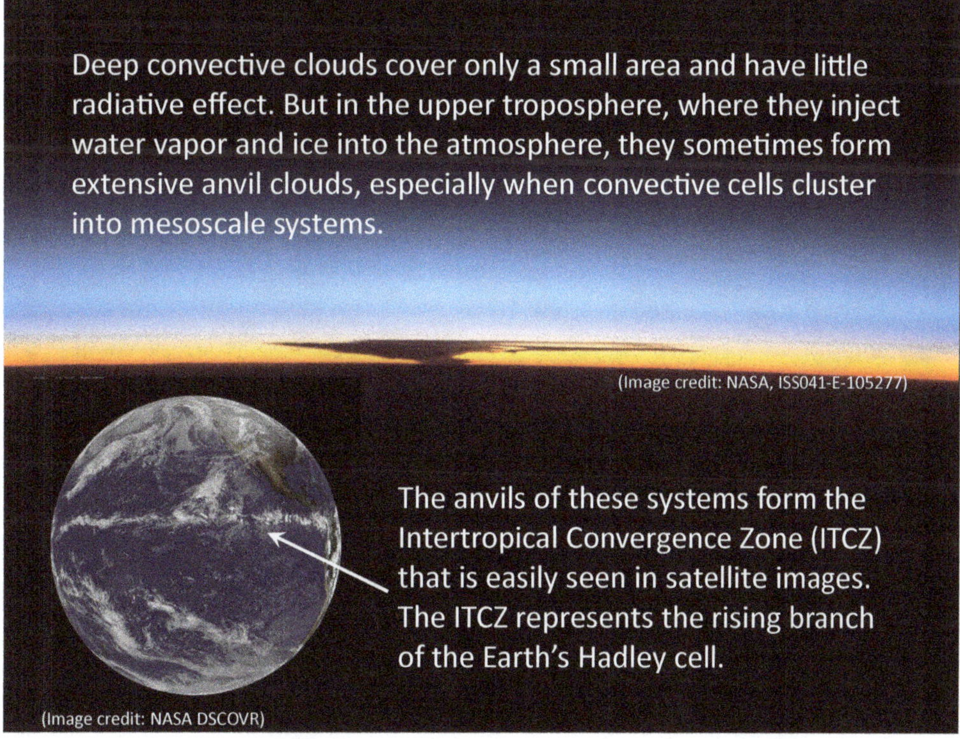
Deep convective clouds cover only a small area and have little radiative effect. But in the upper troposphere, where they inject water vapor and ice into the atmosphere, they sometimes form extensive anvil clouds, especially when convective cells cluster into mesoscale systems.
(Image credit: NASA, ISS041-E-105277)
The anvils of these systems form the Intertropical Convergence Zone (ITCZ) that is easily seen in satellite images. The ITCZ represents the rising branch of the Earth's Hadley cell.
(Image credit: NASA DSCOVR)

Slide 11

detrainment

updraft

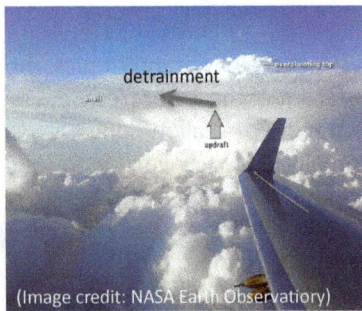

(Image credit: NASA Earth Observatory)

Positive cloud height feedback

- Convection deepens with warming in *all* global climate models, and in cloud resolving models as well

- But cloud top temperature remains nearly constant – the *fixed anvil temperature* (FAT) hypothesis – due to a balance between convective heating and H_2O radiative cooling (Hartmann and Larson, 2002); the result is a positive feedback on surface temperature

- Satellite data confirm that during El Niño, convective anvil cloud tops adjust in altitude in a manner consistent with a modified version of the FAT hypothesis, lending credence to climate model predictions of a positive cloud height feedback (Zelinka and Hartmann, 2011)

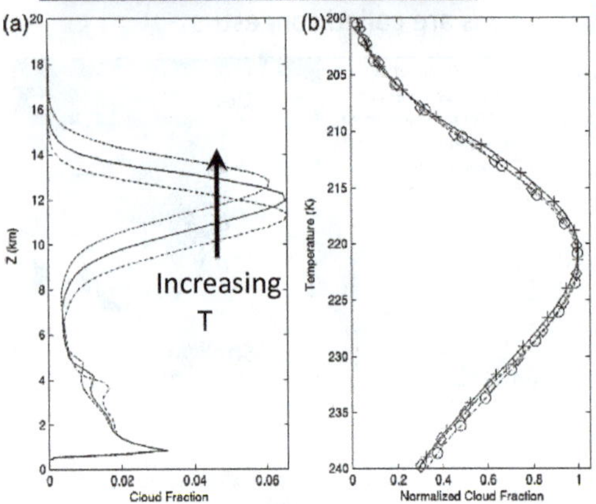

(Kuang and Hartmann © 2007 American Meteorological Society. Used with permission)

Slide 12

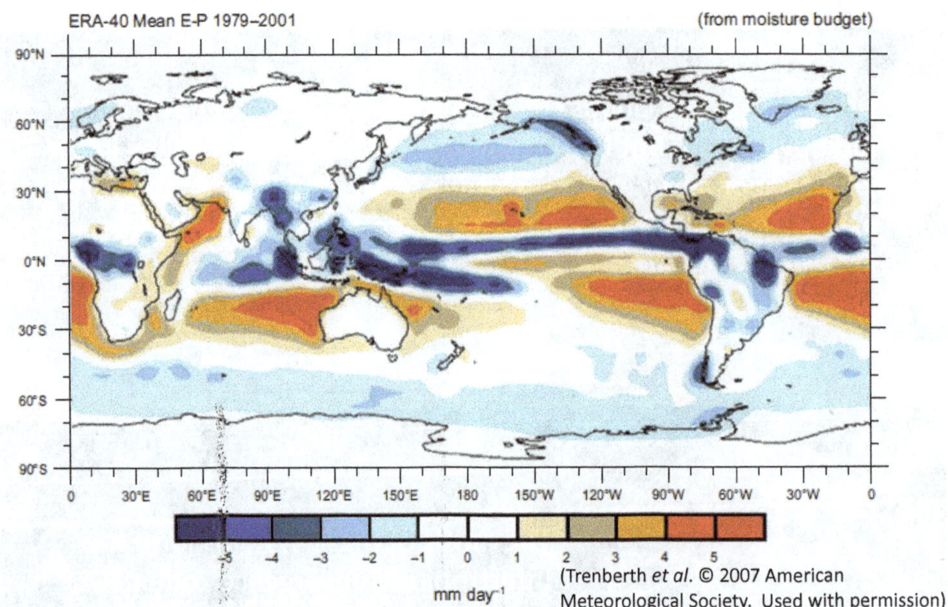

(Trenberth *et al.* © 2007 American Meteorological Society. Used with permission)

Convection determines the sign of the water balance throughout the tropics. In the equatorial region, precipitation (P) from deep convective storms greatly exceeds evaporation (E). In the subtropics, mostly non-precipitating shallow convection exists where E is larger than P.

Slide 13

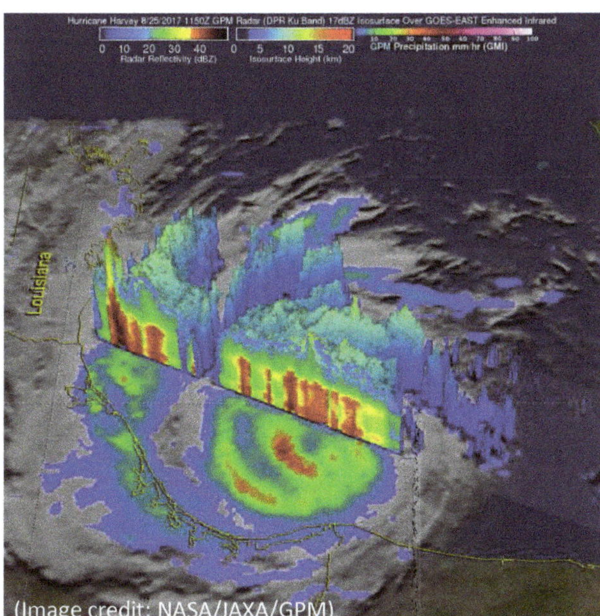
Hurricane Harvey 8/25/2017 1150Z GPM Radar (DPR Ku Band) 17dBZ Isosurface Over GOES-EAST Enhanced Infrared
0 10 20 30 40
Radar Reflectivity (dBZ)
0 5 10 15 20
Isosurface Height (km)
GPM Precipitation mm/hr (GMI)
Louisiana
(Image credit: NASA/JAXA/GPM)

The most intense rain events occur when convection organizes into immense systems with strong winds, called hurricanes or typhoons.

Hurricanes are born over warm tropical oceans, far from land. Satellites are integral to our ability to detect hurricanes' formation and intensification and track their movement toward land. A major question is whether hurricanes will become more intense in a warmer climate.

This satellite image from the Global Precipitation Measurement Mission radar shows Hurricane Harvey bearing down on the Texas coast (at the bottom). The radar sees both the rain at the surface and the vertical structure within the storm.

Slide 14

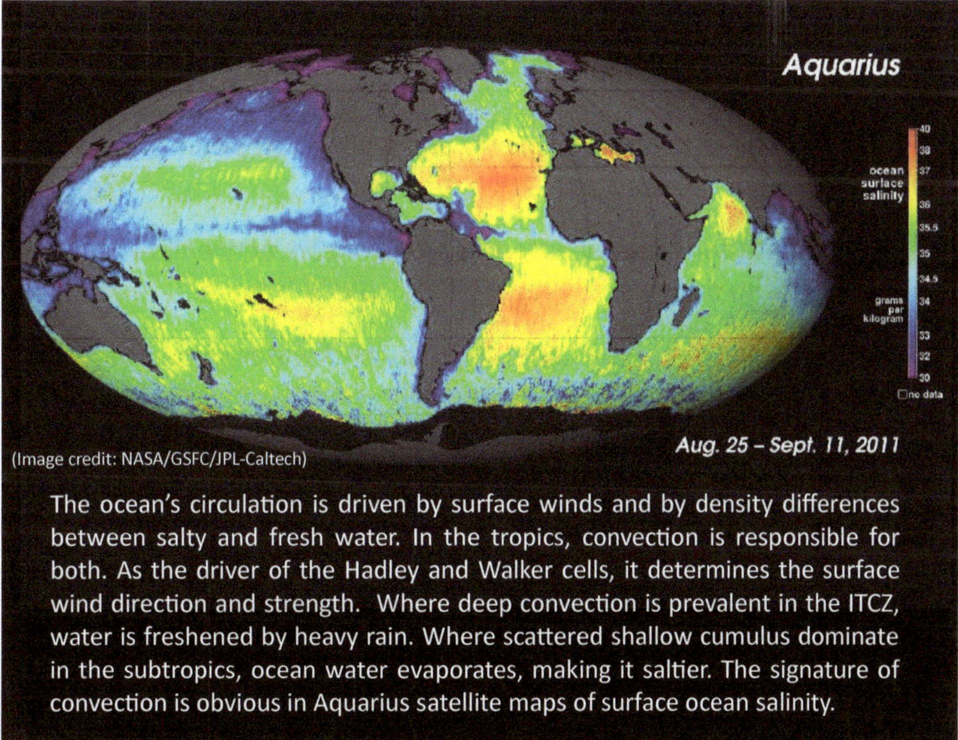
Aquarius
ocean surface salinity
grams per kilogram
40
38
37
36
35.5
35
34.5
34
33
32
30
no data
Aug. 25 – Sept. 11, 2011
(Image credit: NASA/GSFC/JPL-Caltech)
The ocean's circulation is driven by surface winds and by density differences between salty and fresh water. In the tropics, convection is responsible for both. As the driver of the Hadley and Walker cells, it determines the surface wind direction and strength. Where deep convection is prevalent in the ITCZ, water is freshened by heavy rain. Where scattered shallow cumulus dominate in the subtropics, ocean water evaporates, making it saltier. The signature of convection is obvious in Aquarius satellite maps of surface ocean salinity.

Slide 15

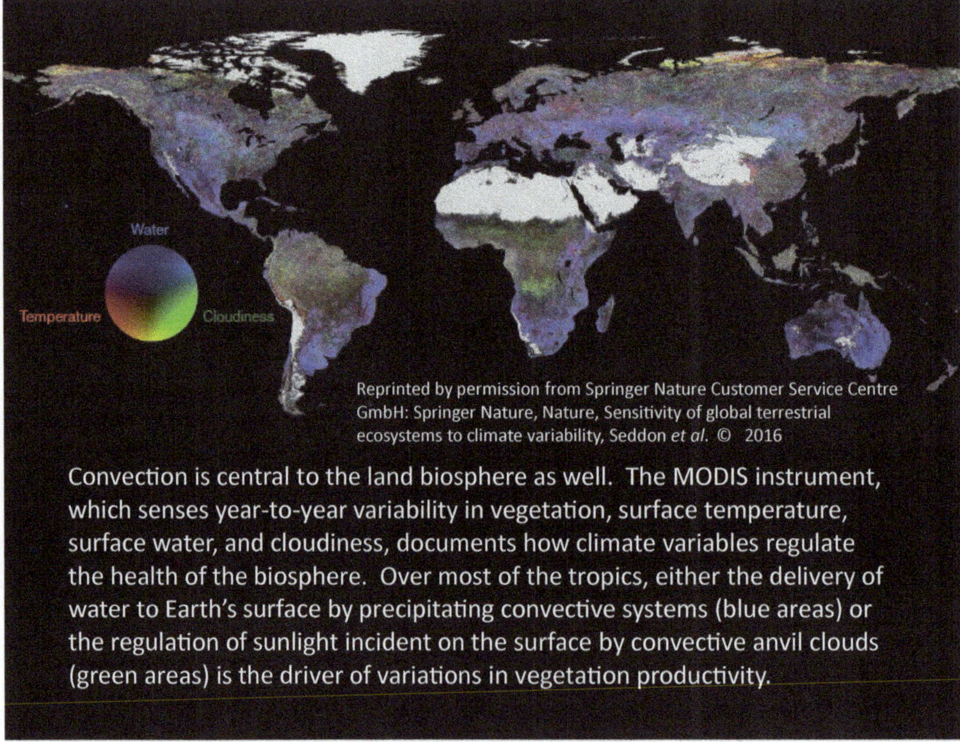

Water
Temperature
Cloudiness
Reprinted by permission from Springer Nature Customer Service Centre GmbH: Springer Nature, Nature, Sensitivity of global terrestrial ecosystems to climate variability, Seddon *et al.* © 2016
Convection is central to the land biosphere as well. The MODIS instrument, which senses year-to-year variability in vegetation, surface temperature, surface water, and cloudiness, documents how climate variables regulate the health of the biosphere. Over most of the tropics, either the delivery of water to Earth's surface by precipitating convective systems (blue areas) or the regulation of sunlight incident on the surface by convective anvil clouds (green areas) is the driver of variations in vegetation productivity.

Slide 16

Summary

- Convection links the global energy and water cycles and thus has an outsized effect on the Earth's climate and climate change.
- Satellite data can identify different types of convective clouds, allowing us to define and quantify their role in climate.
- Shallow convective clouds are the conduct of water vapor from the oceans to the free troposphere and represent the biggest contribution to cloud feedback and its uncertainty.
- Congestus clouds moisten the troposphere and help trigger the transition from shallow to deep convection.
- Deep convection produces extensive anvil clouds that produce the best documented positive cloud feedback; they account for the heaviest rains on Earth, the ocean water balance, and drive year-to-year variability in tropical vegetation productivity.

Slide Captions

Slide 1 Introductory slide.

Slide 2 Clouds affect the energy budget at the top of the atmosphere, at the surface, and within the atmosphere. At the top of the atmosphere, approximately 2/3 of the Earth's ~0.3 planetary albedo is due to the reflection of sunlight by clouds, and outgoing thermal (heat) radiation to space is reduced by ~10% due to the greenhouse effect of clouds. A significant fraction of both these effects is due to convective clouds.

Figure reprinted by permission from Macmillan Publishers, Nature, An update on Earth's energy balance in light of the latest global observations, Stephens *et al.*, *Nature Geoscience* **5**:691–96 © 2012.

Slide 3 Convective clouds are also central to the Earth's water cycle. Water evaporated from the oceans is carried upward into the free troposphere by shallow cumulus clouds. This allows the general circulation to transport it to other parts of the globe, including continental regions where some of it is delivered to the land surface as precipitation, often by convective storms. In heavily precipitating parts of the world, deep convective storms deliver more precipitation to the ocean than is evaporated by the ocean, freshening (i.e., reducing the salinity of) the ocean and affecting its density, helping to drive its circulation.

Figure credit: NASA/GPM.

Slide 4 Convection helps link the global energy and water cycles to each other on climatic time scales (months or longer). At the Earth's surface, radiative heating due to absorbed sunlight and downwelling thermal radiation from the atmosphere is balanced by turbulent fluxes from the surface to the atmosphere (evaporation and sensible heat) that cool the surface. These fluxes create the convective instability either locally in the places where the fluxes occur, or remotely in places to which the resulting water vapor is transported, which produces convective clouds and latent heat release in the atmosphere. The latent heating of the atmosphere by convection must be balanced by the net radiative cooling of the atmosphere. This cooling is produced largely by the thermal radiative flux emitted to space by water vapor carried to the upper troposphere by deep convective clouds, and is modulated by the reduction in emission to space caused by the clouds themselves. At the top of the atmosphere, this thermal cooling to space must be balanced by the absorption of sunlight in order for the Earth's climate to be in equilibrium. The absorption of sunlight is limited by bright convective clouds that help determine the Earth's planetary albedo.

Image credit: NASA.

Slide 5 Diagnosis of differences among the world's climate general circulation models (GCMs) in the sensitivity of climate to external perturbations shows that the models are largely in agreement in the contributions to the sensitivity from changes in clear-sky regions. Most of the differences between more and less sensitive models occur in cloudy regions, and convective clouds play a large role in this disagreement.

Figure from Cess *et al.* (1990) *Journal of Geophysical Research* **95**(D10):16601–15 © American Geophysical Union.

Slide 6 One of Bill Rossow's greatest contributions to Earth science was his innovative and comprehensive approach to the creation of satellite data products. As the leader of the International Satellite Cloud Climatology Project (ISCCP), he combined visible and thermal infrared data from geostationary and polar orbiter satellites to produce the most comprehensive and long-duration cloud climatology. More importantly, he designed innovative data products that went beyond simple parameters such as cloud fraction to provide cloud process insights. The clearest example of this is the cloud top pressure — optical thickness histograms that are produced on a 3-hourly basis worldwide. Bill and his colleagues performed an objective analysis of these histograms to define six cloud classes, or "weather states" (WSs), that explain the major cloud regimes of the tropics. Three of the six WSs identify clouds by how much sunlight they reflect and how large their greenhouse effect is: Organized deep convective cloud systems that occur primarily over warm equatorial oceans and continents in the rising branch of the Hadley cell; disorganized convective cells, including mid-level congestus clouds, that occur in the same places but under different meteorological conditions; and shallow cumulus clouds, which are most frequent over the subtropical oceans in the descending branch of the Hadley cell.

Figure credits: NASA.

Slide 7 Shallow cumulus is common over the subtropical oceans, where the boundary layer of the atmosphere is convective, but the subsiding branch of the Hadley cell maintains an inversion and a dry free troposphere that prevents convection from rising to great height. Shallow cumulus updrafts carry air that has been moistened by ocean evaporation upward into the inversion layer. The trade winds then carry the moistened air toward the Equator, setting the stage for it to be condensed later in the deep convective updrafts of the Intertropical Convergence Zone (ITCZ).

Image credits: NASA.

Slide 8 Global climate models predict that in a warmer climate, the area covered by shallow cumulus clouds will decrease, causing less sunlight to be reflected. This is positive feedback that amplifies the warming due to greenhouse gases. Since these clouds occupy such a large area of the open ocean in the subtropics and mid-latitudes, this is the most important contributor to the shortwave component of cloud feedback. Unfortunately, climate models disagree on the extent of the predicted decrease, explaining a significant amount of the spread in estimates of global climate sensitivity.

Zelinka MD, Klein SA, Hartmann DL. (2012) Computing and partitioning cloud feedbacks using cloud property histograms. Part II: Attribution to changes in cloud amount, altitude, and optical depth. *Journal of Climate* **25**:3736–54 © 2012 American Meteorological Society. Used with permission.

Slide 9 In warmer regions of the tropics, the moistening of the lower free troposphere by shallow cumulus eventually allows convection to penetrate somewhat deeper without losing its buoyancy by the mixing of dry air into the cloud — the resulting clouds with mid-level tops are called congestus. Congestus continues the moistening of the troposphere at higher levels. They also precipitate and bring cold air back to the surface, where it spreads in cold pools. These processes set the stage for the eventual breakout, persistence, and organization of deep convection. The progression from shallow to congestus to deep is clear in CloudSat-CALIPSO and AMSR-E satellite data.

Image credit: NASA.

Figure credit: Del Genio AD, Chen Y, Kim D, Yao M-S. (2012) The MJO transition from shallow to deep convection in CloudSat/CALIPSO data and GISS GCM simulations. *Journal of Climate* **25**:3755–70 © 2012 American Meteorological Society. Used with permission.

Slide 10 Deep convective clouds cover only a small area and have little radiative effect. But in the upper troposphere, where they inject water vapor and ice into the atmosphere, they sometimes form extensive anvil clouds, especially when convective cells cluster into mesoscale systems. The anvils of these systems form the ITCZ that is easily seen as a band of clouds encircling the equatorial region in satellite images. The ITCZ represents the rising branch of the Earth's Hadley cell.

Image credits: NASA, ISS041-E-105277; NASA DSCOVR.

Slide 11 Positive cloud height feedback occurs for the following reason: Convection deepens with warming in *all* global climate models and in cloud-resolving models as well. But cloud top temperature remains nearly constant (this is called the *fixed anvil temperature* [FAT] hypothesis) due to a balance between convective heating and H_2O vapor radiative cooling (Hartmann and Larson, 2002); the result is positive feedback on surface temperature. Satellite data confirm that during El Niño, convective anvil cloud tops adjust in altitude in a manner consistent with a modified version of the FAT hypothesis, lending credence to climate model predictions of a positive cloud height feedback (Zelinka and Hartmann, 2011).

Image credit: NASA Earth Observatory.

Figure credit: Kuang Z, Hartmann DL. (2007) Testing the fixed anvil temperature hypothesis in a cloud-resolving model. *Journal of Climate* **20**:2051–57 © 2007 American Meteorological Society. Used with permission.

Slide 12 Convection determines the sign of the water balance throughout the tropics. In the equatorial region, precipitation (P) from deep convective storms greatly exceeds evaporation (E). In the subtropics, mostly non-precipitating shallow convection exists where E is larger than P. The implication is that water vapor evaporated into the non-precipitating or lightly precipitating regions of the subtropics is transported by the atmospheric circulation into the equatorial region, where it supplies the water that is rained out in the deep convective clouds of the ITCZ.

Figure credit: Trenberth KE, Smith L, Qian T, *et al.* (2007) Estimates of the global water budget and its annual cycle using observations and model data. *Journal of Hydrometeorology* **8**:758–69 © 2007 American Meteorological Society. Used with permission.

Slide 13 The most intense rain events occur when convection organizes into immense systems with strong winds rotating around a central low-pressure area, called hurricanes or typhoons. Hurricanes are born over warm tropical oceans, far from land. Satellites are integral to our ability to detect hurricanes' formation and intensification and track their movement toward land. A major question is whether hurricanes will become more intense in a warmer climate. This satellite image from the Global Precipitation Measurement Mission radar shows Hurricane Harvey bearing down on the Texas coast (at the bottom of the image). The radar sees both the rain at the surface and the vertical structure within the storm.

Image credit: NASA/JAXA/GPM.

Slide 14 The ocean's circulation is driven by surface winds and density differences between salty and fresh water. In the tropics, convection is responsible for both. As the driver of the Hadley and Walker cells, it determines the surface wind direction and strength. Where deep convection is prevalent in the ITCZ, water is freshened by heavy rain. Where scattered shallow cumulus dominates in the subtropics, ocean water evaporates, making it saltier. The signature of convection is obvious in Aquarius satellite maps of surface ocean salinity. This map can be compared to the earlier map of P-E to better see how the convective contribution to the water balance influences the ocean.

Image credit: NASA/GSFC/JPL-Caltech.

Slide 15 Convection is central to the land biosphere as well. The MODIS instrument, which senses year-to-year variability in vegetation, surface temperature, surface water, and cloudiness, documents how climate variables regulate the health of the biosphere. Over most of the tropics, either the delivery of water to the Earth's surface by precipitating convective systems (blue areas) or the regulation of sunlight incident on the surface by convective anvil clouds (green areas) is the driver of variations in vegetation productivity.

Figure reprinted by permission from Springer Nature, Nature, Sensitivity of global terrestrial ecosystems to climate variability, Seddon *et al.*, *Nature* **531**:229–32 © 2016.

Slide 16 Summary

References

Cess RD, Potter GL, Blanchet JP, *et al.* (1990) Intercomparison and interpretation of climate feedback processes in 19 atmospheric general circulation models. *Journal of Geophysical Research* **95**:16601–15.

Del Genio AD, Chen Y, Kim D, Yao M-S. (2012) The MJO transition from shallow to deep convection in CloudSat/CALIPSO data and GISS GCM simulations. *Journal of Climate* **25**:3755–70.

Kuang Z, Hartmann DL. (2007) Testing the fixed anvil temperature hypothesis in a cloud-resolving model. *Journal of Climate* **20**:2051–57.

Seddon AWR, Macias-Fauria M, Long PR, Benz D, Willis KJ (2016) Sensitivity of global terrestrial ecosystems to climate variability. *Nature* **531**:229–32.

Stephens GL, Li J, Wild M, Clayson CA, Loeb N, Kato S, L'Ecuyer T, Stackhouse Jr. PW, Lebsock M, Andrews T. (2012) An update on Earth's energy balance in light of the latest global observations. *Nature Geoscience* **5**:691–6.

Trenberth KE, Smith L, Qian T, *et al.* (2007) Estimates of the global water budget and its annual cycle using observations and model data. *Journal of Hydrometeorology* **8**:758–69

Zelinka MD, Hartmann DL. (2011) The observed sensitivity of high clouds to mean surface temperature anomalies in the tropics. *Journal of Geophysical Research* **116**:D23103. doi:10.1029/2011JD016459

Zelinka MD, Klein SA, Hartmann DL. (2012) Computing and partitioning cloud feedbacks using cloud property histograms. Part II: Attribution to changes in cloud amount, altitude, and optical depth. *Journal of Climate* **25**:3736–54.

Lecture 2

A Satellite-based Estimate of Convective Mass Flux: Methodology, Validation and Applications

Zhengzhao Johnny Luo

Department of Earth and Atmospheric Sciences
City University of New York, City College
New York, NY, USA

Zhengzhao Johnny Luo is a Professor at the Department of Earth and Atmospheric Sciences of City College of the City University of New York (CUNY). He specializes in developing satellite-based methods for studying cloud and convective dynamics.

Introduction

Best known for his leadership in the World Climate Research Programme's (WCRP) International Satellite Cloud Climatology Project (ISCCP) and Global Energy and Water Exchanges (GEWEX) project, William (Bill) Rossow is a leading figure in the study of clouds and their impact on climate. A defining characteristic of Bill's research, which has been accurately summarized in the quote of the American Meteorological Society's (AMS) 2015 Verner E. Suomi Award that he won, is his "tireless efforts using multi-satellite observations" in addressing cloud-climate questions. Bill truly believes that advancement in understanding the complicated processes in the atmosphere and climate should come from bringing together all relevant observations from different platforms. The subject presented in this chapter, which discusses a global estimation of convective mass flux through synergistic analysis of a suite of satellite observations, together with ambient sounding and a plume model, is a direct response to Bill Rossow's call for the multi-platform approach.

Moist convection is a crucial element of atmospheric circulation, especially in the tropics. To properly represent moist convection and its interaction with the large-scale circulation is a central task in global models. Despite decades of research and development, it remains a major challenge in the modeling community (Arakawa, 2004; Moncrieff *et al.*, 2017). Part of the difficulties can be attributed to the lack of global observations that can be used to effectively evaluate existing theories and parameterizations used in the models. At the heart of the convection parameterization problem is the determination of convective mass flux (defined as the product of air density, convective cloud coverage, and vertical velocity), which controls the collective effects of convection on heat, moisture, momentum, and constituent budgets. Field measurements using aircraft (e.g., [LeMone and Zipser, 1980]) and ground-based radar and wind profilers (e.g., [Kumar *et al.*, 2015; Giangrande *et al.*, 2016]) provide important information and insights, but they are often limited in space and time, making it difficult to generalize the findings for developing and evaluating cumulus parameterizations in global models. Long-term, global observations from satellites offer a potential solution to the problem. However, no satellite instruments currently in orbit are designed for measuring convective vertical velocity or convective mass fluxes. Even spaceborne Doppler radars that are being planned for future missions (e.g., ESA/JAXA's EarthCARE mission) will still find it difficult to measure the vertical velocity profiles inside convective cores due to attenuation by heavy precipitation. To get around the difficulty, a new approach has been presented that brings together multiple satellite observations, as well as ambient sounding and a plume model, to estimate convective vertical velocity and convective mass flux (Masunaga and Luo, 2016; Jeyaratnam *et al.*, 2020).

A large number of satellite measurements were used and analyzed in developing the new method. They include temperature and moisture profiles from the Atmospheric Infrared Sounder and Advanced Microwave Sounder Unit (AIRS/AMSU), cloud-top heights and cloud profiling information from CloudSat and Cloud-Aerosol Lidar and Infrared Pathfinder Satellite Observation (CALIPSO), cloud-top temperature from Moderate Resolution Imaging Spectroradiometer (MODIS), and in one application, even precipitation from the Tropical Rainfall Measuring Mission (TRMM) Precipitation Radar (PR) and ocean surface wind from Quick Scatterometer (QuickSCAT). These observations are combined with a convective plume model driven by the ambient sounding in a Bayesian manner to produce the best estimate of convective vertical velocity and mass flux.

The new global estimate of convective mass flux enables a number of insightful scientific investigations of convective dynamics. Here we present a few examples. The first one uses the *k*-means cluster analysis to identify global patterns related to the structures and distribution of convective vertical velocity, revealing the distinct differences among extratropical convection, tropical land convection, and tropical oceanic convection (Jeyaratnam *et al.*, 2020). The second example examines the life cycle and temporal evolution of convective and large-scale mass fluxes (Masunaga and Luo, 2016). Finally, we show some preliminary results concerning the use of the satellite-based estimate of convective mass flux to evaluate cumulus parameterization of a Global Climate Model (GCM). These studies exemplify how a deeper understanding of cloud and convective problems can be obtained through analysis of multiple satellite data — a defining character of Bill Rossow's career.

References

Arakawa A. (2004) The cumulus parameterization problem: Past, present, and future. *Journal of Climate* **17**:2493–525.

Giangrande SE, Toto T, Jensen MP, *et al*. (2016) Convective cloud vertical velocity and mass-flux characteristics From radar wind profiler observations during GoAmazon2014/5. *Journal of Geophysical Research: Atmospheres* **121**:12891–913. https://doi.org/10.1002/2016JD025303

Jeyaratnam J, Luo ZJ, Giangrande SE, *et al*. (2020) A satellite-based estimate of convective vertical velocity and convective mass flux: Global survey and comparison with radar wind profiler observations, *Geophysical Research Letters*. doi:10.1029/2020GL090675

Kumar VV, Jakob C, Protat A, *et al.* (2015) Mass-flux characteristics of tropical cumulus clouds from wind profiler observations at Darwin, Australia. *Journal of the Atmospheric Sciences* **72**:1837–55. doi:10.1175/JAS-D-14-0259.1

LeMone MA, Zipser EJ. (1980) Cumulonimbus vertical velocity events in GATE. Part I: Diameter, intensity, and mass flux. *Journal of the Atmospheric Sciences* **37**:2444–57.

Masunaga H, Luo ZJ. (2016) Convective and large-scale mass flux profiles over tropical oceans determined from synergistic analysis of a suite of satellite observations. *Journal of Geophysical Research: Atmospheres* **121**. doi:10.1002/2016JD024753

Moncrief MW, Liu C, Bogenschutz P. (2017) Simulation, modeling, and dynamically based parameterization of organized tropical convection for global climate models. *Journal of the Atmospheric Sciences* **74**:1363–80. doi:10.1175/JAS-D-16-0166.1

Slide 1

A Satellite-Based Estimate of Convective Mass Flux: Methodology, Validation, and Applications

Zhengzhao Johnny Luo

City University of New York, City College, New York

With contributions from: Hirohiko Masunaga and Jeyavinoth Jeyaratnam

Studies of Cloud, Convection and Precipitation Processes through Satellite Observations

Lectures in Climate Change, Vol. 3 2021

Slide 2

1. Motivation 2. Methodology 3. Validation 4. Applications

Convective mass flux (M_c) is a key parameter controlling the effects of cumulus convection on large-scale heat (Q_1) and moisture (Q_2) budgets.

Heat budget:

$$Q_1 \equiv \frac{\partial \bar{s}}{\partial t} + \overline{\nabla \cdot s\mathbf{V}} + \frac{\partial \bar{s}\bar{\omega}}{\partial p} = Q_R + L(c-e) - \frac{\partial}{\partial p}\overline{s'\omega'},$$

Moisture budget:

$$Q_2 \equiv -L\left(\frac{\partial \bar{q}}{\partial t} + \overline{\nabla \cdot q\mathbf{V}} + \frac{\partial \bar{q}\bar{\omega}}{\partial p}\right) = L(c-e) + L\frac{\partial}{\partial p}\overline{q'\omega'},$$

(Yanai *et al.*, 1973)

A cumulus ensemble model

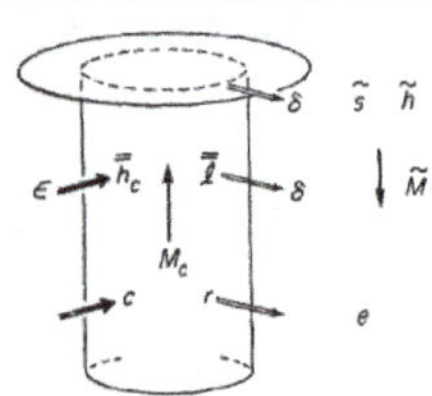

$$Q_1 - Q_R = -\boxed{M_c}\frac{\partial \tilde{s}}{\partial p} - Le,$$

$$Q_2 = \boxed{LM_c}\frac{\partial \tilde{q}}{\partial p} - L\delta(\tilde{q}^* - \tilde{q}) - Le.$$

$$\boxed{M_c = \sigma \rho w_c}$$

w_c: *vertical velocity*
σ: *cloud coverage*
ρ: *air density*

Slide 3

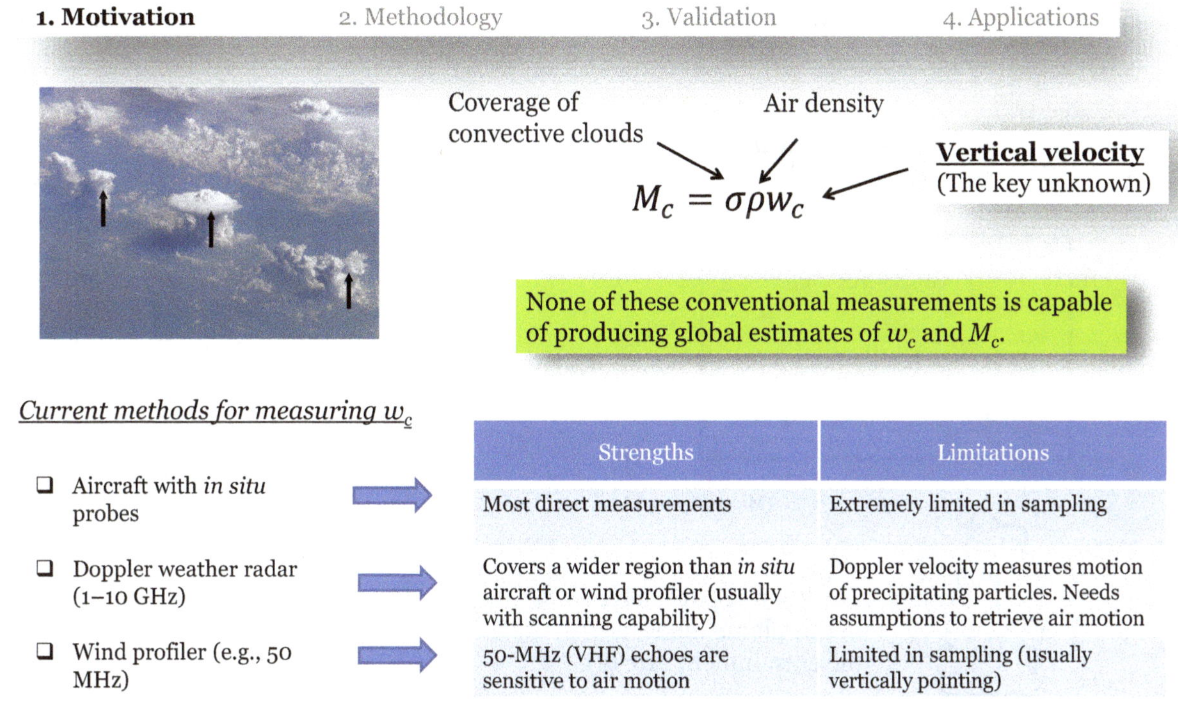

1. Motivation
2. Methodology
3. Validation
4. Applications
Coverage of convective clouds
Air density
$M_c = \sigma \rho w_c$
Vertical velocity
(The key unknown)
None of these conventional measurements is capable of producing global estimates of w_c and M_c.
Current methods for measuring w_c
Aircraft with in situ probes
Doppler weather radar (1–10 GHz)
Wind profiler (e.g., 50 MHz)
Strengths
Limitations
Most direct measurements
Extremely limited in sampling
Covers a wider region than in situ aircraft or wind profiler (usually with scanning capability)
Doppler velocity measures motion of precipitating particles. Needs assumptions to retrieve air motion
50-MHz (VHF) echoes are sensitive to air motion
Limited in sampling (usually vertically pointing)

Slide 4

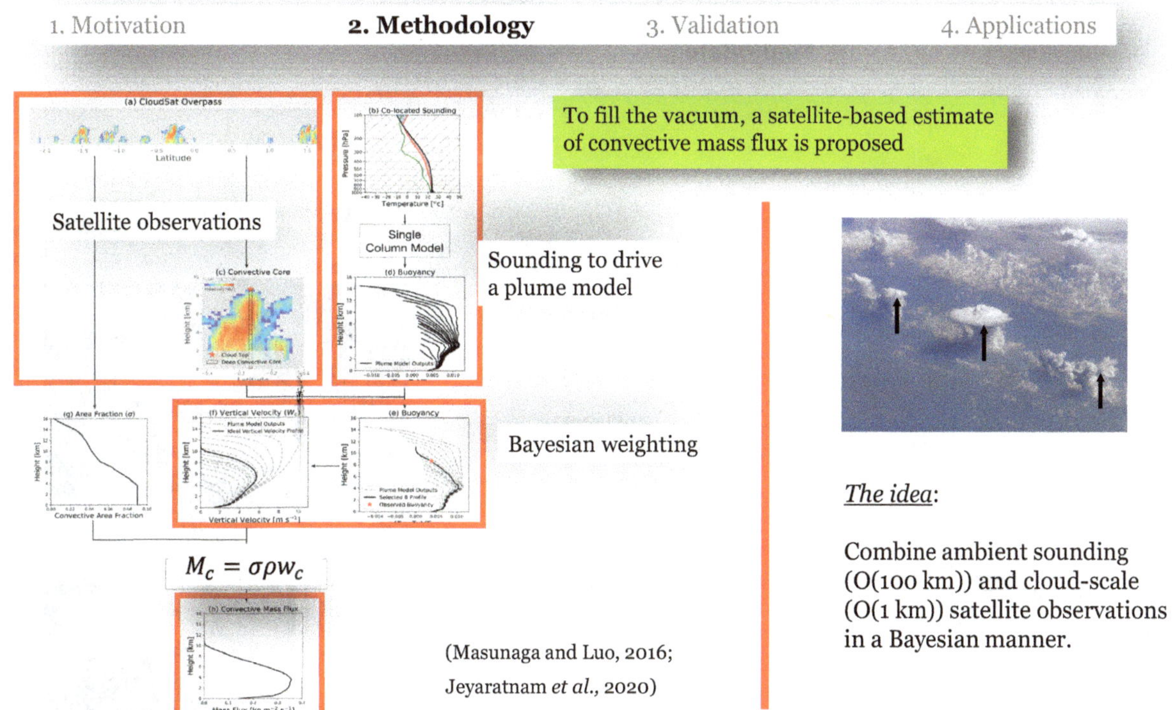

1. Motivation
2. Methodology
3. Validation
4. Applications
To fill the vacuum, a satellite-based estimate of convective mass flux is proposed
(a) CloudSat Overpass
Satellite observations
(b) Co-located Sounding
Single Column Model
Sounding to drive a plume model
Bayesian weighting
$M_c = \sigma \rho w_c$
(Masunaga and Luo, 2016; Jeyaratnam et al., 2020)
The idea:
Combine ambient sounding (O(100 km)) and cloud-scale (O(1 km)) satellite observations in a Bayesian manner.

Slide 5

1. Motivation **2. Methodology** 3. Validation 4. Applications

Convective plume model
(driven by ambient sounding)

$$\frac{1}{2}\frac{\partial w_c^2}{\partial z} = a_B B - \epsilon w_c^2 - c_D w_c^2,$$

$$\frac{\partial (h_c - L_i q_i)}{\partial z} = -\epsilon (h_c - L_i q_i - h_a),$$

$$\frac{\partial q_w}{\partial z} = -\epsilon q_w + \frac{1}{w_c}(\dot{q}_{\rm cond} - \dot{q}_{\rm auto}),$$

(Masunaga and Luo, 2016)

w_c
h_c
q_w

The key free parameter in this plume model is the entrainment rate (ε).

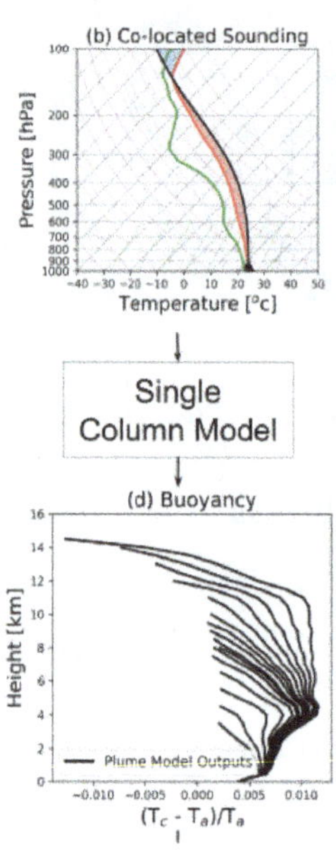

Slide 6

1. Motivation **2. Methodology** 3. Validation 4. Applications

A-Train satellite measurements (such as cloud-top vertical velocity and buoyancy) provide observational *constraints* on the plume model simulations

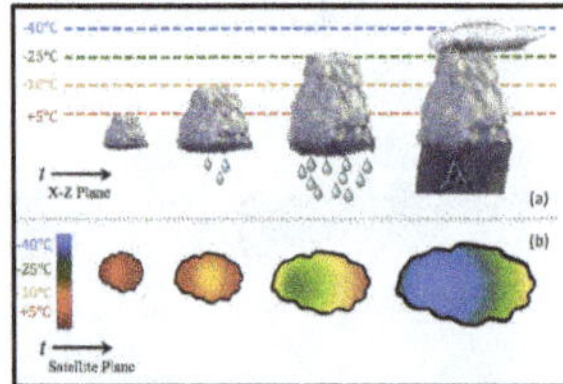

(Luo *et al.*, 2014)

$$w = \left(\frac{\partial T}{\partial z}\right)^{-1} \frac{dT_{BB}}{dt}$$

Convective cloud-top w

$$B = g\frac{T_{parcel} - T_{env}}{T_{env}}$$

Convective cloud-top buoyancy

(Luo *et al.*, 2010)

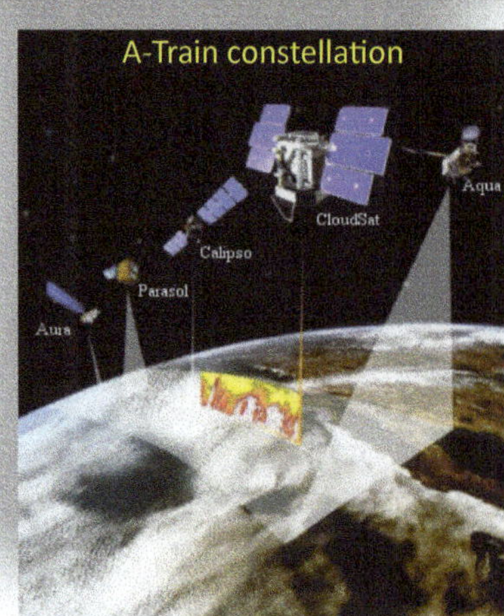

Slide 7

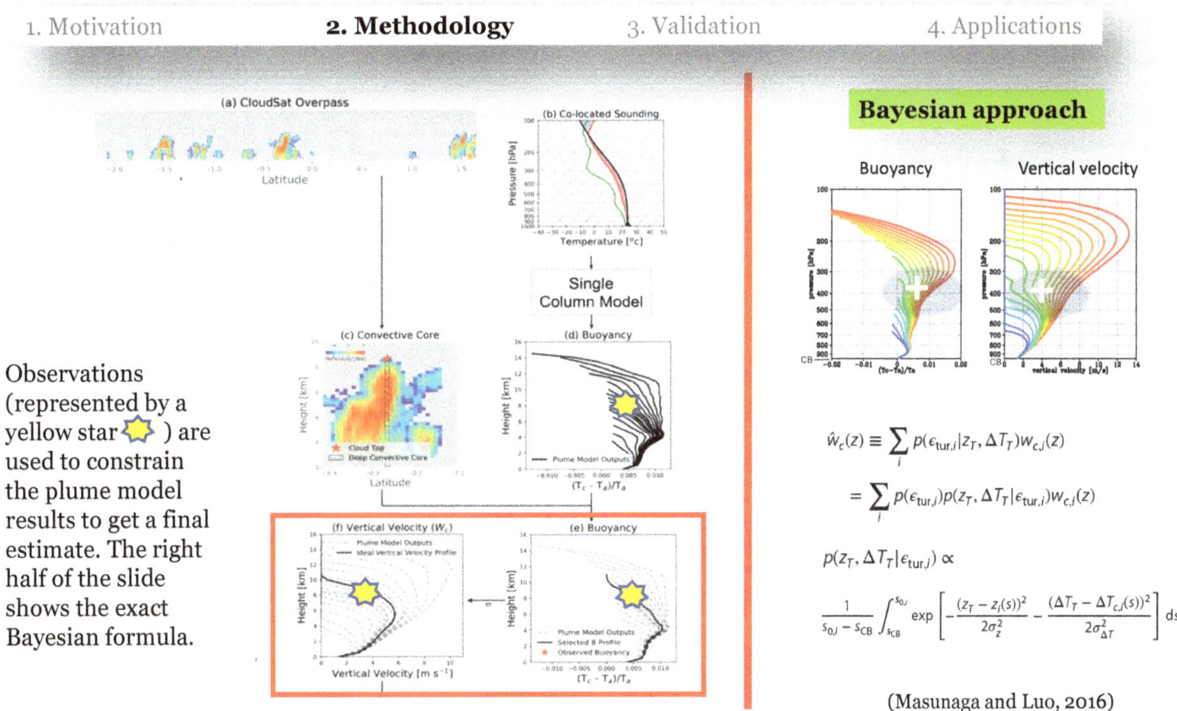

Observations (represented by a yellow star) are used to constrain the plume model results to get a final estimate. The right half of the slide shows the exact Bayesian formula.

Slide 8

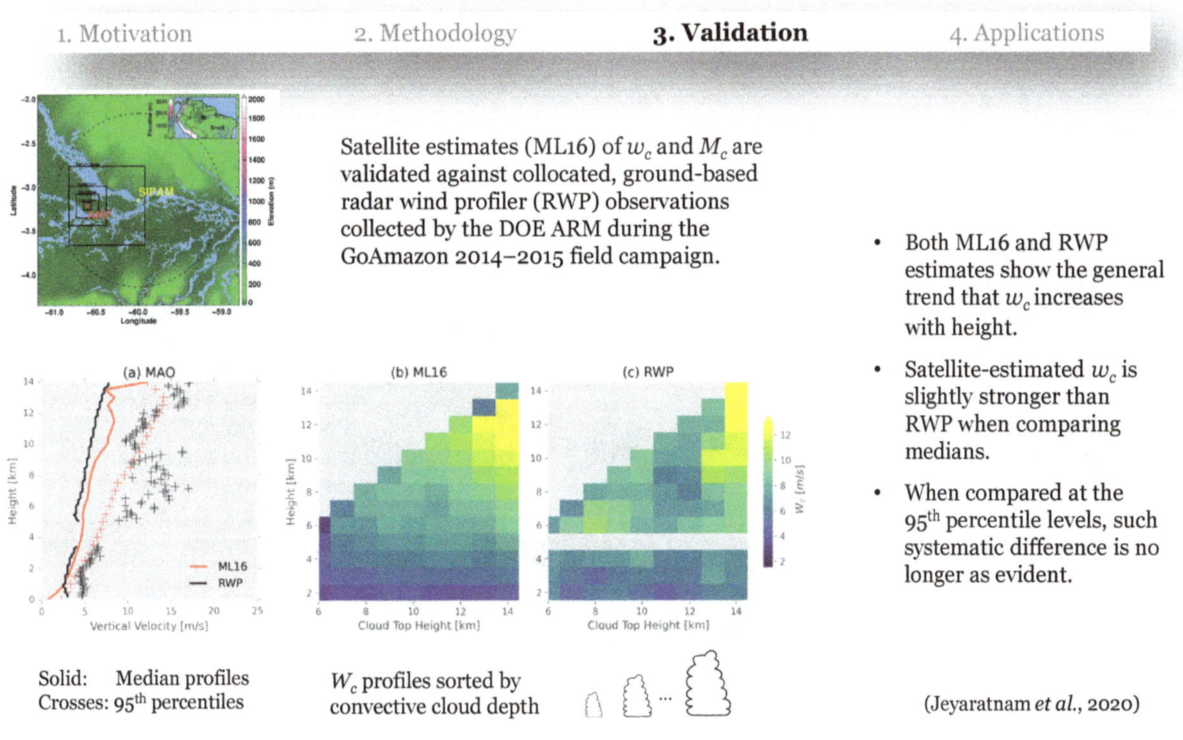

1. Motivation
2. Methodology
3. Validation
4. Applications
SIPAM
Latitude
Longitude
Elevation (m)
Satellite estimates (ML16) of w_c and M_c are validated against collocated, ground-based radar wind profiler (RWP) observations collected by the DOE ARM during the GoAmazon 2014–2015 field campaign.
• Both ML16 and RWP estimates show the general trend that w_c increases with height.
• Satellite-estimated w_c is slightly stronger than RWP when comparing medians.
• When compared at the 95th percentile levels, such systematic difference is no longer as evident.
(a) MAO
Height [km]
ML16
RWP
Vertical Velocity [m/s]
$\hat{w}_c(z) \equiv \sum_i p(\epsilon_{\mathrm{tur},i}|z_T, \Delta T_T) w_{c,i}(z)$
$= \sum_i p(\epsilon_{\mathrm{tur},i}) p(z_T, \Delta T_T|\epsilon_{\mathrm{tur},i}) w_{c,i}(z)$
$p(z_T, \Delta T_T|\epsilon_{\mathrm{tur},i}) \propto$
$\frac{1}{s_{0,i} - s_{\mathrm{CB}}} \int_{s_{\mathrm{CB}}}^{s_{0,i}} \exp\left[-\frac{(z_T - z_i(s))^2}{2\sigma_z^2} - \frac{(\Delta T_T - \Delta T_{c,i}(s))^2}{2\sigma_{\Delta T}^2}\right] \mathrm{d}s$
(b) ML16
Height [km]
Cloud Top Height [km]
(c) RWP
Cloud Top Height [km]
W_c [m/s]
Solid: Median profiles
Crosses: 95th percentiles
W_c profiles sorted by convective cloud depth
(Jeyaratnam et al., 2020)

Slide 9

1. Motivation 2. Methodology **3. Validation** 4. Applications

M_c and its three constituents

$$M_c = \sigma \rho w_c$$

Satellite estimates of M_c are compared with collocated, ground-based radar wind profiler (RWP) observations. Also, the constituents of M_c are analyzed individually for the wet and dry seasons of Amazon.

(a) Convective Mass Flux

(b) Convective Vertical Velocity

(c) Convective Area Fraction

(d) Convective Mass Flux

- M_c estimates are broadly comparable in the lower and middle troposphere, with some differences in the upper troposphere.
- Analysis of different constituents of convective mass flux (M_c) shows that M_c is primarily controlled by $\boldsymbol{\sigma}$, with w_c playing a secondary role.

(Jeyaratnam *et al.*, 2020)

Slide 10

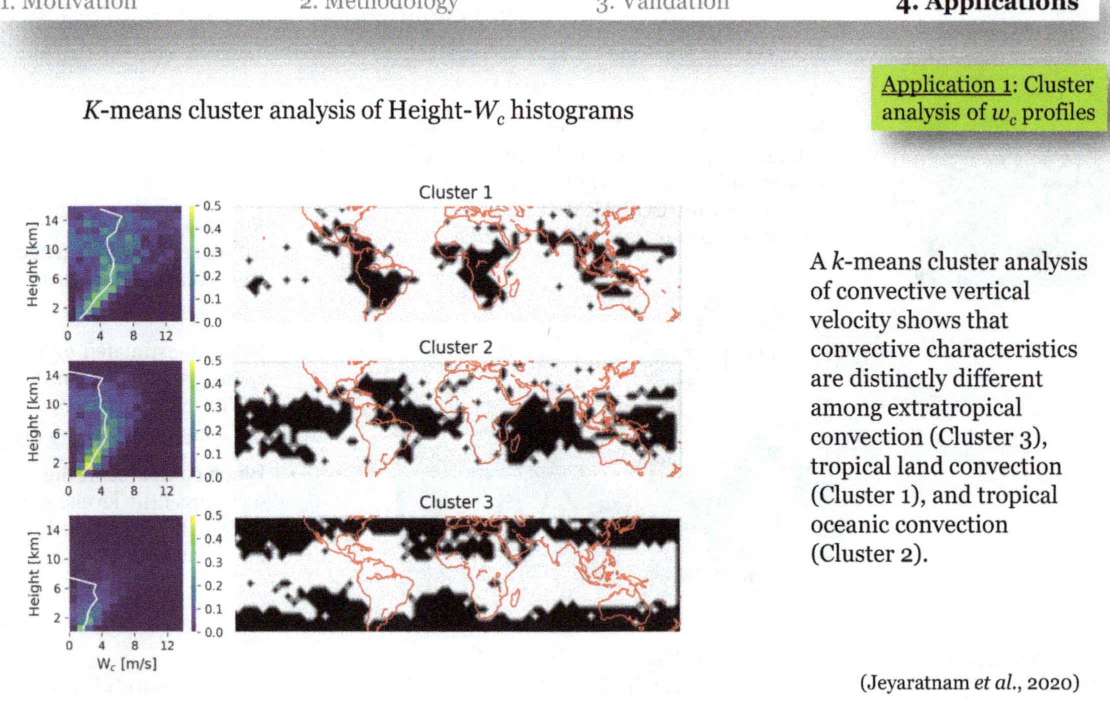

1. Motivation
2. Methodology
3. Validation
4. Applications
Application 1: Cluster analysis of w_c profiles
K-means cluster analysis of Height-W_c histograms
Cluster 1
Cluster 2
Cluster 3
Height [km]
W_c [m/s]
A k-means cluster analysis of convective vertical velocity shows that convective characteristics are distinctly different among extratropical convection (Cluster 3), tropical land convection (Cluster 1), and tropical oceanic convection (Cluster 2).
(Jeyaratnam et al., 2020)

Slide 11

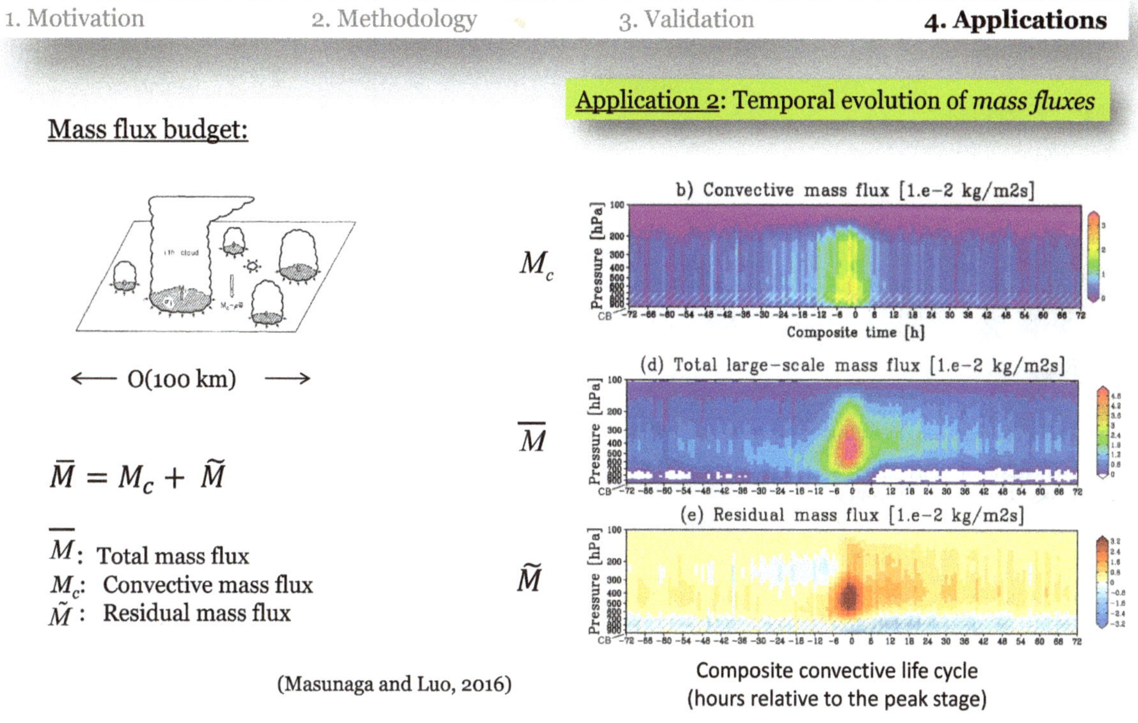

Slide 12

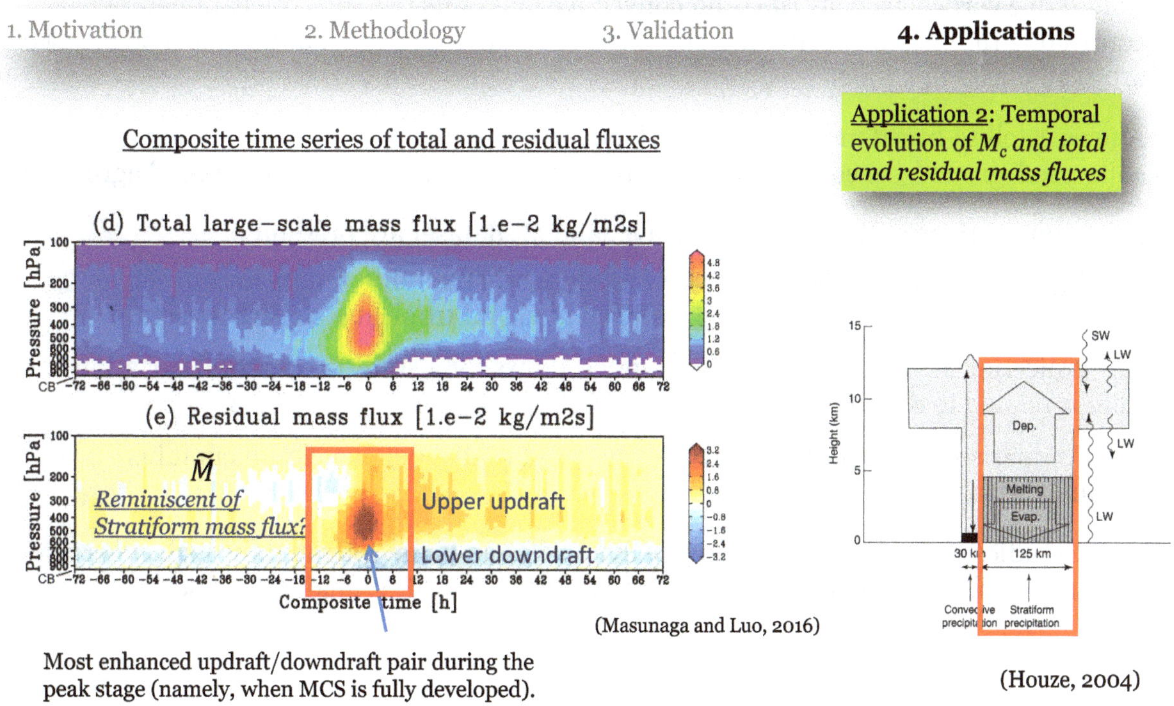

Slide 13

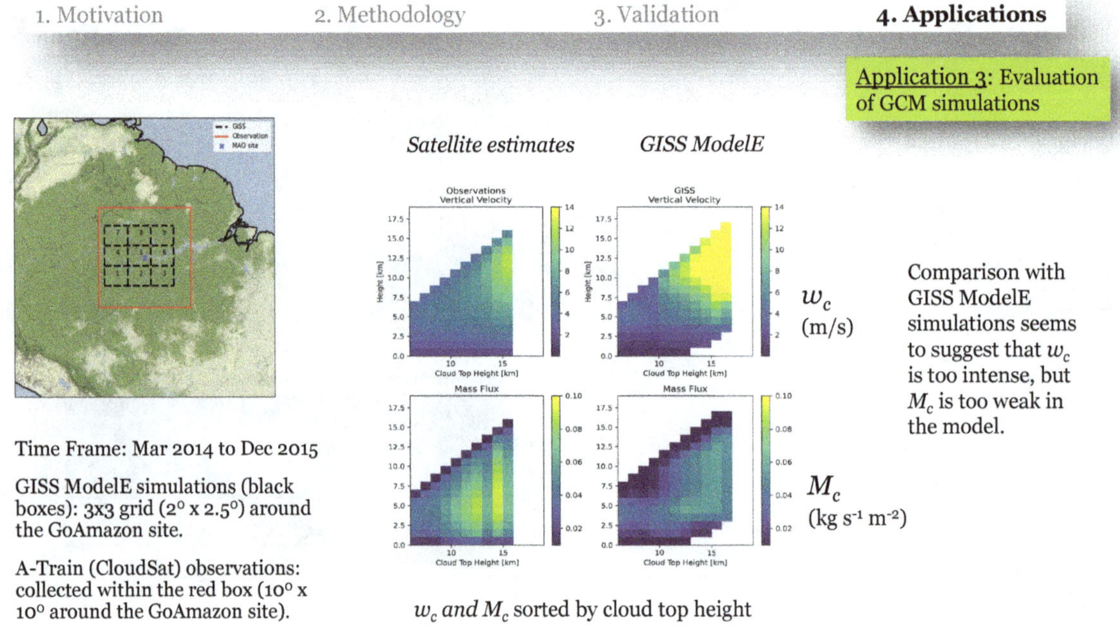

1. Motivation
2. Methodology
3. Validation
4. Applications
Application 3: Evaluation of GCM simulations
Satellite estimates
GISS ModelE
Observations Vertical Velocity
GISS Vertical Velocity
Mass Flux
Cloud Top Height [km]
Height [km]
w_c (m/s)
M_c (kg s^{-1} m^{-2})
Comparison with GISS ModelE simulations seems to suggest that w_c is too intense, but M_c is too weak in the model.
Time Frame: Mar 2014 to Dec 2015
GISS ModelE simulations (black boxes): 3x3 grid (2° x 2.5°) around the GoAmazon site.
A-Train (CloudSat) observations: collected within the red box (10° x 10° around the GoAmazon site).
w_c and M_c sorted by cloud top height

Slide 14

Summary

- Convective vertical velocity (w_c) and convective mass flux (M_c) are key parameters controlling the effects of convection on large-scale environment, but no global observations are available at this time.
- A new, satellite-based method was developed to estimate w_c and M_c. Validation against collocated, ground-based radar wind profiler observations show generally good agreements.
- A few applications of the satellite-based w_c and M_c dataset are presented:

  1) A global *k*-means cluster analysis of w_c shows distinct differences in convective characteristics among extratropical, tropical land and tropical oceanic convection.

  2) Projecting M_c onto composite time series allows for a study of temporal variation of M_c that is related to the convective life cycle.

  3) Comparison with GCM simulations reveals potential deficiencies in cumulus parameterization.

- These studies exemplify how convective dynamics can be investigated through synergistic use of multiple satellite data — a defining character of Prof. William B. Rossow's career.

Slide Captions

Slide 1 Title slide

Slide 2 Convective mass flux (M_c) is a central parameter for understanding the interaction between convection and the environment. As an example, we show that the impacts of cumulus convection on the large-scale environment (represented by Q1 and Q2 as introduced by Yanai *et al.* [1973]) are largely controlled by M_c. Because of this, a large number of current Global Climate Model (GCM) cumulus parameterization schemes are based on the concept of convective mass flux (Arakawa, 2004).

Slide 3 Of the three terms in convective mass flux (M_c), convective vertical velocity (w_c) is the most difficult to measure. A few conventional methods for measuring w_c are reviewed with respect to their strengths and limitations. However, none of them is capable of producing global estimates of w_c. Long-term, global observations from satellites offer a potential solution to the problem. Slides 4–7 describe a satellite-based approach.

Slide 4 An overview (flowchart) of the satellite-based method. The central idea is to combine multiple information contents and combine them in a Bayesian manner. Slides 5–7 explain each component, including ambient sounding and plume model (Slide 5), satellite observations (Slide 6), and the Bayesian approach (Slide 7).

Slide 5 *Convective plume model and ambient sounding*: A plume model driven by ambient sounding is used to compute a range of candidate profiles of w_c and buoyancy under different entrainment rates (ε). Then, satellite observations are used to constrain the plume model solutions and select the final solution that best matches the observations in a Bayesian manner.

Slide 6 *Satellite observations*: (1) *Convective cloud-top vertical velocity.* Time-delayed measurements of cloud-top temperature allow for an estimate of cloud-top vertical velocity. Here, we implemented the idea using a pair of infrared (IR) sensors in the A-Train constellation, namely, IIR onboard CALIPSO and MODIS onboard Aqua (Luo *et al.*, 2014). As part of the A-Train constellation, the two IR sensors fly in close formation and are separated by 1–2 minutes, (2) *Convective cloud-top buoyancy.* Here, CloudSat Cloud Profiling Radar (CPR) is used to identify convective cores. MODIS provides an estimate of cloud-top temperature (T_{parcel}). Ambient sounding, together with knowledge of cloud-top height from CPR, provides an estimate of the environmental temperature (T_{env}). Then, the difference between T_{parcel} and T_{env} determines cloud-top buoyancy (Luo *et al.*, 2010). Both cloud-top vertical velocity and cloud-top buoyancy are closely related to the vertical profiles of convective vertical velocity and convective mass flux.

Slide 7 A Bayesian approach is used to combine satellite observations and plume model predictions. See (Masunaga and Luo, 2016) for details.

Slide 8 Validations were made against collocated, ground-based radar wind profiler (RWP) observations collected by the DOE ARM during the GoAmazon 2014–2015 field campaign. The upper panel shows the location of the GoAmazon field campaign. The three lower panels are adopted from (Jeyaratnam *et al.*, 2020): (a) Red color represents satellite estimates ([Masunaga and Luo, 2016] or satellites [ML16]), and the black color represents RWP observations. Solid lines are the median profiles, and crosses are the 95th percentiles (i.e., intense updrafts). (b) ML16-retrieved vertical velocity profiles sorted by convective depths. Only the 95th percentiles are shown. (c) The same as (b), except for RWP observations. The main conclusion from the comparison study is that satellite-estimated vertical velocity is slightly stronger than that retrieved by RWPs (this difference is attributed to the differences in the definitions for allowable convective regions), but when compared by vertical velocity properties at the 95th percentile levels, such systematic difference is no longer evident except at a 6–8-km height range (see [Jeyaratnam *et al.*, 2020] for a discussion).

Slide 9 Validation of convective mass flux M_c and examination of the variations in its two constituents (namely, vertical velocity w_c and convective area fraction σ) between the wet and dry seasons. The main conclusion is that variation in M_c is primarily controlled by that of σ.

Slide 10 *Application 1*: With a global dataset from satellite observations, one could investigate general patterns related to the structures and distributions of convective vertical velocity and mass flux. Here, we run a *k*-means cluster analysis through w_c profiles saved in each 0.5°×0.5° grid within the latitude range of 45°S to 45°N (higher latitudes are excluded because convection is less frequent there). In each grid, the retrieved w_c profiles are compiled in a height-w_c histogram to summarize the characteristics of their vertical structures. The *k*-mean cluster analysis is performed on these height-w_c histograms. Three clusters emerge from the analysis: the left three panels show the centroids of the clusters in terms of the height-*wc* histograms, and the right panels show the corresponding geographical distributions.

Slide 11 *Application 2*: Breakdown of the total mass flux budget into contributions from the convective mass flux and residual mass flux. Results are further projected on a composite convective life cycle. A-Train measurements are instantaneous snapshots. To obtain the temporal dimension, all snapshots are sorted into composite time series that delineate a convective lifecycle using precipitation from TRMM PR as the anchor, following Masunaga (2012).

Slide 12 *Application 2 (continued)*: Results show that M_c dominates the total mass flux during the early hours of the convective evolution. But when a convective system matures and reaches its peak, a residual mass flux reminiscent of the stratiform dynamics builds up.

Slide 13 *Application 3*: Comparisons of convective vertical velocity and convective mass flux between satellite estimates and a GCM simulation. Results are still preliminary, but it illustrates an important application of the new dataset, that is, a detailed and direct evaluation of GCM cumulus parameterizations.

Slide 14 Summary

References

Arakawa A. (2004) The cumulus parameterization problem: Past, present, and future. *Journal of Climate* **17**:2493–525.

Houze RA, Jr. (2004) Mesoscale convective systems, *Rev. Geophys.* **42**:RG4003. doi:10.1029/2004RG000150

Jeyaratnam J, Luo ZJ, Giangrande SE, *et al.* (2020) A satellite-based estimate of convective vertical velocity and convective mass flux: Global survey and comparison with radar wind profiler observations, *Geophysical Research Letters*. doi:10.1029/2020GL090675

Luo ZJ, Liu GY, Stephens GL. (2010) Use of A-Train data to estimate convective buoyancy and entrainment rate. *Geophysical Research Letters* **37**:L09804. doi:10.1029/2010GL042904

Luo ZJ, Jeyaratnam J, Iwasaki S, *et al.* (2014) Convective vertical velocity and cloud internal vertical structure: An A-Train perspective. *Geophysical Research Letters* **41**. doi:10.1002/2013GL058922

Masunaga H. (2012) A satellite study of the atmospheric forcing and response to moist convection over tropical and subtropical oceans. *Journal of the Atmospheric Sciences* **69**:150–67.

Masunaga H, Luo ZJ. (2016) Convective and large-scale mass flux profiles over tropical oceans determined from synergistic analysis of a suite of satellite observations. *Journal of Geophysical Research: Atmospheres* **121**. doi:10.1002/2016JD024753

Yanai M, Esbensen S, Chu J-H. (1973) Determination of bulk properties of tropical cloud clusters from large scale heat and moisture budgets. *Journal of the Atmospheric Sciences* **30**:611–27.

Lecture 3

Convective Core, Detrainment Level, and Convective Dilution: New Perspectives Based on Five Years of CloudSat Data

Hanii Takahashi

Joint Institute for Regional Earth System Science and Engineering (JIFRESSE), University of California Los Angeles, and NASA Jet Propulsion Laboratory, California, USA

Hanii Takahashi is a JIFRESSE Assistant Researcher at the NASA Jet Propulsion Laboratory. She has worked extensively on satellite data analysis, especially using CloudSat and other A-Train observations, focusing on convective dynamics in tropical deep convective clouds. She has developed metrics that uniquely capture important processes pertaining to convective dynamics, convective evolution, and cloud microphysics.

Introduction

Among the many convective processes in deep convection, *two* are of key importance in affecting the climate through controlling the energy budget, hydrological cycle, and radiative-convective feedbacks: *vertical mass transport* by convective updrafts and *horizontal mass transport* by deep convective outflow layers that transition from such updrafts into widespread cirrus anvils (**Slide 2**). The vertical transport is like an express elevator transferring the near-surface air directly into the upper troposphere or lower stratosphere, which effectively controls the energy budgets and hydrological cycles (e.g., [Riehl and Malkus, 1958]). The vertical transport will eventually make a transition to horizontal outflows where cirrus anvils develop. These cirrus anvils cause a large amount of radiative warming in the upper troposphere, reinforce their internal latent heating structure through their radiative effects (e.g., [Machado and Rossow, 1993]), and thus play an important role in radiative-convective feedbacks (e.g., [Stephens *et al.*, 2008]). Therefore, advancing knowledge of tropical deep convection requires, in part, ways of observing these fundamental convective processes — the vertical and horizontal mass transports.

Deep convective clouds can be identified from various satellite observations using a number of methods. Previous studies widely used satellite-borne infrared (IR) sensors to identify cold cloud tops to find deep convective cores. Later, the Tropical Rainfall Measuring Mission (TRMM) precipitation radar (PR) was used to illustrate the vertical structure of overshooting cores or penetrating convective cores, and it was found that the distribution of deep convective cores is quite different between IR and PR-based observations. TRMM is an active sensing system and provides profiling capability; however, TRMM is mostly sensitive to heavier precipitation since TRMM PR operates at 13.8 GHz with a sensitivity at ~17 dBZ (**Slide 3, left**). CloudSat, on the other hand, carries a 94-GHz Cloud Profiling Radar (CPR) with a sensitivity at ~–30 dBZ, which is sensitive to both cloud and precipitation-size particles (**Slide 3, right**). Therefore, CloudSat can provide us a more complete sampling of deep convective clouds and adds many insights to our knowledge of deep convective cores and anvil outflows (Takahashi and Luo, 2012; 2014; Takahashi *et al.*, 2017; 2018; Luo *et al.*, 2017).

Although CloudSat has a great profiling capability, exploring vertical and horizontal mass transports is still a challenge. For the vertical transport, a key convective property is convective cores (**Slide 2**), which are strongly affected by entrainment because entrainment dilutes the strength of convective updrafts and weakens the intensity of vertical transport (Houze, 2014). For the horizontal transport, a key convective property is the convective detrainment levels where anvil outflow occurs, which can be controlled by the strength of convective cores (**Slide 2**). However, no existing satellite instruments can directly measure the updraft velocity or amount of entrainment/detrainment. Therefore, to overcome this challenge, it is important to develop innovative ways to estimate the key convective properties and to investigate the relationships between them. Recent publications by Takahashi and Luo (2012; 2014) and Takahashi *et al.* (2017; 2018) have developed methods that allow us to examine convective dynamics, structure, and environment using CloudSat observations. How can we detect convective cores and convective detrainment levels? How can we estimate convective strength and the amount of entrainment dilution? How are convective cores, entrainment, and convective detrainment levels linked? This chapter answers these questions.

How can we detect convective cores and convective detrainment levels?

Slide 4 illustrates an example of a cloud object, embedded convective cores, and attached anvils. First, using the 2B-GEOPROF CloudSat data product, we define deep cloud objects as the continuous area enclosed by a cloud mask value ≥ 20 (Riley and Mapes, 2009). Second, we find deep convective cores in each cloud object as a CloudSat CPR profile that has a continuous radar echo from the cloud top to within 2 km of the surface with an Echo Top Height (ETH) of 10 dBZ at ≥10 km. ETHs of 10 dBZ are the highest altitudes that 10 dBZ (e.g., larger particles) radar echoes reach. Once a deep convective core is identified, we finally find attached anvils by searching on both sides of the cloud object; attached anvil clouds are defined by a CPR profile that has a cloud base higher than or equal to 5 km (Yuan and Houze, 2010). These attached anvil outflows, the convective detrainment levels, are tightly linked to the concept of level of neutral buoyancy (LNB), where convection stops ascending and starts to detrain mass. To describe the full range over which convective detrainment develops, three forms of LNB are estimated based on the attached anvils (LNB_observation, **Slide 5**). LNB_CTH is the cloud top height of the anvils, and LNB_CBH is the cloud base height of anvils. LNB_maxMass is the height of the maximum radar reflectivity within the anvil, which is well correlated with the maximum mass detrainment level (Mullendore *et al.*, 2009) and thus is most relevant to convective mass transport. LNB_observation is calculated over only "fresh" anvils (20 km of a horizontal anvil on each side of the convective core) in order to mitigate the bias owing to particle sedimentation with longer residence time in the anvil. Key findings for convective cores and LNB_observation are summarized in Takahashi and Luo (2012) and **Slide 6**. These have important implications on how continental and oceanic deep convective clouds are fundamentally different from each other, which is rarely captured in climate models.

How can we estimate convective strength and the amount of entrainment dilution?

Without direct measurement of vertical velocity, there are several ways to analyze the intensity of deep convection. For example, CPR ETH can be a good proxy for convective strength (**Slide 7**). ETHs of 0 and 10 dBZ are the highest altitudes that 0 (e.g., smaller particles) and 10 (e.g., larger particles) dBZ radar echoes reach. Since strong updrafts can loft both smaller and larger particles to greater altitudes, an ETH tends to be high over strong updrafts (Luo *et al.*, 2008). Smaller distances between ETH and CTH (cloud top height — roughly corresponds to the ETH of about −30 dBZ) are also indicative of stronger convective intensity since both small and large particles can similarly be lofted high into the troposphere. On the other hand, larger particles will fall out over weak updrafts; thus, the distance between ETH and CTH tends to be large. We call the distance between ETH and CTH the cloud top-echo top distance (CTETD). Results of proxies of convective strength based on CPR ETH are summarized in Takahashi and Luo (2014) and **Slide 7**.

Another way to understand the intensity of deep convection is to estimate the amount of entrainment dilution. LNB based on the parcel theory (**Slide 8**) is the ideal case of an undiluted convective core (i.e., entrainment effect is assumed to be zero), which can be estimated from sounding datasets (referred to as LNB_sounding, **Slide 9**). However, in reality, convection interacts with the environment, loses buoyancy due to entrainment dilution, and lowers its LNB to be LNB_observation (**Slide 10**). Indeed, LNB_sounding has little direct correlation with LNB_observation (**Slide 11**). Therefore, the difference between LNB_sounding and LNB_observation can be interpreted as a measure of the magnitude of the entrainment effects — the greater the entrainment dilution, the larger the height difference (**Slide 12**). Results of the difference between LNB_sounding and LNB_observation are summarized in Takahashi *et al.* (2017) and **Slide 13**. Continental deep convective clouds tend to have stronger convective cores, and thus their vertical transport is stronger and their horizontal transport is more likely to occur at a higher level than oceanic deep convective clouds.

We can zoom in further and see the same phenomena in the internal structure of convective cores — the shape of CloudSat CPR profile (**Slide 14**). Internal vertical structures of convective cores show that continental convection has stronger cores that have better capabilities of transporting larger particles to a higher level compared to oceanic convection, which is consistent with the previous result. These results show, at a process level, how continental and oceanic deep convective clouds are different.

How are convective cores, entrainment, and convective detrainment levels linked?

Since the 1960s, it has been assumed in cumulus cloud modeling that the entrainment rate is negatively correlated with the size of convective cores (e.g., [Simpson and Wiggert, 1969], **Slide 15**). Why does this make sense? The physical explanation is that larger convective cores are better protected from the environment and thus less diluted by entrainment. However, this relation was never examined with global satellite observations. Considering this, we estimate the bulk entrainment rate for each LNB_observation. The difference between LNB_sounding and LNB_observation can be cast into a simple framework of a 1-D entraining plume model (**Slide 16**). The corresponding bulk entrainment rates (λ; unit: %/km) for LNB_maxMass and the size of convective cores are compared over the Warm Pool and tropical land (Africa and Amazon) in Takahashi *et al.* (2017) and **Slide 17**. This reveals a negative correlation between the size of convective cores and entrainment (core-entrainment relation), which adds observational evidence from global satellite observations to support the longstanding fundamental assumption. Moreover, it also shows a positive correlation between the size of convective cores and convective detrainment levels (core-LNB relation), which suggests that the strength of vertical transport can control horizontal transport.

How do convective cores, entrainment, and convective detrainment levels change during El Niño and La Niña?

It is interesting to investigate how vertical and horizontal mass transports change in a warmer world since it has important implications for understanding current and future climate. Surface water in the equatorial Pacific becomes warmer during El Niño than during the La Niña periods. Therefore, we assess the core-entrainment relation and core-LNB relation during El Niño (2006–2007DJF, 2009–2010DJF) and La Niña periods (2007–2008DJF, 2010–2011DJF), which can expose interannual variations that provide unique insights into how convection may change in a warmer climate. Key findings are summarized in (Takahashi *et al.*, 2018), as well as **Slides 18 and 19**. This has important implications on how deep convective clouds might change in a warmer climate. Although caution needs to be taken, this El Niño/La Niña comparison can be used as observational evidence to guide future climate projections in climate models.

Some limitations are unavoidable when the analysis involves satellite instruments that do not have the capability of directly measuring the updraft velocity and amount of entrainment/detrainment. However, it is still possible to develop metrics that uniquely capture important convective processes to understand deep convective clouds — the vertical and horizontal mass transports. CloudSat reveals a negative core-entrainment relation and a positive core-LNB relation, which both appear with a land-ocean comparison and with an El Niño/La Niña comparison. The physical explanation is that larger convective cores are better protected from the environment and thus less diluted by entrainment. As a result, more intense convective cores are associated with stronger vertical transport and higher horizontal transport. In a warmer world, we expect to see more intense convection.

References

Hartmann DL, Larson K. (2002) An important constraint on tropical cloud–climate feedback. *Geophysical Research Letters* **29**:1951. doi:10.1029/2002GL015835.

Houze Jr R A. (2014) *Cloud Dynamics.* Academic Press.

Luo Z, Liu GY, Stephens GL. (2008) CloudSat adding new insight into tropical penetrating convection. *Geophysical Research Letters* **35**(19).

Luo ZJ, Anderson RC, Rossow WB, Takahashi H. (2017) Tropical cloud and precipitation regimes as seen from near-simultaneous TRMM, CloudSat, and CALIPSO observations and comparison with ISCCP. *Journal of Geophysical Research: Atmospheres* **122**(11):5988–6003. doi:10.1002/2017JD026569

Machado LAT, Rossow WB. (1993) Structural characteristics and radiative properties of tropical cloud clusters. *Monthly Weather Review* **121**(12):3234–60.

Mullendore GL, Homann AJ, Bevers K, Schumacher C. (2009) Radar reflectivity as a proxy for convective mass transport. *Journal of Geophysical Research* **114**(D16103). doi:10.1029/2008JD011431.

Riehl H, Malkus JS. (1958) On the heat balance in the equatorial trough zone. *Geophysics* **6**:503–38.

Riley EM, Mapes BE. (2009) Unexpected peak near −15°C in CloudSat echo top climatology. *Geophysical Research Letters* **36**(9).

Simpson J, Wiggert V. (1969) Models of precipitating cumulus towers. *Monthly Weather Review* **97**(7):471–89.

Stephens GL, Van Den Heever S, Pakula L. (2008) Radiative–convective feedbacks in idealized states of radiative–convective equilibrium. *Journal of the Atmospheric Sciences* **65**(12):3899–916.

Takahashi H, Luo Z. (2012) Where is the level of neutral buoyancy for deep convection? *Geophysical Research Letters* **39**(15). doi:10.1029/2012GL052638

Takahashi H, Luo ZJ. (2014) Characterizing tropical overshooting deep convection from joint analysis of CloudSat and geostationary satellite observations. *Journal of Geophysical Research: Atmospheres* **119**(1):112–21. doi:10.1002/2013JD020972

Takahashi H, Luo ZJ, Stephens GL. (2017) Level of neutral buoyancy, deep convective outflow, and convective core: New perspectives based on 5 years of CloudSat data. *Journal of Geophysical Research: Atmospheres* **122**(5):2958–69. doi:10.1002/2016JD025969

Takahashi H, Luo ZJ, Stephens GL. (2018) Variations of deep convective outflow, entrainment rates and convective cores during El Niño and La Niña: A CloudSat Perspective. *To be submitted.*

Yuan J, Houze Jr RA. (2010) Global variability of mesoscale convective system anvil structure from A-Train satellite data. *Journal of Climate* **23**(21):5864–88.

Slide 1

Slide 2

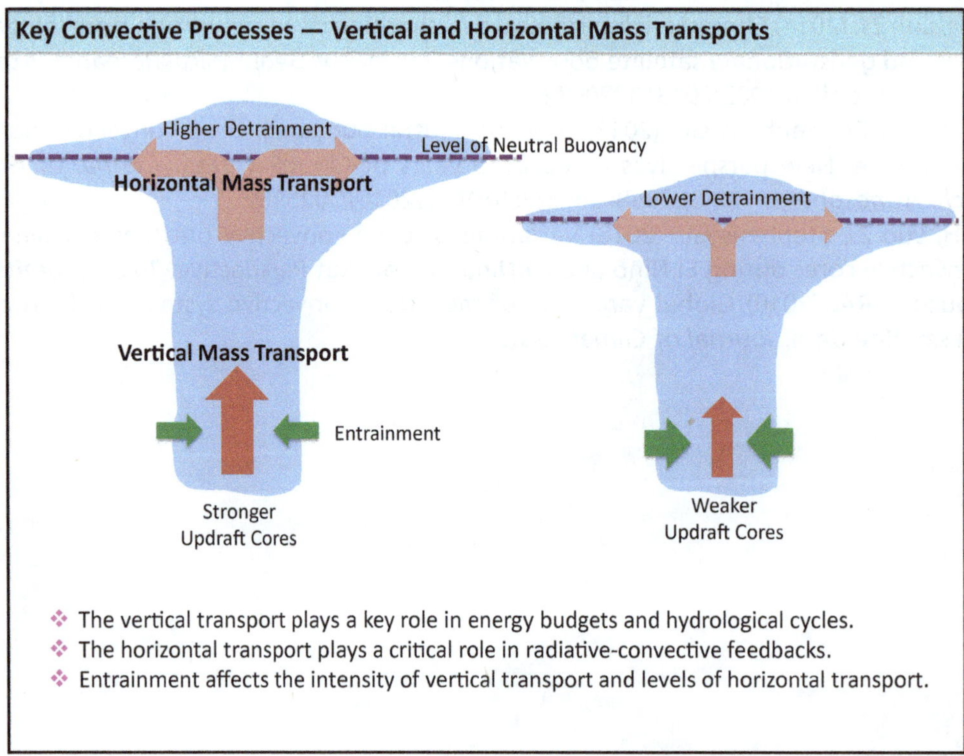

Slide 3

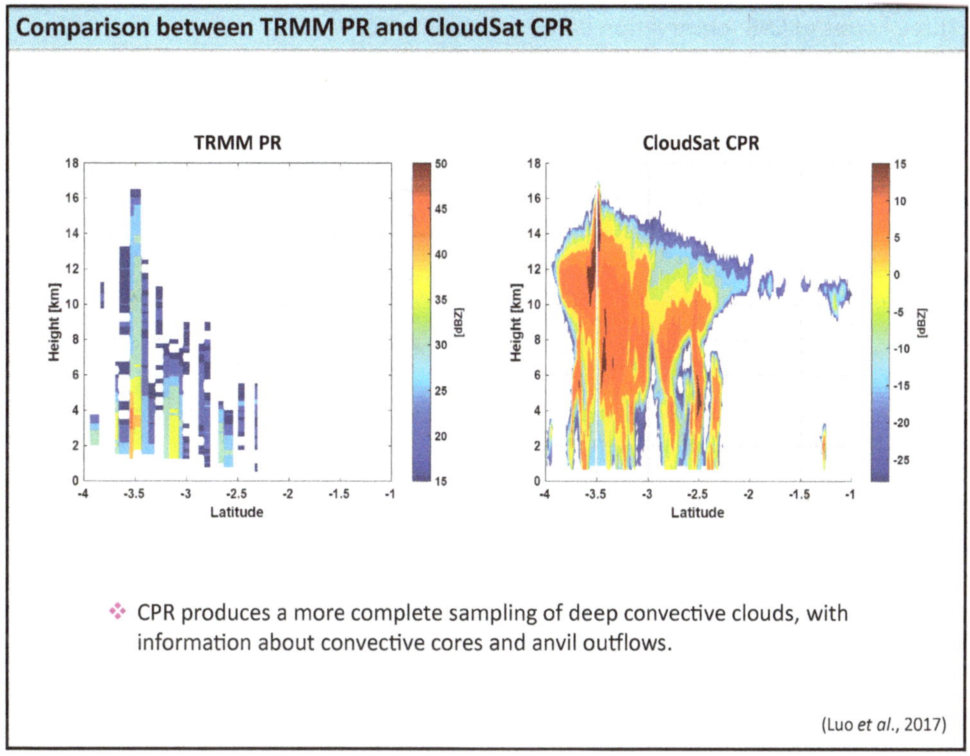
Comparison between TRMM PR and CloudSat CPR
TRMM PR
CloudSat CPR
Height [km]
Latitude
[dBZ]
CPR produces a more complete sampling of deep convective clouds, with information about convective cores and anvil outflows.
(Luo et al., 2017)

Slide 4

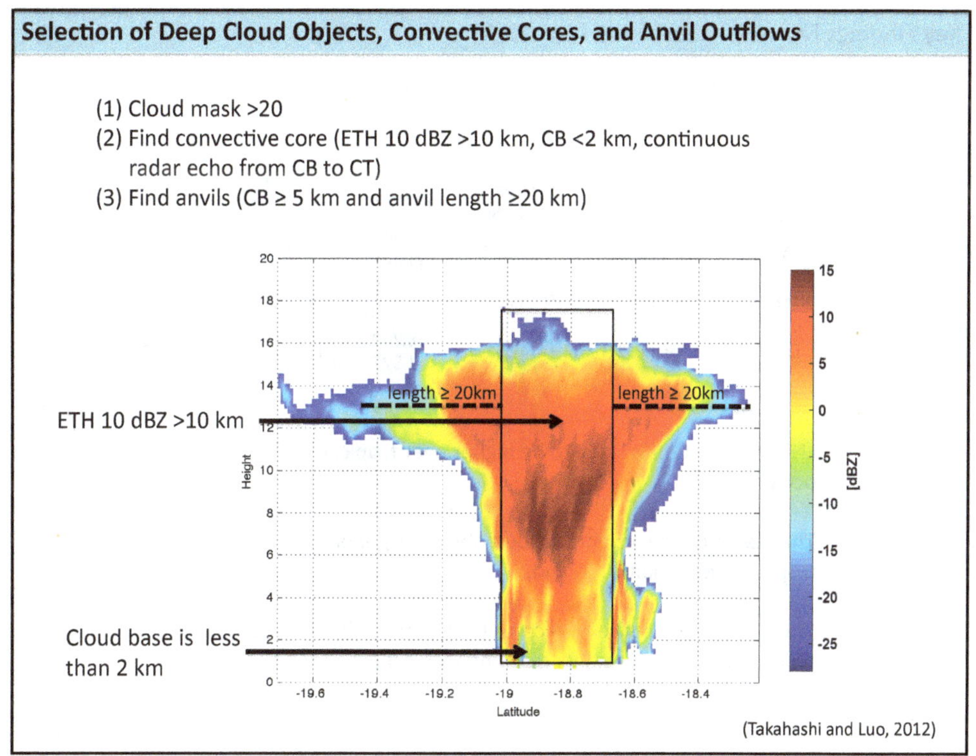
Selection of Deep Cloud Objects, Convective Cores, and Anvil Outflows
(1) Cloud mask >20
(2) Find convective core (ETH 10 dBZ >10 km, CB <2 km, continuous radar echo from CB to CT)
(3) Find anvils (CB ≥ 5 km and anvil length ≥20 km)
length ≥ 20km
length ≥ 20km
ETH 10 dBZ >10 km
Cloud base is less than 2 km
Height
Latitude
[dBZ]
(Takahashi and Luo, 2012)

Slide 5

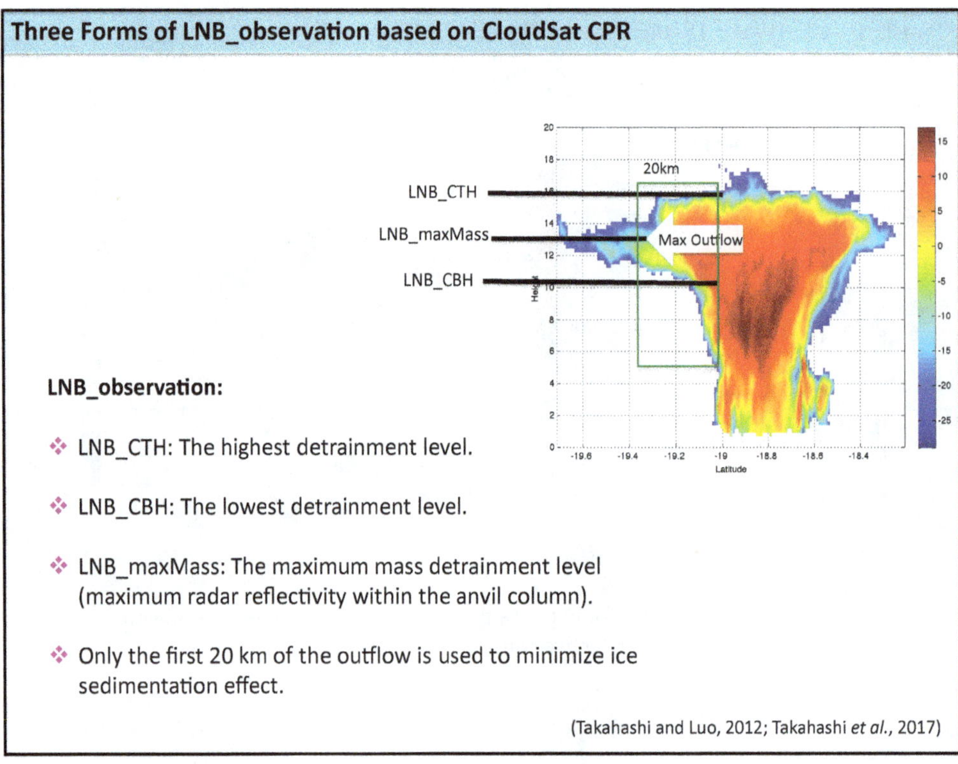

Three Forms of LNB_observation based on CloudSat CPR
LNB_CTH
LNB_maxMass
LNB_CBH
20km
Max Outflow
Latitude
LNB_observation:
LNB_CTH: The highest detrainment level.
LNB_CBH: The lowest detrainment level.
LNB_maxMass: The maximum mass detrainment level (maximum radar reflectivity within the anvil column).
Only the first 20 km of the outflow is used to minimize ice sedimentation effect.
(Takahashi and Luo, 2012; Takahashi et al., 2017)

Slide 6

Key Findings From (Takahashi and Luo, 2012)

Unit [m]

Median (STD)	LNB_CTH	LNB_maxMass	LNB_CBH	System Size	DCC Size
All	13,406 (1,365)	10,680 (1,342)	8,409 (1,495)	159.5 (159.7)	11 (25.7)
Ocean	13,293 (1,358)	10,548 (1,315)	8,272 (1,473)	167.2 (165.1)	9.9 (25.1)
Land	13,756 (1,327)	11,141 (1,337)	8,783 (1,527)	139.7 (136.5)	14.3 (27.5)

- LNB_observation is likely to be higher over tropical land than over the ocean.
- The size of deep convective cloud objects is generally larger over the tropical ocean than over land.
- The size of deep convective cores (DCCs) embedded in these cloud objects tends to be smaller over the tropical ocean than over land.

(Takahashi and Luo, 2012)

Slide 7

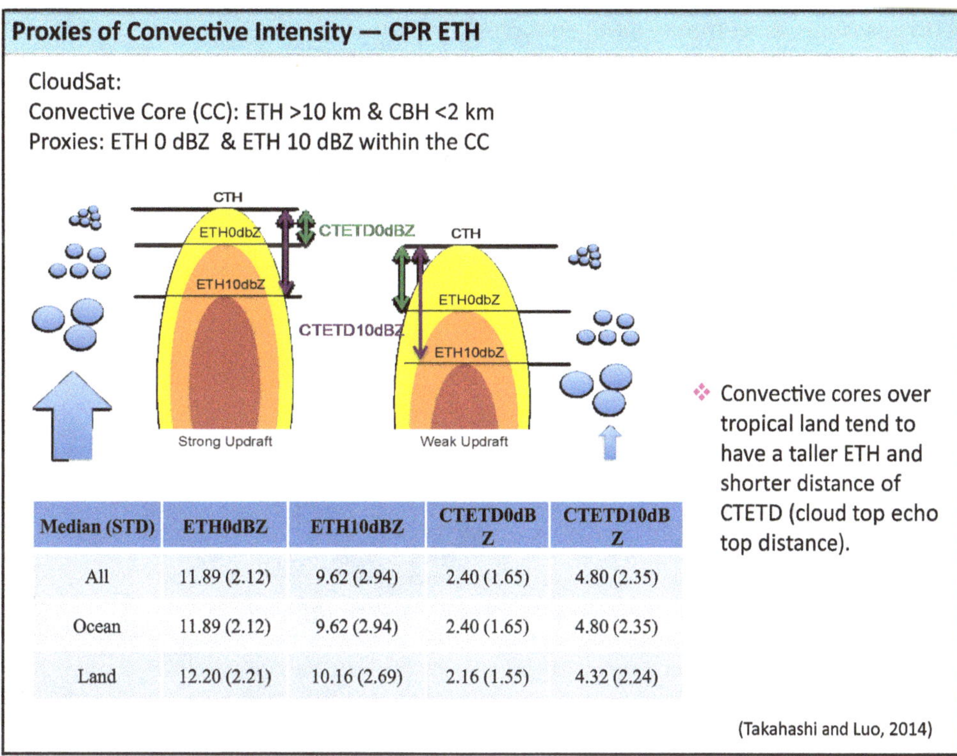

Median (STD)	ETH0dBZ	ETH10dBZ	CTETD0dBZ	CTETD10dBZ
All	11.89 (2.12)	9.62 (2.94)	2.40 (1.65)	4.80 (2.35)
Ocean	11.89 (2.12)	9.62 (2.94)	2.40 (1.65)	4.80 (2.35)
Land	12.20 (2.21)	10.16 (2.69)	2.16 (1.55)	4.32 (2.24)

Slide 8

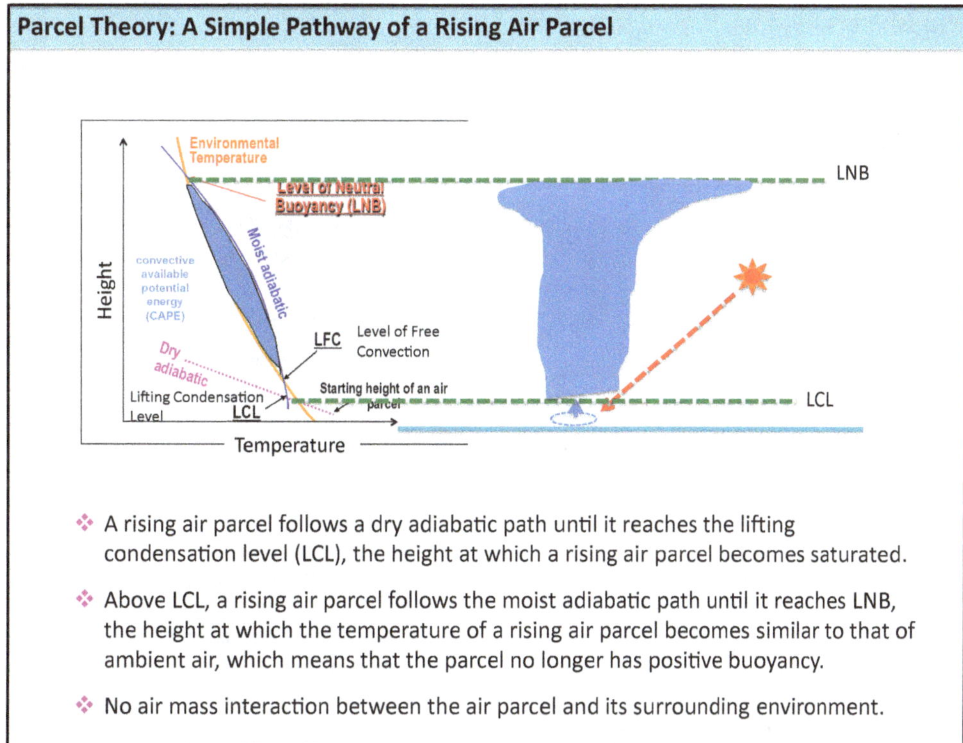

Parcel Theory: A Simple Pathway of a Rising Air Parcel
Environmental Temperature
Level of Neutral Buoyancy (LNB)
Moist adiabatic
convective available potential energy (CAPE)
Height
LFC
Level of Free Convection
Dry adiabatic
Lifting Condensation Level
LCL
Starting height of an air parcel
Temperature
LNB
LCL
A rising air parcel follows a dry adiabatic path until it reaches the lifting condensation level (LCL), the height at which a rising air parcel becomes saturated.
Above LCL, a rising air parcel follows the moist adiabatic path until it reaches LNB, the height at which the temperature of a rising air parcel becomes similar to that of ambient air, which means that the parcel no longer has positive buoyancy.
No air mass interaction between the air parcel and its surrounding environment.

Slide 9

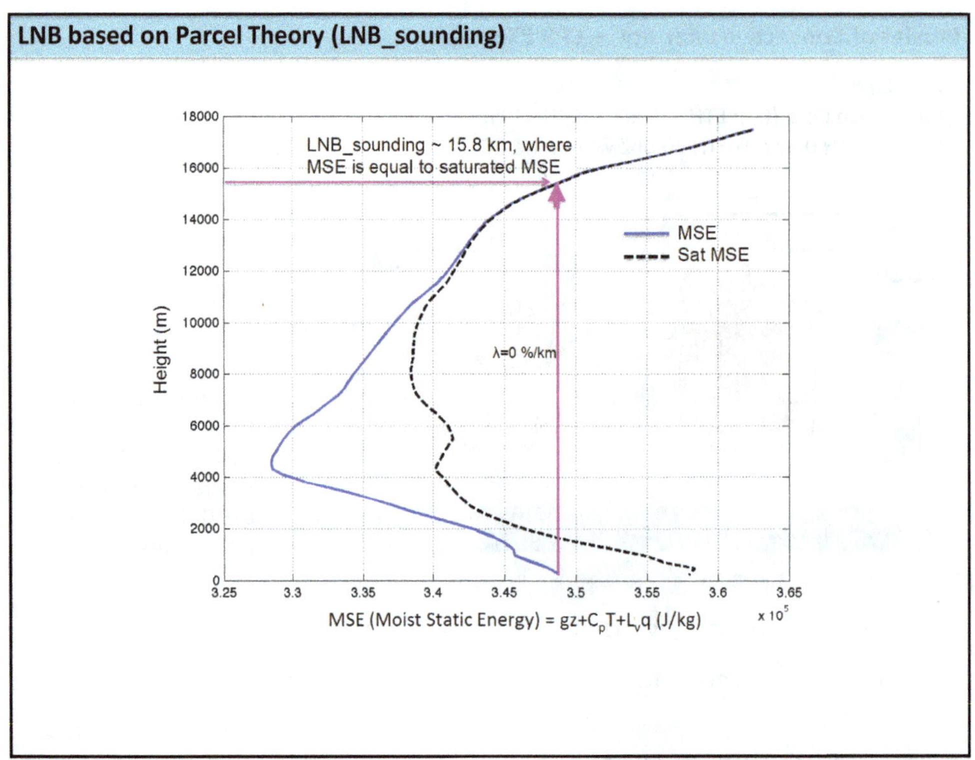

Slide 10

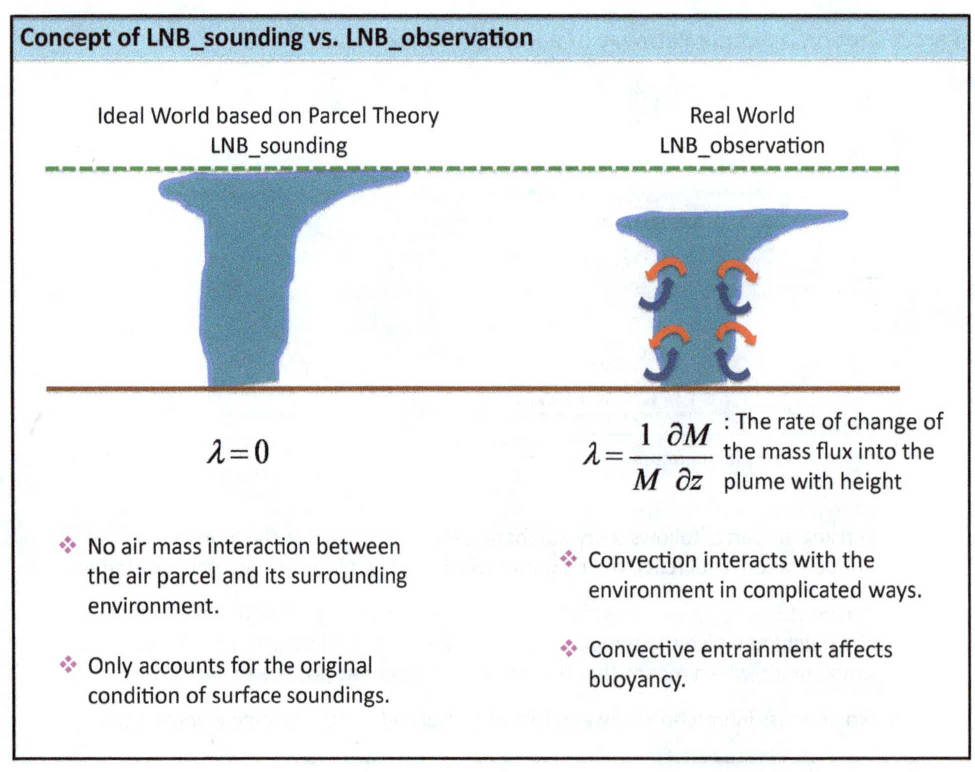

Concept of LNB_sounding vs. LNB_observation
Ideal World based on Parcel Theory
LNB_sounding
Real World
LNB_observation
$\lambda = 0$
$\lambda = \frac{1}{M}\frac{\partial M}{\partial z}$
: The rate of change of the mass flux into the plume with height
No air mass interaction between the air parcel and its surrounding environment.
Only accounts for the original condition of surface soundings.
Convection interacts with the environment in complicated ways.
Convective entrainment affects buoyancy.

Slide 11

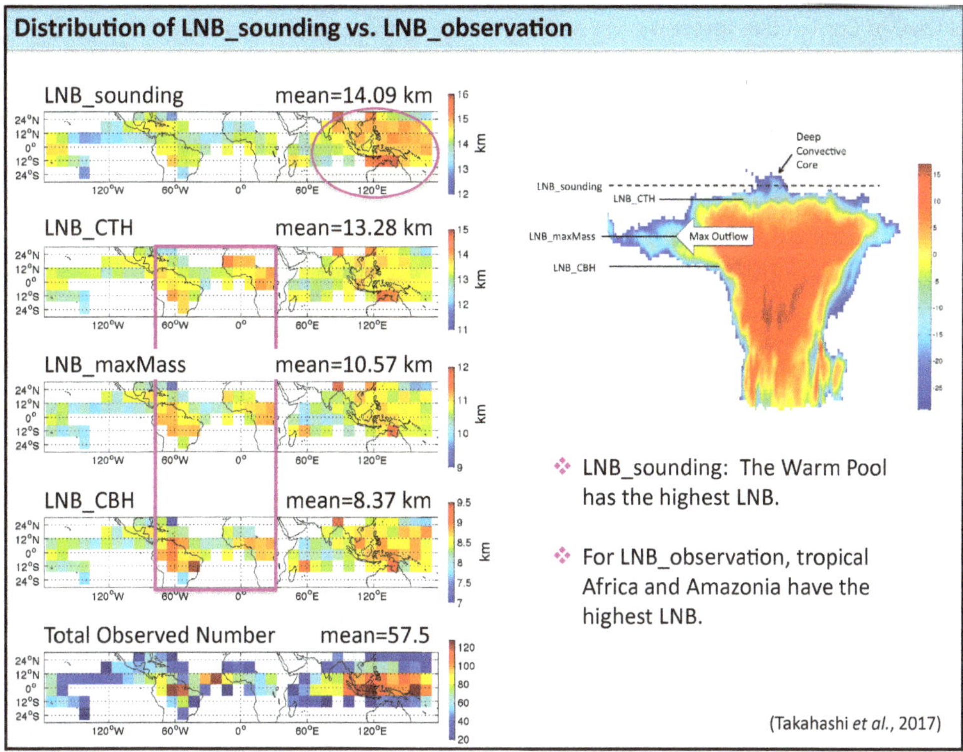

Slide 12

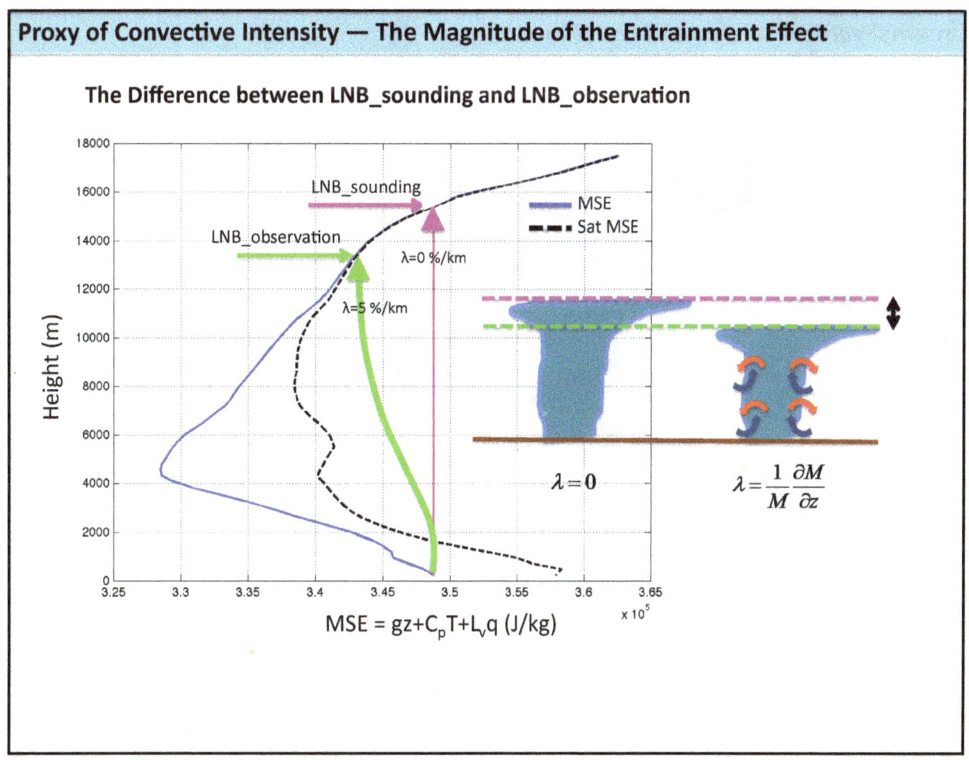

Slide 13

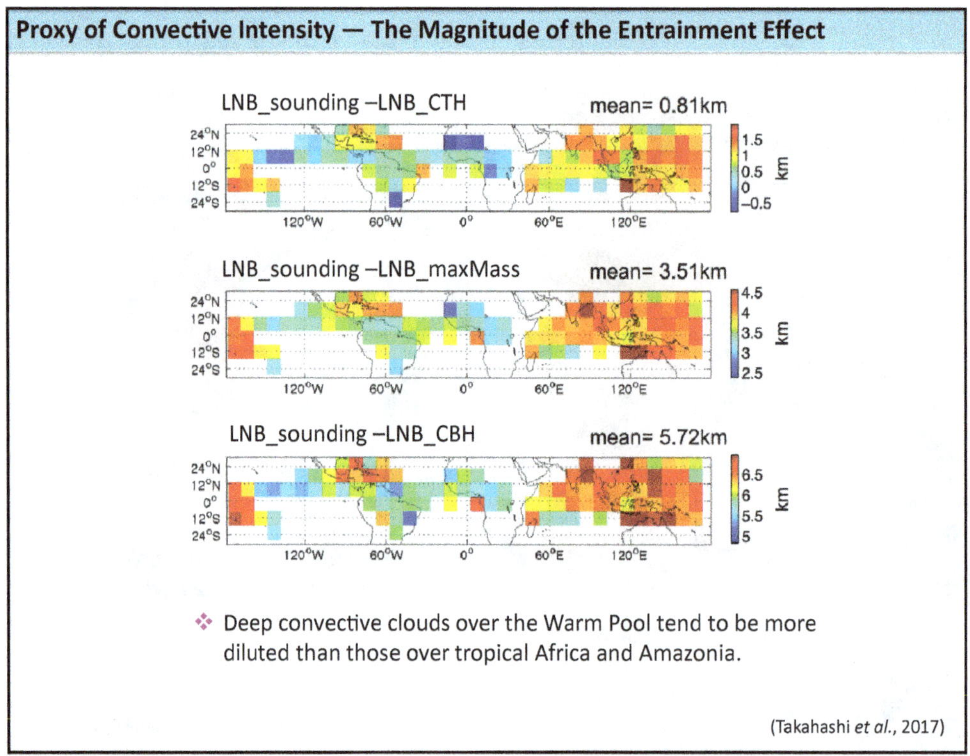

Proxy of Convective Intensity — The Magnitude of the Entrainment Effect
LNB_sounding –LNB_CTH
mean= 0.81km
LNB_sounding –LNB_maxMass
mean= 3.51km
LNB_sounding –LNB_CBH
mean= 5.72km
Deep convective clouds over the Warm Pool tend to be more diluted than those over tropical Africa and Amazonia.
Attenuation due to heavy precipitation is more severe in tropical land than in the Warm Pool.
(Takahashi et al., 2017)

Slide 14

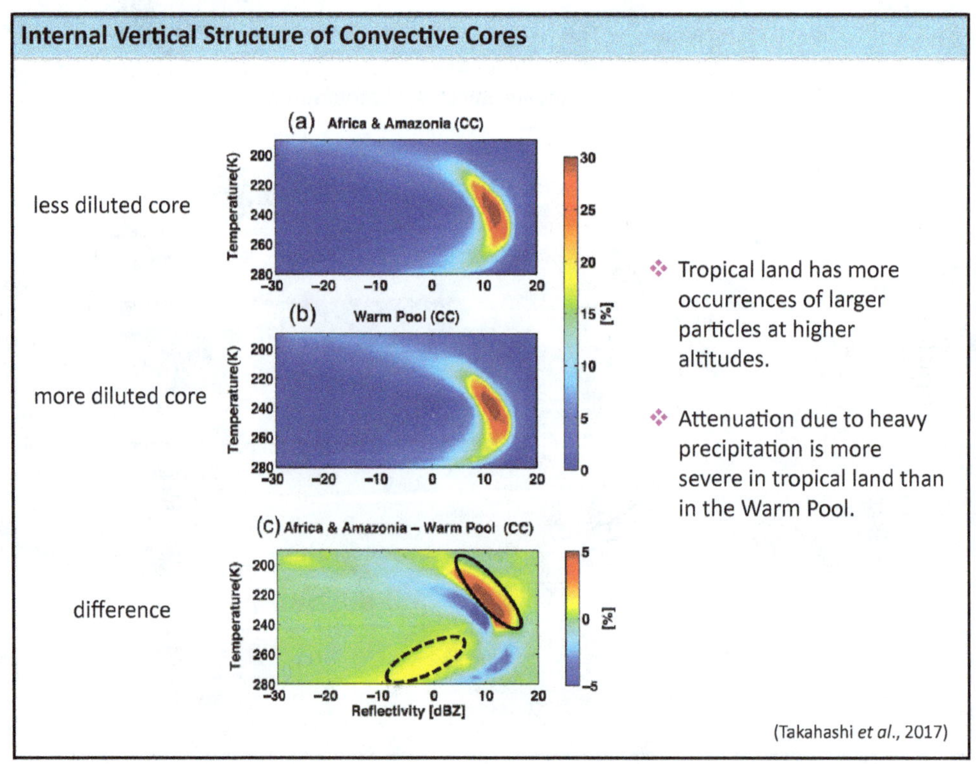

Temperature(K)

Slide 15

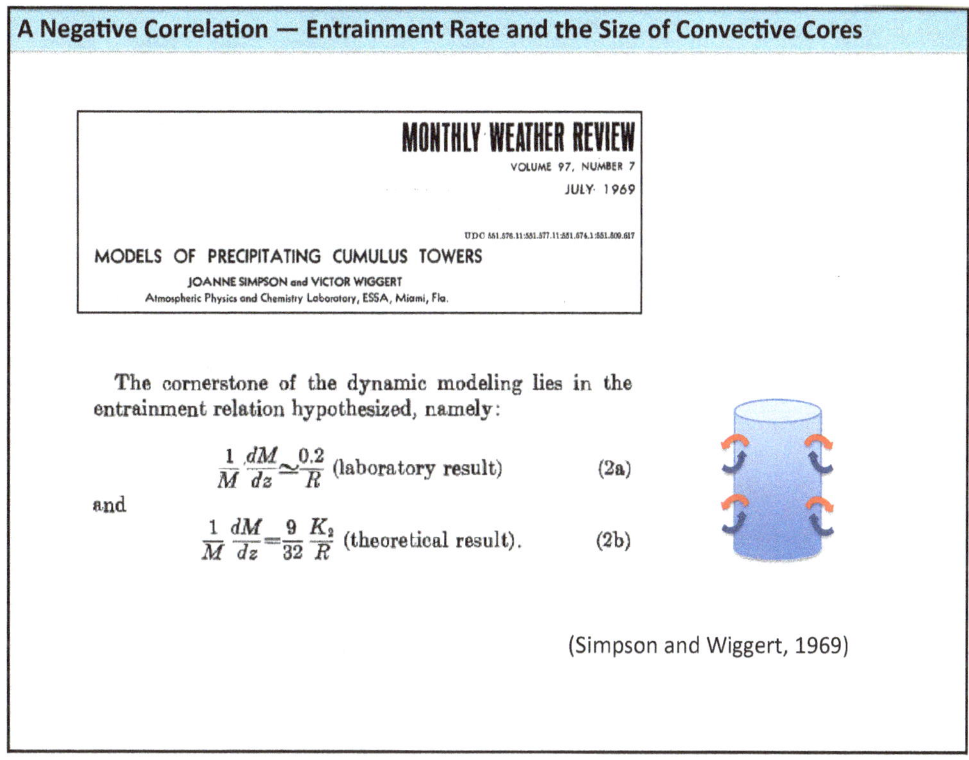

A Negative Correlation — Entrainment Rate and the Size of Convective Cores
MONTHLY WEATHER REVIEW
VOLUME 97, NUMBER 7
JULY 1969
MODELS OF PRECIPITATING CUMULUS TOWERS
JOANNE SIMPSON and VICTOR WIGGERT
Atmospheric Physics and Chemistry Laboratory, ESSA, Miami, Fla.
The cornerstone of the dynamic modeling lies in the entrainment relation hypothesized, namely:
$\frac{1}{M}\frac{dM}{dz}\simeq\frac{0.2}{R}$ (laboratory result) (2a)
and
$\frac{1}{M}\frac{dM}{dz}=\frac{9}{32}\frac{K_2}{R}$ (theoretical result). (2b)
(Simpson and Wiggert, 1969)

Slide 16

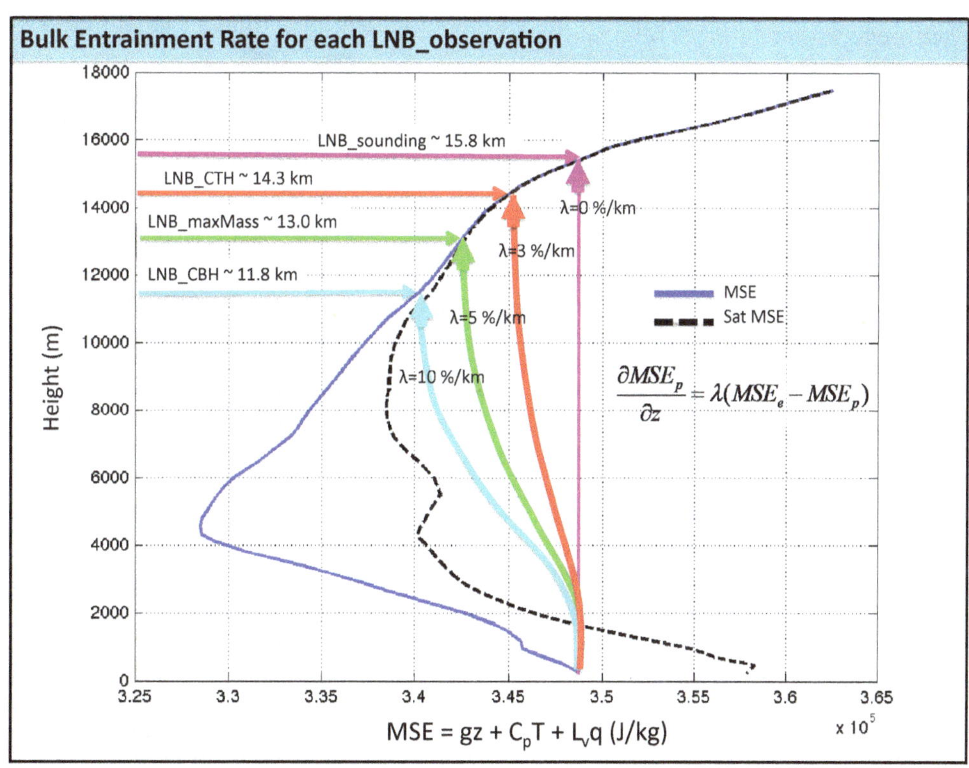

Slide 17

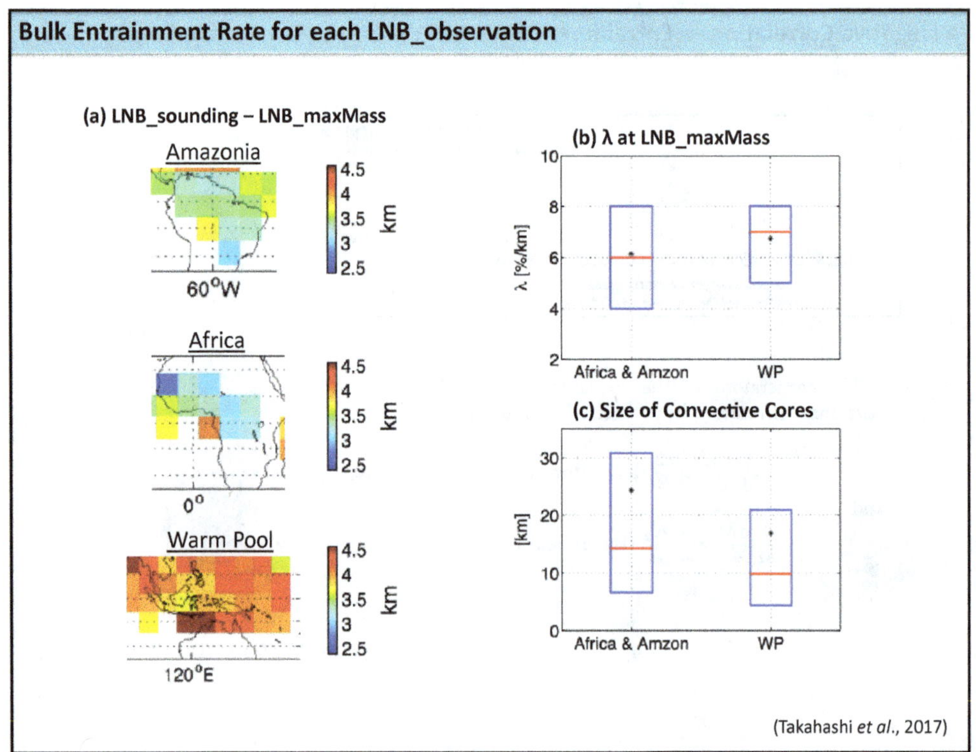
Bulk Entrainment Rate for each LNB_observation
(a) LNB_sounding – LNB_maxMass
Amazonia
60°W
Africa
0°
Warm Pool
120°E
4.5
4
3.5
3
2.5
km
(b) λ at LNB_maxMass
λ [%/km]
Africa & Amzon
WP
(c) Size of Convective Cores
[km]
Africa & Amzon
WP
(Takahashi et al., 2017)

Slide 18

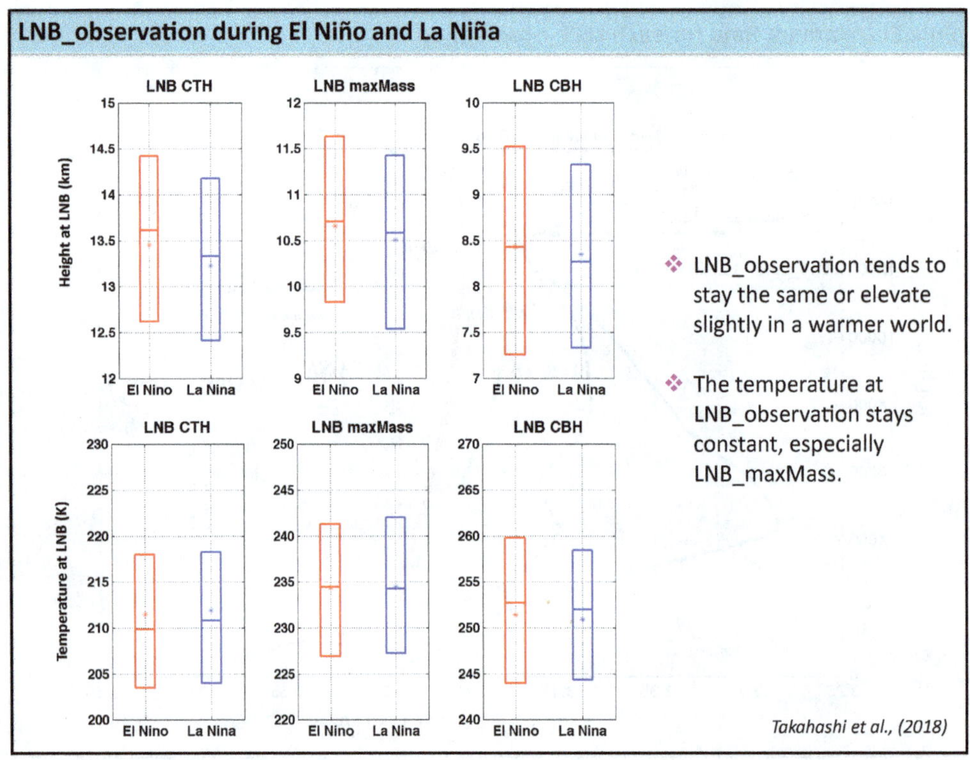
LNB_observation during El Niño and La Niña
LNB CTH
LNB maxMass
LNB CBH
Height at LNB (km)
Temperature at LNB (K)
El Nino
La Nina
LNB_observation tends to stay the same or elevate slightly in a warmer world.
The temperature at LNB_observation stays constant, especially LNB_maxMass.
Takahashi et al., (2018)

Slide 19

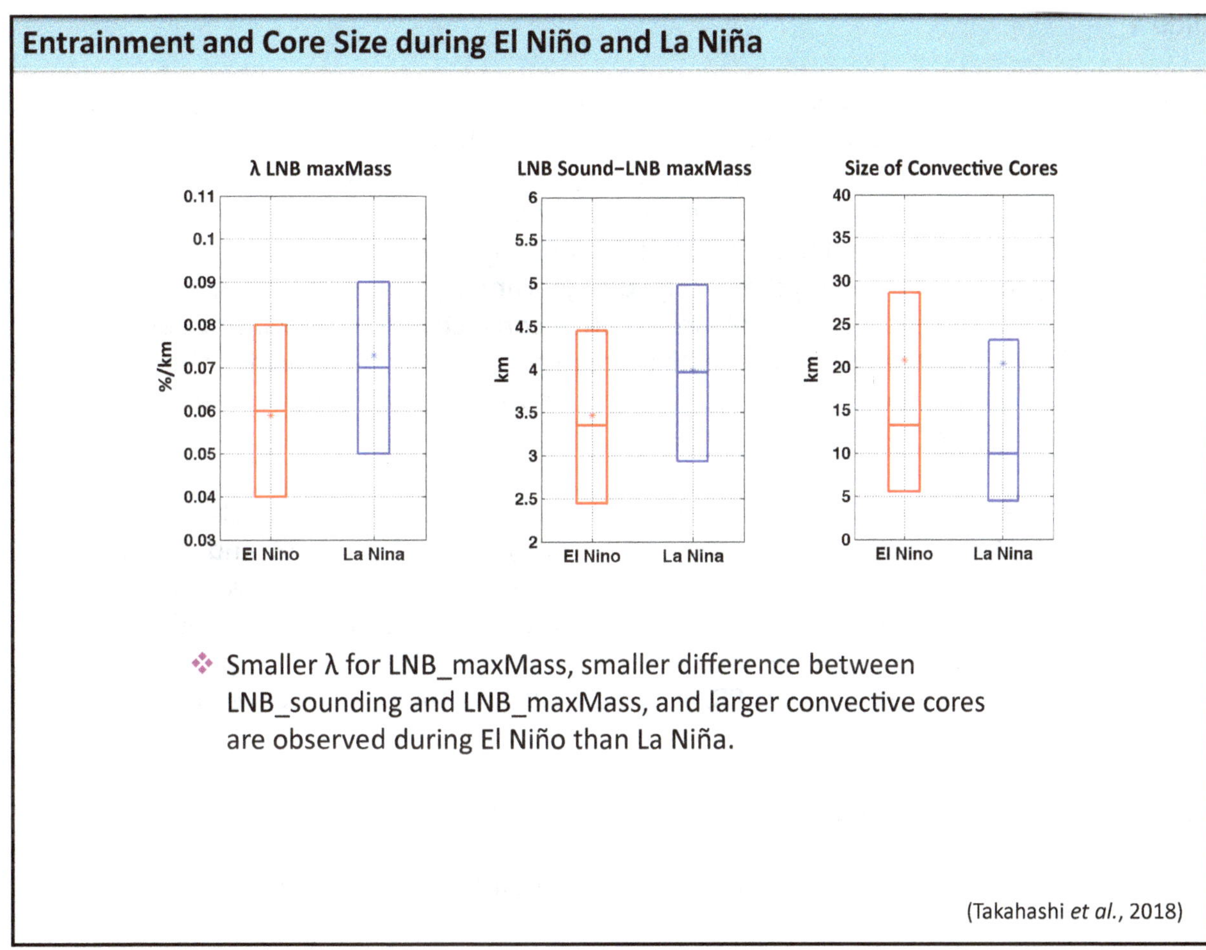
Entrainment and Core Size during El Niño and La Niña
λ LNB maxMass
%/km
El Nino
La Nina
LNB Sound-LNB maxMass
km
El Nino
La Nina
Size of Convective Cores
km
El Nino
La Nina
Smaller λ for LNB_maxMass, smaller difference between LNB_sounding and LNB_maxMass, and larger convective cores are observed during El Niño than La Niña.
(Takahashi et al., 2018)

Slide Captions

Slide 1

Slide 2 Schematic images of *two* key convective processes over stronger convective updrafts (left) and weaker convective updrafts (right): *vertical mass transport* by convective updraft cores that interact with entrainment and its transition to *horizontal mass transport* by deep convective outflows at a level of neutral buoyancy where convective detrainment occurs and anvils forms. Vertical transport plays a key role in controlling the energy budgets and hydrological cycles, and horizontal transport plays a critical role in radiative-convective feedbacks. Since entrainment dilutes the strength of convective updrafts and reduces buoyancy, the intensity of vertical transports by convective cores and the convective detrainment levels are strongly affected by entrainment.

Slide 3 A deep convective cloud off the coast of tropical Africa was observed by both TRMM PR (left) and CloudSat CPR (right). Both TRMM PR and CloudSat CPR have profiling capabilities. However, CloudSat CPR produces a more complete sampling of deep convective clouds since CPR is sensitive to both cloud and precipitation-size particles while PR is only sensitive to large precipitation-size particles. Based only on PR, it is hard to find deep convective cores and anvil outflows.

Slide 4 An example of a selected deep convective cloud object based on CloudSat CPR that was observed on 24 February 2007, over tropical Amazonia. The size of the system is about 140 km, and the highest point is about 17 km.

Slide 5 An example of three forms of LNB_observation (LNB_CTH, LNB_maxMass, and LNB_CBH) based on the same image in Slide 4. LNB_CTH is the average of anvil tops, and LNB_CBH is the average of anvil bottoms. LNB_CTH and LNB_CBH represent the destiny of parcels that are relatively less and more diluted during their ascent, respectively. It is also possible that parcels that reach LNB_CTH and LNB_CBH are the ones that have the larger and the lower moist static energy in the planetary boundary layer, respectively. Between an LNB_CTH and LNB_CBH, there is a level where the average maximum radar reflectivity is detected, which is called LNB_maxMass. LNB_maxMass represents the maximum mass detrainment level (Mullendore *et al.*, 2009). For all the three forms of LNB_observation, estimation is made profile-by-profile first, then averaged over the first 20 km of horizontal anvils on each side of the convective core in order to minimize random noise and the effect of ice sedimentation.

Slide 6 Median values (standard deviations, STDs) of LNB_CTH, LNB_maxMass, and LNB_CBH, together with the convective system size and the size of convective cores over topical ocean and land. Some land-ocean differences are worth pointing out. All three forms of LNB_observation are higher over land than the ocean; the median values for land (ocean) LNB_CTH, LNB_maxMass, and LNB_CBH are, respectively,

13.76 km (13.29 km), 11.14 km (10.55 km), and 8.78 km (8.27 km). The Student's *t*-test confirms that all the differences are statistically significant at a 0.95 confidence level. This land-ocean difference can be explained by the land-ocean difference in convective intensity, as illustrated in Slide 2. Stronger convective updrafts in land convection may bring the ascending air parcels to a higher altitude and detrain mass at a higher level. Moreover, oceanic convective systems are bigger (the horizontal span of the CloudSat profiles including both anvils and cores), but the embedded convective cores are narrower than continental convective systems: the median size of the land convective systems is 140 km and that of the ocean is 167 km, while the median size of land convective cores is 14 km and that of the ocean is 10 km. Both differences are again statistically significant at a 0.95 confidence level.

Slide 7 Median values (STDs) of ETH0dBZ, ETH10dBZ, CTETD0dBZ, and CTETD10dBZ over topical ocean and land. Convective cores embedded in continental convective systems have higher ETH and shorter CTETD than those in oceanic convective systems; the median values for land (ocean) ETH0dBZ, ETH10dBZ, CTETD0dBZ, and CTETD10dBZ are, respectively, 12.20 km (11.89 km), 10.16 km (9.62 km), 2.16 km (2.40 km), and 4.32 km (4.80 km). These land-ocean differences indicate that continental convective systems tend to have stronger updraft cores than oceanic convective systems, which is consistent with the results in Slide 6.

Slide 8 A schematic illustration to explain the parcel theory, a simple pathway of a rising air parcel. It shows which path — dry adiabatic or moist adiabatic path — an air parcel follows during its ascent, when it becomes saturated (i.e., LCL), and when it stops rising and forms anvils (i.e., LNB). It is important to point out that a rising air parcel based on the parcel theory is an ideal case of an air parcel that does not have the effect of entrainment dilution since there are no air mass interactions between the air parcel and its surrounding environment.

Slide 9 Since the parcel theory does not allow convection to interact with the environment, the final destination of an air parcel is defined only by the originating moist static energy (MSE) in the planetary boundary layer, which can be estimated from sounding datasets. This is a demonstration of how to estimate LNB_sounding for a deep convective cloud object. MSE and saturated MSE are estimated based on ambient soundings from the ECMWF-AUX product containing temperature and moisture profiles from the European Centre for Medium-Range Weather Forecast (ECMWF) operational analysis interpolated in space and time to the CloudSat track. In the parcel theory, since there is no air mass interaction between the air parcel and its surrounding environment, the surface MSE is conserved during the ascent. Therefore, LNB_sounding can be defined as the level where MSE is equal to the surface MSE or is equal to saturated MSE. In this case, MSE is equal to the surface MSE (~3.48×10^5 J/kg) or saturated MSE at the level of ~15.8 km and, thus, this is the level of LNB_sounding.

Slide 10 As demonstrated in Slides 8 and 9, LNB_sounding is the theoretical LNB, which is the highest possible level that an air parcel can ever reach. However, in reality, convection interacts with the environment in complicated ways, and convective entrainment weakens buoyancy. Therefore, the LNB cannot reach as high as LNB_sounding. LNB_observation is lower than LNB_sounding as a result of entrainment dilution.

Slide 11 Tropical (30°S to 30°N) distribution of LNB_sounding, LNB_observation, and sample size of cloud objects in 10°×10° boxes. To ensure enough statistics, the sample size in those 10°×10° boxes with less than 20 cloud objects are left blank. The most interesting difference between LNB_sounding and LNB_observation is the different ranking of the highest to lowest LNB over different regions. For LNB_sounding, the Warm Pool (14.9 km) is the highest, while LNB_CTH is the highest over tropical Africa (13.9 km), followed by Amazonia (13.7 km). LNB_maxMass also shows the highest values in tropical Africa (11.3 km) and Amazonia (11.1 km). A cloud object that has a higher LNB_sounding than other cloud objects does not mean that the cloud object also has a higher LNB_observation, which suggests LNB_sounding has little direct correlation with LNB_observation. Indeed, the correlation between LNB_sounding and LNB_CTH is 0.29 (0.30 over ocean and 0.28 over land), between LNB_sounding and LNB_maxMass is 0.28 (0.30 over ocean and 0.25 over land), and between LNB_sounding and LNB_CBH is 0.20 (0.21 over ocean and 0.16 over land), which is demonstrated in (Takahashi and Luo, 2012). This points to the limited value of using LNB_sounding to predict LNB_observation.

Slide 12 As illustrated in Slide 10, LNB_sounding is the ideal case of LNB with no entrainment dilution, and LNB_observation is the actual case of LNB with some entrainment dilution (e.g., λ = 5%/km). The vertical transports by undiluted (pink) and diluted (green) convective cores whose convective outflows are, respectively, associated with LNB_sounding and LNB_observation are illustrated. The undiluted convective core transports surface MSE up to LNB_sounding with no entrainment dilution (e.g., λ = 0%/km), and the diluted convective core transports surface MSE up to LNB_observation with some entrainment dilution (e.g., λ = 5%/km). Since LNB_sounding and LNB_observation have little one-to-one correlation, as seen in Slide 11, the difference between LNB_sounding and LNB_observation can be interpreted as a measure of the magnitude of the entrainment effects — the greater the entrainment dilution, the larger the height difference.

Slide 13 The mean difference between LNB_sounding and LNB_observation in 10° × 10° boxes over the whole tropics (30°S to 30°N). The sample size in those 10° × 10° boxes with less than 20 cloud objects are again left blank. A larger difference between LNB_sounding and LNB_observation is observed over the Warm Pool, and a smaller difference is seen over tropical Africa and Amazonia. Results suggest that deep convection over the tropical ocean (e.g., Warm Pool) tends to have larger entrainment dilution and thus has weaker convective cores than that over tropical land counterparts (e.g., tropical Africa and Amazonia).

Slide 14 Contoured frequency by temperature diagrams (CFTDs) for deep convective cores over (a) tropical land (Africa and Amazonia), (b) the Warm Pool, and (c) the difference between them (i.e., (a) and (b)). In CFTDs, the probability density function of the radar reflectivity is normalized at each bin of temperature. Results show that continental convective cores have more occurrences of larger radar echoes (reflectivities > 10 dBZ) at higher altitudes, as highlighted by the black solid oval. This indicates that continental convection has stronger cores that can transport larger particles to a higher level compared to oceanic convection, suggesting the capability of stronger vertical transports and higher horizontal transports by continental convective cores than that by oceanic convective cores. Moreover, radar attenuation due to intense precipitation (i.e., decreasing reflectivities below the melting layer) is more evident over the continental convection than oceanic convection, which is highlighted by the black dashed oval. Continental convection produces heavier rainfall than oceanic convection, which is also additional evidence of stronger continental convection than oceanic convection.

Slide 15 The fractional entrainment rate per unit height was documented in (Simpson and Wiggert, 1969) as $\frac{1}{M}\frac{dM}{dz} = \frac{9}{32}\frac{K}{R}$, where M was the mass in the rising tower, K was the entrainment coefficient, and R was the radius of the cumulus tower (i.e., convective core). Based on their theoretical results, K varied between 0.55 and 0.65, and laboratory results showed that K was 0.71 (i.e., $\frac{1}{M}\frac{dM}{dz} \cong \frac{0.2}{R}$). This negative core-entrainment relation was supported by several airborne measurements in the 1960s and has been a longstanding assumption in cumulus cloud modeling. The physical explanation is that larger convective cores are better protected from the environment and thus less diluted by entrainment. However, this relation has never been systematically explored with global satellite observations.

Slide 16 The corresponding bulk entrainment rates (λ; unit: %/km) for LNB_observation can be estimated as $\frac{\partial MSE_p}{\partial z} = \lambda(\partial MSE_e - \partial MSE_p)$, where $MSE = C_pT + gz + L_vq$ (T, z, and q, are temperature, height, and specific humidity, respectively, Cp, g, and Lv are the specific heat of dry air, gravitational acceleration, and the latent heat of condensing, respectively, and subscripts p and e refer to properties of the in-cloud air parcel and environment, respectively. As demonstrated in Slides 9 and 12, MSE is estimated based on ambient soundings from the ECMWF-AUX product. Here, different λ (0 to 10%/km) are used to demonstrate how convective MSE decreases during vertical transport due to entrainment dilution. For example, when λ = 3%/km, a convective core transports MSE from a surface (3.48×10^5 J/kg) up to LNB_CTH of 14.3 km (3.45×10^5 J/kg), where convective MSE (red line) is similar to environmental MSE (blue). Therefore, LNB_observation can be used to find its corresponding λ: LNB_maxMass corresponds to λ = 5%/km (green path), and LNB_CBH corresponds to λ = 10%/km (cyan path).

Slide 17 Results compare (a) the difference between LNB_sounding and LNB_maxMass, (b) λ at LNB_maxMass, and (c) the size of convective cores over tropical land

(Africa and Amazonia) and ocean (Warm Pool). For (b) and (c), the bottom and top of the blue boxes show, respectively, the 25 and 75% percentile, the central lines show the median, and the stars inside the box show the mean. It is clear that the effect of dilution (the difference between LNB_sounding and LNB_maxMass) is smaller, λ for LNB_maxMass is smaller, and the size of the convective core is larger over tropical land than the Warm Pool. In addition (see Slide 11), it is known that LNB_maxMass is higher over tropical land than the Warm Pool. Therefore, results indicate that larger convective cores are associated with smaller entrainment effects and higher convective detrainment levels.

Slide 18 Box-scatter plots of LNB_observation height (top panels) and temperature (bottom panels) during El Niño (red) and La Niña (blue). The bottom and top of the blue boxes show, respectively, the 25 and 75% percentile, the central lines show the median, and the stars inside the box show the mean. LNB_observation is more likely to stay the same or elevate slightly in a warmer world, but the temperature at LNB_observation stays constant, which supports the Fixed Anvil Temperature (FAT) hypothesis (Hartmann and Larson, 2002). The Student's *t*-test suggests that these results are statistically significant at a 0.95 confidence level. The explanation of the FAT hypothesis is based on the heat balance at the upper troposphere. The radiative cooling over clear-sky regions and convective heating by latent heat release over deep convective regions are roughly balanced in the tropical troposphere. Therefore, to maintain the heat balance at the upper troposphere, convective outflows in the tropics must occur at the altitudes where the largest radiative cooling happens (e.g., around 200 hPa). Since radiative emission from water vapor becomes significantly low above 200 hPa and the temperature at convective outflow levels is only controlled by the Clausius Clapeyron relation through the clear-sky radiative cooling rate, the temperature at convective outflow levels should not be affected by the surface temperature and thus it would stay constant during climate changes. This FAT hypothesis is most evident at LNB_maxMass, possibly because this level is most relevant to convective mass outflow and convective transport.

Slide 19 Box-scatter plots of λ at LNB_maxMass, the difference between LNB_sounding and LNB_maxMass, and the size of convective cores during El Niño (red) and La Niña (blue). Again, the bottom and top of the blue boxes show, respectively, the 25 and 75% percentile, the central lines show the median, and the stars inside the box show the mean. The difference between El Niño and La Niña is clear — smaller λ for LNB_maxMass, a smaller difference between LNB_maxMass and LNB_observation, and larger convective cores are observed during El Niño than La Niña. The student's *t*-test suggests that these differences are statistically significant at a 0.95 confidence level. Results reveal that convective cores during La Niña are weaker and narrower since they experience more intense convective dilution than those during El Niño. The implication is that convection tends to be stronger in a warmer world.

References

Hartmann DL, Larson K. (2002) An important constraint on tropical cloud–climate feedback. *Geophysical Research Letters* **29**:1951. doi:10.1029/2002GL015835

Luo ZJ, Anderson RC, Rossow WB, Takahashi H. (2017) Tropical cloud and precipitation regimes as seen from near-simultaneous TRMM, CloudSat, and CALIPSO observations and comparison with ISCCP. *Journal of Geophysical Research: Atmospheres* **122**(11):5988–6003. doi:10.1002/2017JD026569

Mullendore GL, Homann AJ, Bevers K, Schumacher C. (2009) Radar reflectivity as a proxy for convective mass transport. *Journal of Geophysical Research* **114**(D16103). doi:10.1029/2008JD011431

Simpson J, Wiggert V. (1969) Models of precipitating cumulus towers. *Monthly Weather Review* **97**(7):471–89.

Takahashi H, Luo Z. (2012) Where is the level of neutral buoyancy for deep convection? *Geophysical Research Letters* **39**(15). doi:10.1029/2012GL052638

Takahashi H, Luo ZJ. (2014) Characterizing tropical overshooting deep convection from joint analysis of CloudSat and geostationary satellite observations. *Journal of Geophysical Research: Atmospheres* **119**(1):112–21. doi:10.1002/2013JD020972

Takahashi H, Luo ZJ, Stephens GL. (2017) Level of neutral buoyancy, deep convective outflow, and convective core: New perspectives based on 5 years of CloudSat data. *Journal of Geophysical Research: Atmospheres* **122**(5):2958–69. doi:10.1002/2016JD025969

Takahashi H, Luo ZJ, Stephens GL. (2018) Variations of deep convective outflow, entrainment rates and convective cores during El Niño and La Niña: A CloudSat Perspective. *To be submitted.*

Lecture 4

Cloud Life Cycle and Space-Time Organization

Luiz A. T. Machado

Multiphase Chemistry Department, Max Planck Institute for Chemistry, Mainz, Germany
Instituto de Física, Universidade de São Paulo, São Paulo, Brazil

Luiz Machado has a BSc (1981) and Masters (1984) in Meteorology from the University of São Paulo (1981), Diplôme d' Etudes Approfondies Oceanologie Meteorologie (1989) and PhD in Meteorology and Oceanography from the University of Paris VI (1992). Today, he is a senior researcher at São Paulo University, Visiting Researcher at Max Planck Institute for Chemistry and Professor of the PhD Program at Instituto Nacional de Pesquisas Espaciais. He has also served as Head of Satellite Division, Centro de Previsão do Tempo e Estudos Climáticos and acted as its Coordinator from 2008 to 2010.

Introduction

Clouds are organized in a large range of space-time scales. Since clouds vary in size from a warm cumulus cloud of a few meters to cloud organizations of thousands of kilometers, such as supercloud clusters observed in the Pacific by Nakasawa (1988). Each of these scales is associated with a specific dynamic and directly interacts with the radiative, water and energy balance. Some interesting characteristics are observed in how these range of scales clouds are organized: (1) The effective radius of a cloud cluster or a rain cell (cell obtained using radar) is nearly linearly related to its lifetime duration, (2) Cloud life cycle can be classified in the following phases: initiation, mature and dissipation, (3) Clouds in the initiation–mature phase have the largest area expansion, rainfall and lightning occurrence. In the dissipation phase, convection is reduced, but the cloud deck continues to increase in size until it dissipates. There is a typical reflectivity profile in each component of the cloud cluster. The convective updrafts is the driver to develop the ice hydrometer above the melting layer, the ice water path is related to the rain rate and lightning. In the dissipation phase, at the melting layer, ice melts to generate a specific signature in radar called a bright band.

In general, when all synoptic patterns are analyzed together, the cloud cluster's size distribution can be approximated by a power law of the radius with an exponent of around $N = A.r^{-2}$. This gives an almost equal contribution of each cloud cluster size to the total cloud cover (for the specific threshold employed). Clouds are organized such that cloud cover contribution is independent of size; each size range gives nearly the same contribution to the total cloud cover. However, fewer large cloud clusters give the largest contribution to the amount of rainfall and lighting. This size distribution is observed up to the breaking radius, a cloud cluster size when the probability is very low, and only few cloud organizations are found. The interannual variability of cloud cluster size distribution is very small (though it is very stable); this includes the breaking radius, even among years with very different cloud cover and rainfall patterns, such as in 1983 and 1988 in Sahel Africa.

Reproducing cloud organization in numerical atmospheric models is still a challenge, mainly in the upscale and convective initiation. The knowledge about cloud clusters organization provides a benchmark for models. Machado and Charboreau (2015), using a cloud-resolving model and satellite-radar data, showed that the model creates too many small clouds compared to the observation. Changes in turbulence parameterization indicate that cloud entrainment should be improved to fix the cloud size distribution. There are several studies trying to understand this and new parametrizations are being proposed to improve the cloud size distribution.

The cloud size distribution varies with the synoptic pattern as the diurnal cycle. A specific synoptic pattern has a specific cloud space-time scale, or the diurnal march of the convection modulates the cloud size distribution. This modulation is composed of the cloud life cycle. When a cloud is in the initiation process, the cloud area expansion (or the increase in the size of the cloud) is mainly due to the condensation process, followed by the upper-level wind divergence as a response to the mass flux. The maximum upper-level wind divergence occurs just before the maturation phase (see Machado and Laurent, 2004 for a detailed description).

The area expansion in the initiation phase determines lifetime duration — the larger the area expansion, the longer the life cycle's duration. Normalized area expansion is the response to the mass flux, whose intensity in the initiation increases the duration of the system. The normalized area expansion varies nearly linearly during the life cycle; thus, it is possible to estimate its size in the near future by following the system during its life cycle and knowing the normalized area expansion.

Cloud cluster electrification processes can be characterized by the ice-effective radius on the cloud top, and only systems that have ice larger than 1 mm and a minimum brightness temperature colder than 215 K have cloud-to-ground lightning. The larger the ice and the colder the top is, the higher the number of cloud-to-ground lightning. The ice water path is linearly related to cloud ground lightning. In addition, the larger the cloud cluster is, the greater the frequency of the lightning, with maximum activity occurring just before the mature phase.

A cloud cluster comprises the convective, transition, and anvil cloud deck, each of which has specific radiative properties and a specific radiative effect on the heating-cooling rate profile and net radiative effect. Cloud clusters heat the middle atmosphere and cool the cloud top. The net radiative effect over land or ocean cools the atmosphere (during the day at 1200 LST, it is around 35 to 64 W.m^{-2}).

These are some of the cloud cluster properties and characteristics of cloud organization. However, global circulation models still lack accuracy in describing cloud organization. More efforts should be put into understanding how dynamics and thermodynamics fields can describe organization properties. This presentation intends to introduce cloud cluster organization properties, from its size distribution to its life cycle.

References

Machado LAT, Chaboureau J-P. (2015) Effect of turbulence parameterization on assessment of cloud organization. *Monthly Weather Review* **143**:3246–62.

Machado LAT, Laurent H. (2004) The convective system area expansion over Amazonia and its relationships with convective system life duration and high-level wind divergence. *Monthly Weather Review* **132**:714–25.

Slide 1

Slide 2

Cloud Cluster and Lifetime Observed by Satellite

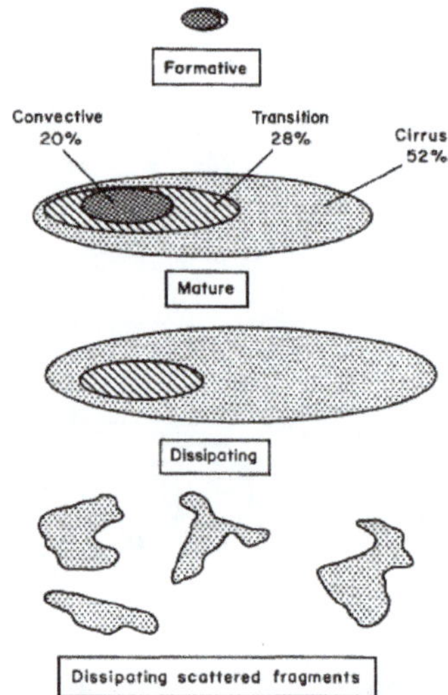

Cloud Clusters can be defined in the range of Brightness Temperature smaller of around 255K, mostly of the studies employ 235K.

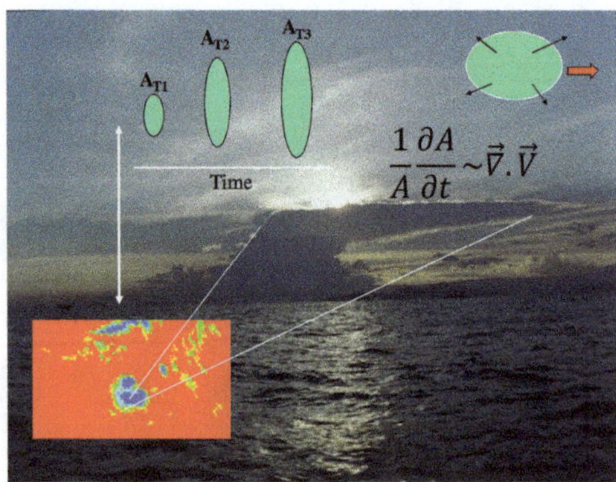

Machado, L. A. T., and W. B. Rossow, 1993: Structural characteristics and radiative properties of tropical cloud clusters. Mon. Wea. Rev., 121, 3234–3260.

Slide 3

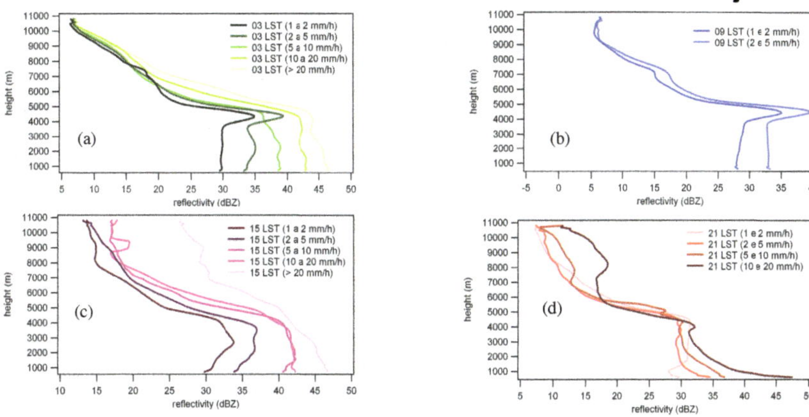

Amazonas Vertical Pointing Radar profiles corresponding to different rainfall intensities and time intervals: (a) 03:00 LST, (b) 09:00 LST, (c) 15:00 LST and (d) 21:00 LST

These reflectivity vertical profiles of Amazonas convective clouds separated by rain rate and local time describe the cloud life cycle. At 15:00 LST convective cloud starts do develop and these are the typical profile of convective cloud initiation (except for the rain rate above 20.0 mm/h). Very few ice formation and maximum liquid water at around the melting layer. The profile with rainfall above 20.0 mm/h or the one later at 21:00 LST (rain rate above 5 mm/h) the ice layer is well developed, this is the mature stage. It is interesting to see the increase of ice content and rainfall, this is the base of passive microwave rainfall estimation over land. At night or early morning the profiles are typical of the stratiform cloud deck in the dissipation phase. The maximum reflectivity pic around the melting layer is a result of the melting ice and is called bright band.

*Martins, R.C.G., Machado, L.A.T., Costa, A.A. Characterization of the microphysics of precipitation over Amazon region using radar and disdrometer data. (2010) Atmospheric Research, **96**(2-3), pp. 388–394.*

Slide 4

The Cloud Size Distribution

Clouds are organized in different scales from few meters to thousands of kilometers. Cloud organization´s have an interesting general feature, they follow a nearly power law distribution with N(r)= $A.r^{-2}$, where r is the effective cloud cluster radii and N(r) is the number of cloud cluster with size r and A is a constant. This size parameterization implies in the nearly same contribution to the cloud cover by all cloud cluster sizes, up to a break size where only rare systems are organize on this scale. One example of this width and rare scale organization are the super cloud clusters found in Pacific as described by Nakazawa (1998).

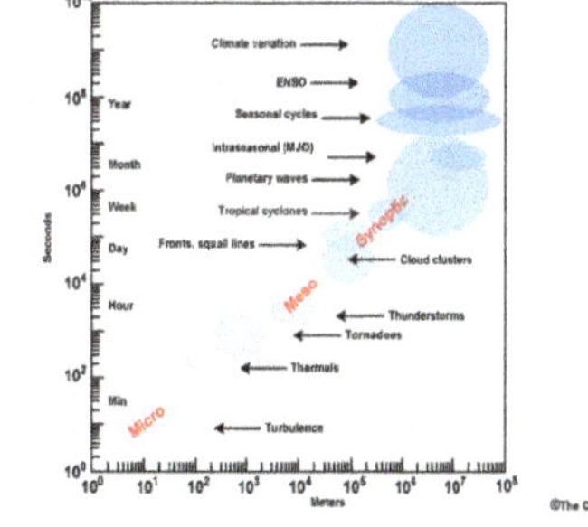

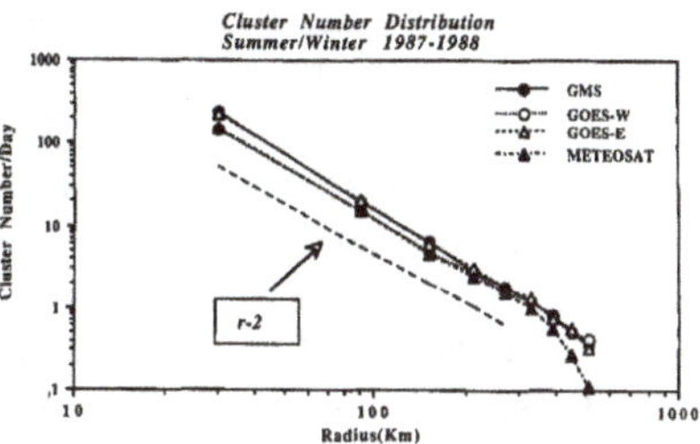

Machado, L. A. T., and W. B. Rossow, 1993: Structural characteristics and radiative properties of tropical cloud clusters. Mon. Wea. Rev., 121, 3234–3260.

Slide 5

Thresholds and the Cloud Cluster Size Distribution

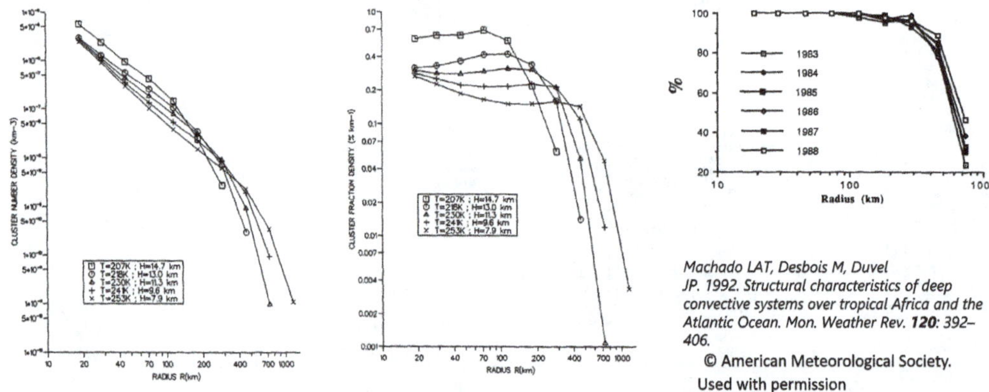

Machado LAT, Desbois M, Duvel JP. 1992. Structural characteristics of deep convective systems over tropical Africa and the Atlantic Ocean. Mon. Weather Rev. ***120****: 392–406.*

Cloud Cluster Number Density and fraction density, number density describes the number of cloud cluster in a given area for each size range and fraction density describes the area fraction covered by each size range. The thresholds change the absolute value, however, its do not change the shape of the distribution. Iti s clear that size distribution follow a specific distribution up to a breaking radius. This breaking radius defines the typical sizes of cloud clusters for a specific threshold. Colder thresholds are associated to smaller cloud organization. The breaking radius is related to the temporal probability of observing cloud clusters with larger radius in a specific area and threshold. This breaking radius is very stable from year to year in despite of large total cloud cover variability and set up the scale of the cloud clusters for a specific threshold, region and season.

Slide 6

The Modulation of the Cloud Size Distribution by Diurnal Cycle

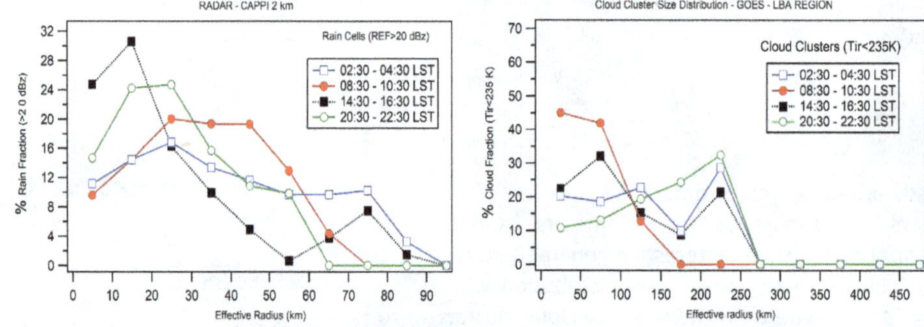

Diurnal cycle change the size distribution from the climatology distribution from the standard N(r)= $A.r^{-2}$. Figures above are for Amazonas collected during TRMM-LBA field campaign. On the left side are the rain cell size distribution, the rainfall at 2 km height and on the right side are the cloud cluster size distribution for a 235K threshold. Iti s clear the increase in size from early morning to the night. Also one can note the differences between cloud and rain. The rainfall in the early morning are a result of clouds of reduced size. From the other side at 2030–2230 LST almost the entire high cloud cover is organized into large convective systems, butt he rain cells are smaller. There is a preferential size organization for cloud and rainfall.

Slide 7

The Modulation of the Cloud Size Distribution by Synoptic Systems

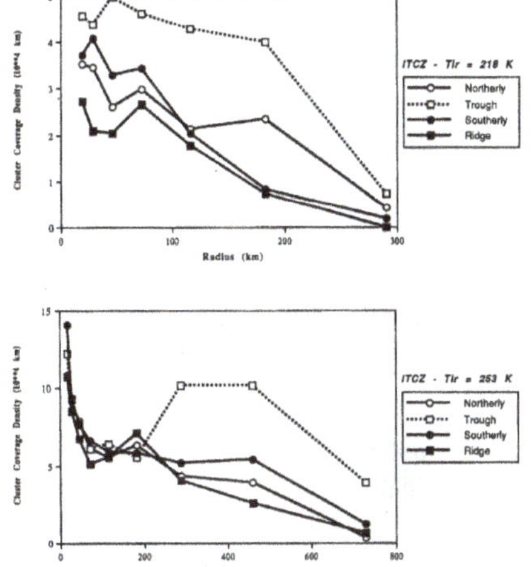

FIG. 10. Easterly wave modulation of the cluster coverage density $S(\Delta R)$ versus cluster radius over the ITCZ region (a) for threshold T_{IR} = 218 K, and (b) for threshold T_{IR} = 253 K.

Easterly waves, in Africa, modulates de cloud cluster number distribution, the system with effective radius around 200–400 km are typical in the trough (system is defined by 253 K – cloud organization in middle to high levels clouds). At colder thresholds (218K), associated to the more intense parts of the cloud cluster easterly waves modulated the number of cluster at all sizes, in the trough there is a larger number of these more convective cells.

The same is observed by looking cold fronts penetration and others synoptic systems.

Machado, L. A. T., J. P. Duvel, and M. Desbois, 1993: Diurnal variations and modulation by easterly waves of the size distribution of convective cloud clusters over West Africa and the Atlantic Ocean. Monthly Weather Review, ***121****,37–49*

Slide 8

Deep Convective System Evolution over Land and Ocean

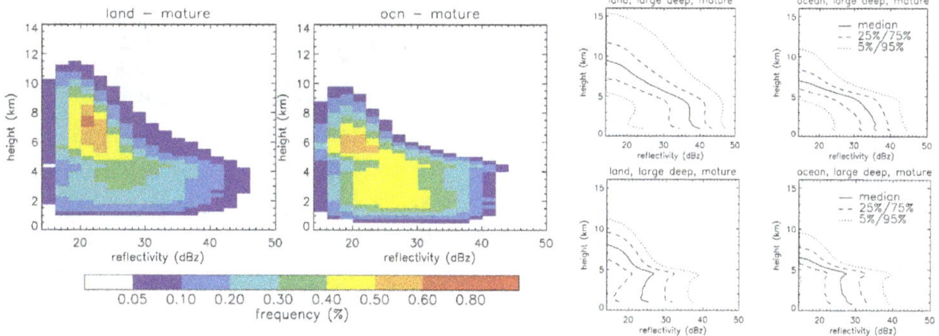

Frequency distribution of the radar reflectivity (left), and median and quartile profiles for Convective and Stratiform (right) regions of mature African and Atlantic convective systems. Reflectivity values measured by a TRMM-PR radar.

African storms evolve from convectively active systems with frequent lightning in their developing stages to more stratiform conditions as they dissipate. Over the Atlantic, the convective fraction remains essentially constant into the dissipating stages, and lightning occurrence peaks late in the life cycle.

Slide 9

The Cloud Lifetime and Size Relationship

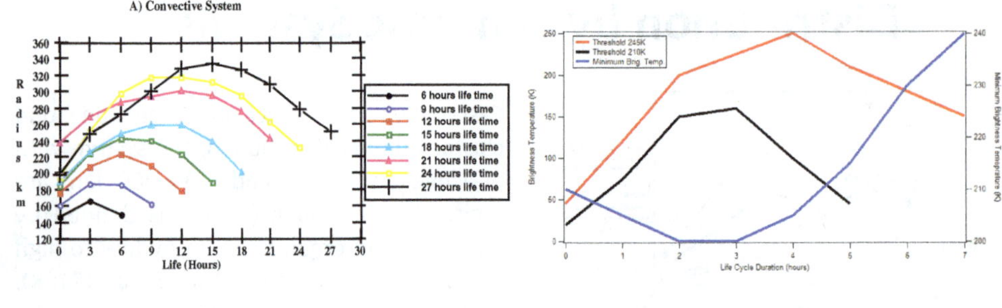

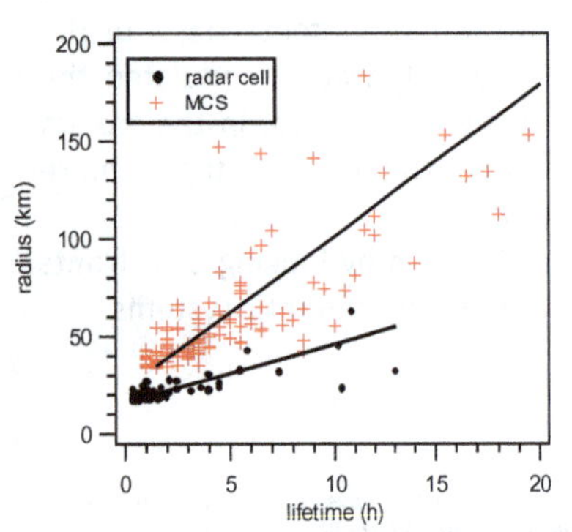

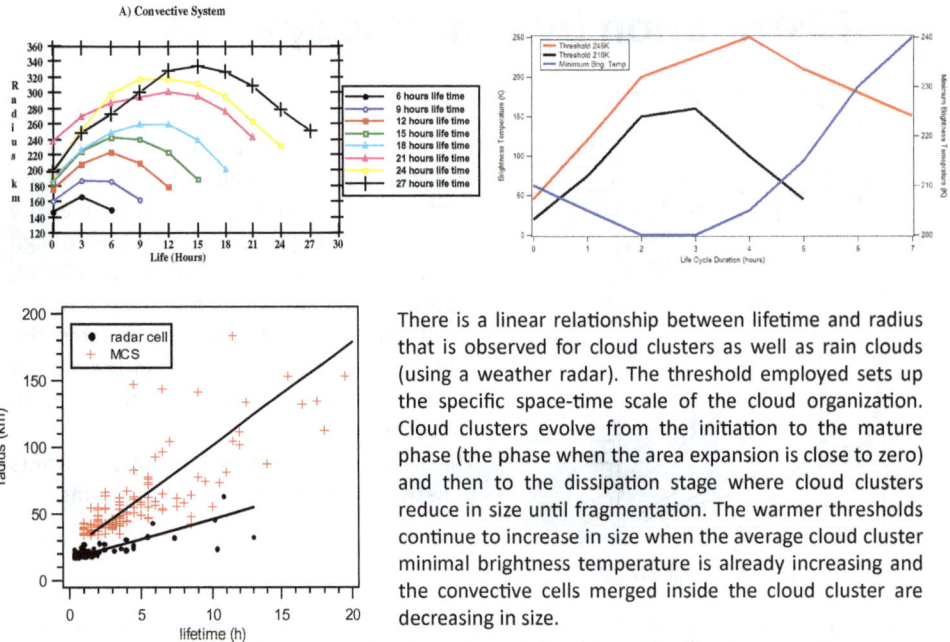

There is a linear relationship between lifetime and radius that is observed for cloud clusters as well as rain clouds (using a weather radar). The threshold employed sets up the specific space-time scale of the cloud organization. Cloud clusters evolve from the initiation to the mature phase (the phase when the area expansion is close to zero) and then to the dissipation stage where cloud clusters reduce in size until fragmentation. The warmer thresholds continue to increase in size when the average cloud cluster minimal brightness temperature is already increasing and the convective cells merged inside the cloud cluster are decreasing in size.

Slide 10

The Cloud Area Expansion

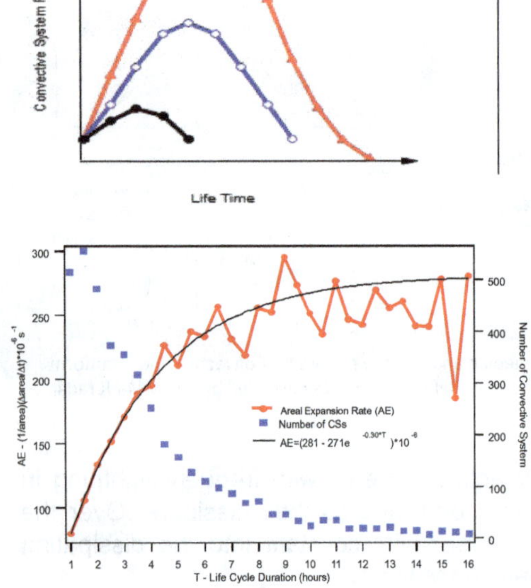

Cloud clusters with larger size stay for longer than smaller one as showed before. Larger cloud clusters show in the initiation phase a larger Normalized Area Expansion (NAE) than system with shorter duration. The NAE in the initiation is related to the cloud cluster lifetime. The NAE is related to the mass flux of the convective process, larger mass flux longer and larger the system will be.

Slide 11

The Cloud Area Expansion

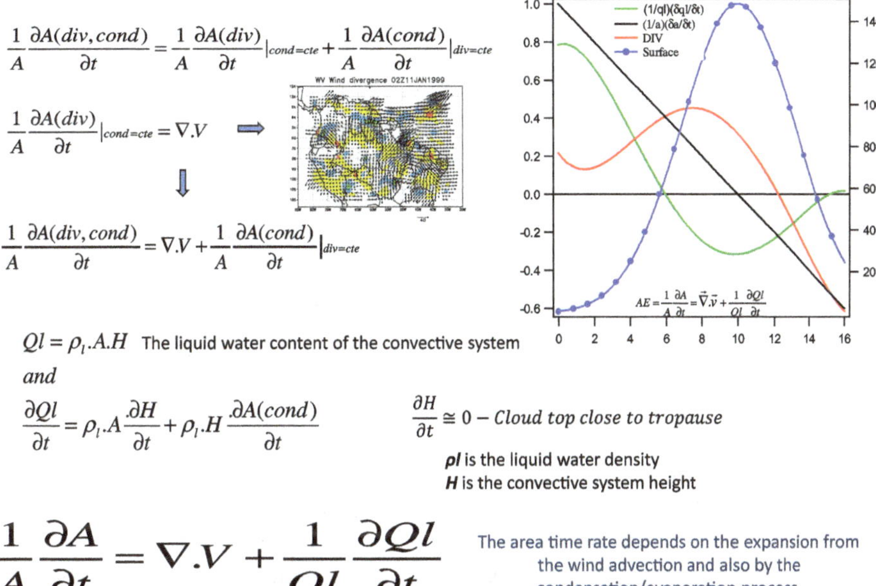

The area time rate depends on the expansion from the wind advection and also by the condensation/evaporation process.

Slide 12

The Cloud Area Expansion

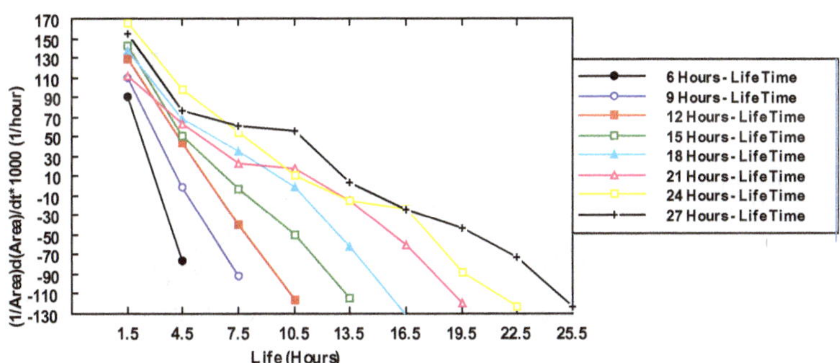

The Normalized Area Expansion change nearly linearly during the Cloud cluster lifecycle. Stronger the NAE in the initiation, longer the system lifetime. The system, in the beginning, expand mainly by the condensation process, the upper-level wind divergence become more and more important as the system evolve to the Maturation. In the dissipation the system reduce area mainly by upper-level wind convergence, evaporation (sublimation) and fragmentation. If the NAE is followed one can roughly forecast the change in size of the cloud cluster.

Slide 13

The Cloud Cluster and Rain Cell Lifetime

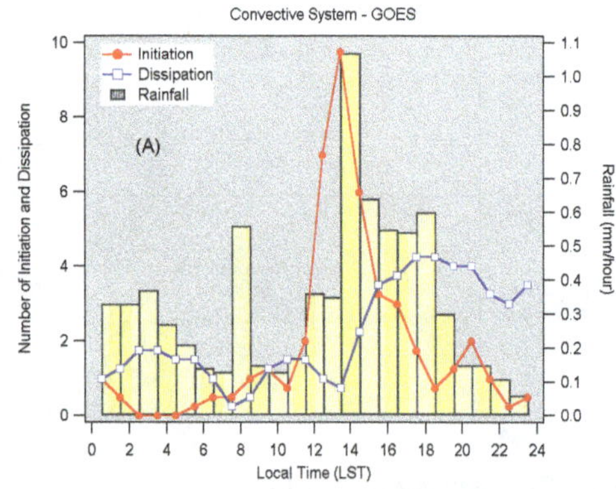

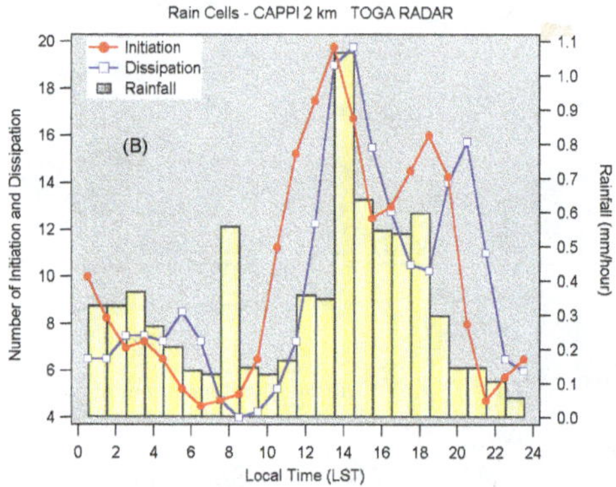

The examples above are for Amazonas convection. Cloud clusters are mainly initiated at 1400 LST, but dissipation varies. The majority of the lifetime duration is around 4–6 hours. The maximum rain rate is close to the initiation when the system is growing. Rain cells, from another side, are also initiated at the same time, but the lifetime is much smaller around one hour. The maximum convective activity occurs in the phase between initiation and the mature time step.

Slide 14

The Cloud Cluster and the Electrification Process

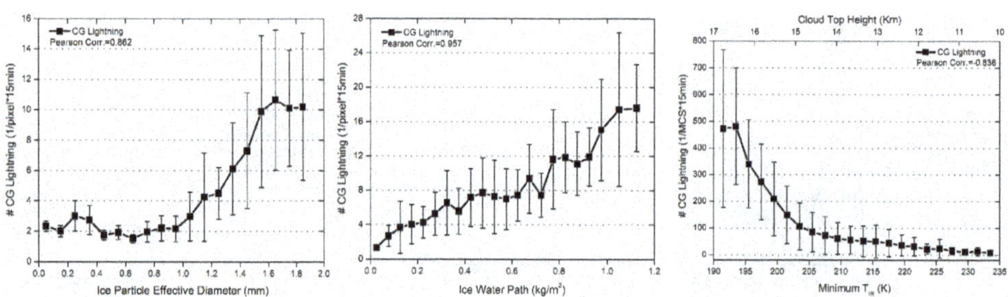

The Cloud cluster electrification processes depends on the ice water path, cloud top (Minimum Brightness Temperature) and ice particles size. Is interesting to see that for effective ice diameter smaller than 1 mm very few cloud to ground lighting is observed (CG). CG increases nearly linearly with the ice water path and exponentially with the minimum brightness temperature.

Mattos, E. V. and L.A.T. Machado. Cloud-to-ground lightning and mesoscale convective systems. Atmospheric Research., ***99****(2011), pp.377–390*

Slide 15

The Cloud Cluster and the Electrification Process

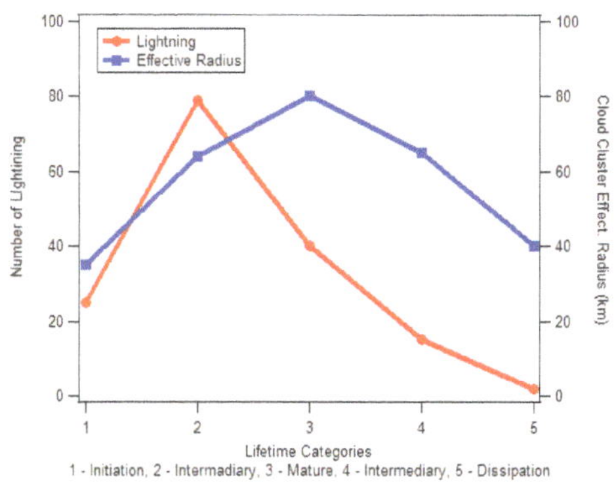

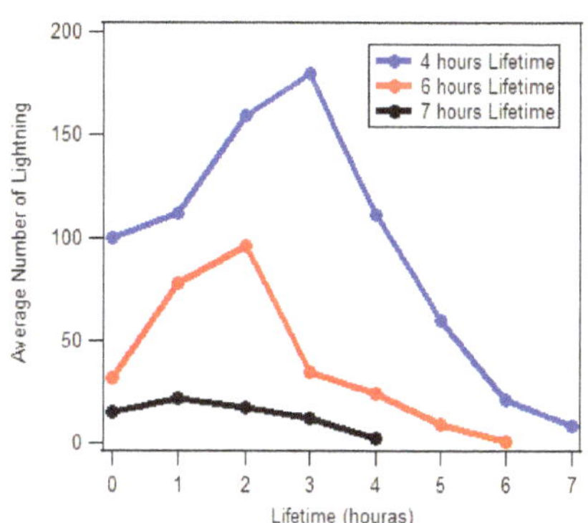

Maximum lightning activity occurs between the initiation and mature stages. This is also the time of maximum convective activity. The larger the system is, the more lightning is expected. However, regardless of the size of the system, the maximum lightning activity will always occur between the initiation and mature stages.

Slide 16

The Cloud Cluster and the Electrification Process

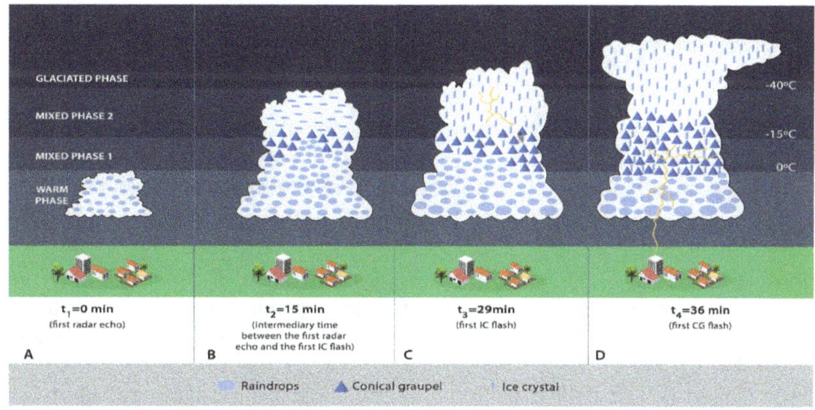

Conceptual model of the thunderstorm electrification life cycle. It shows the evolution from the first radar echo up to the time of the first cloud-to-ground flash: (a) the time of the first radar echo (#1Echo, $t_1 = 0$ min), (b) the intermediate time between the first echo radar and first intracloud lightning flash (Int., $t_2 = 15$ min), (c) the time of the first intracloud lightning flash (#1IC, $t_3 = 29$ min), and the (d) time of the first cloud-to-ground lightning flash (#1CG, $t_4 = 36$ min).

Mattos EV, Machado LAT, Williams ER, *et al.* (2017) Electrification life cycle of incipient thunderstorms. *Journal of Geophysical Research: Atmospheres* **122**:4670–97

Slide 17

The Cloud Cluster Radiative Effect

Change in net radiative fluxes (W m^{-2})				
Region	Cloud type	Surface	Atmosphere	Top
Land	CS	−133	+98	−35
	Convective	−188	+71	−118
	Mesoscale anvil	−119	+104	−14
	Transition anvil	−164	+92	−71
	Cirrus anvil	−95	+111	+16
Ocean	CS	−160	+96	−64
	Convective	−218	+65	−153
	Mesoscale anvil	−146	+104	−41
	Transition anvil	−197	+92	+105
	Cirrus anvil	−118	+111	−7

Based on the tropical cloud cluster mean properties it was computed, for 1200 LST, the heating-cooling rates of the longwave, short wave and the net effect. The change in net radiative flux, for land and ocean and for the different components of the cloud cluster is also presented in the Table above. Cloud cluster has a net cooling effect, mainly due to the convective parts of around 35–64 W m^{-2}.

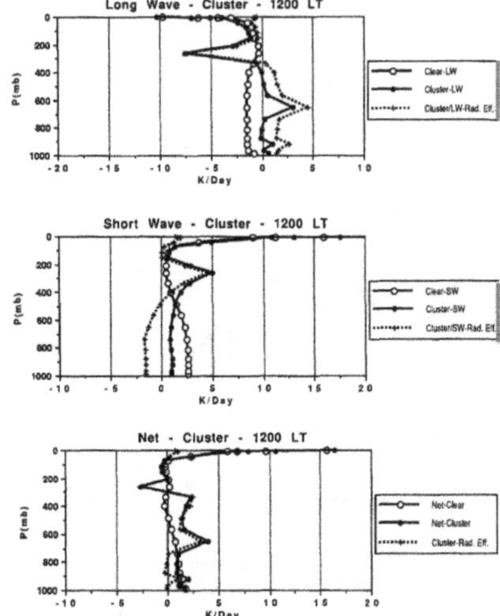

Vertical profile heating-cooling rates for a typical cloud cluster of around 180–360 km radius, based on average cloud properties. The heating-cooling rates were computed for noon time over tropical region.

Machado, L. A. T., and W. B. Rossow, 1993: Structural characteristics and radiative properties of tropical cloud clusters. Monthly Weather Review., **121**, *3234–60*

Slide 18

Take-away messages

- Clouds are organized in a large range of space-time scales, since a warm cumulus cloud of few meters to thousands of kilometers as the super cloud clusters
- Cloud size distribution varies with the synoptic pattern or forcing as the diurnal cycle.
- A cloud cluster's life cycle can be separated into phases as the initiation, mostly only convective clouds, is followed by a fast increase in size up to the mature stage. The next phase is dissipation when the system is composed of the stratiform cloud deck.
- A cloud cluster or rain cells presents a nearly linear relationship between effective radius and the lifetime duration.
- The stretching of the convection can be evaluated by the cloud cluster area expansion. The area expansion is related to the mass flux, and cloud clusters with a strong mass flux will be large and have a long life cycle.
- The cloud cluster electrification processes are related to the lifetime phase. The maximum electrical activity occurs between the initiation and mature stage.
- A cloud cluster changes the radiation budget and the heating/cooling profiles. The net effect is heating the middle atmosphere and cooling the upper level.

Slide Captions

Slide 1

Slide 2 Cloud, a 3-D object, is projected in two dimensions in satellite images. One way to select a cloud cluster in a satellite image is to define a threshold in infrared brightness temperature. There are several definitions and thresholds; a specific definition as mesoscale convective complexes includes two different brightness temperature threshold and a threshold for cloud organization eccentricity. A general description normally employs 1–2 thresholds to define the convective cores merged inside the cloud clusters. Values between 255 and 210 K are normally used; the most common value is 235 K, where clouds in the tropics reaching these levels are normally associated with convective processes. In addition, this value is largely employed in rainfall estimation. Several studies employ 235 K to define the cloud cluster and 210 K to define the most intense core merged inside the cloud cluster. Defined as a cloud cluster structure in satellite images, the life cycle definition depends on classification in a sequence of images of the same cloud clusters. The tracking of a cloud cluster normally uses the overlapping condition to set up the same cloud cluster in the first and subsequent images. The cloud cluster life cycle can be separated in phases as the initiation; mostly, only convective clouds are observed, followed by a fast increase in size up to the mature stage, where cloud clusters usually do not change in size and composition: 20% (area of convective clouds), 28% of the transition (the region with characteristics of stratiform and advected hidrometeors from convective clouds), and 52% (the cloud deck basic composed of stratiforms clouds). The next life cycle phase is dissipation, where the system is basically composed of the stratiform cloud deck, followed by fragmentation and the collapse of high clouds.

Slide 3 The typical cloud cluster reflectivity profile for each life cycle phase is described. Three important features describe the typical profile of each phase of the cloud cluster's lifetime. Reflectivity is proportional to the sixth power of drop diameter and raindrop number concentration. The figure shows the reflectivity profiles with different rain rates. At 15:00 LST, from 1 to 20 mm/h, the profiles describe the convective initiation and evolution toward the mature stage. The main difference is below the melting layer, where larger droplets and larger concentration occur in the warm phase. When the rain rate is larger than 20 mm/h, note the increase of reflectivity above the melting layer, showing that the ice phase is well developed. This is the time when clouds are building the glaciation layer, increasing the reflectivity. The same is observed at 21 LST, where it is clear how ice increases (reflectivity above 8,000 m) as the rain rate increases; this typically happens around the mature stage. This is the base consideration for passive microwave rainfall estimation, a relationship between the the ice water path and rain rate. Later in the night, the typical profile describes stratiform clouds — the picture at around the melting layer is due to the ice melting that creates a liquid water layer over the ice, and reflectivity is very high because the refraction index of liquid water is much higher than the one for ice. This is the dissipation stage where the convective processes have finished and the ice buffer is collapsing.

Slide 4 Clouds can be organized in different scales, from meters to thousands of kilometers. Each scale is related to a specific type of cloud or cloud cluster. However, cloud organization, when integrated over a long period of time, considering all forces acting on the region, the cloud cluster number distribution of density, so-named by some authors, follows a radius size dependency of a power law distribution of around –2. This feature is observed in several parts of the globe and by using different thresholds. There is a larger number of small systems, and the distribution of the number of systems as a function of the effective radius (suppose all clouds as circles) follows the relationship $N(r) = A.r^{-2}$. To compute the contribution to the total cloud area by size range, consider $S(r) = \pi.r^2 A.r^{-2}$ = cte. It means that each spatial scale equally contributes to the total cloud cover.

Slide 5 The figures in these slides show the number and area size distribution, confirming what was discussed in the previous slides. In addition, the size distribution for different thresholds shows the same pattern where only the absolute value is different because the colder the threshold is, the smaller the total area covered. Another feature described in these figures is the breaking radius. After a specific size, the number of systems decreases very fast. This behavior defines the spatial scale of the cloud cluster (for each threshold) characterizing the maximum size of cloud organization. Finally, another interesting behavior is the unchanged size distribution pattern, even between two very different years — 1983 and 1988. The cloud cover between these two years was very different, though the size distribution was similar. The main difference was that the breaking radius in 1988 (a rainy year in Africa Sahel) had a larger number of systems than in 1983 (a drier year in Africa Sahel).

Slide 6 These slides show the size distribution of the Amazon's convective systems observed during the Tropical Rainfall Measuring Mission – Large Scale Biosphere-Atmosphere Experiment in Amazonia (TRMM-LBA) experiment. If the size distribution is analyzed for specific cases, the distribution is far from $N(r) = A.r^{-2}$. This slide shows the size distribution as a function of the diurnal cycle for cloud clusters (threshold 235 K) and rain cells (from radar Constant Altitude Plan Position Indicator [CAPPI] 2 km, threshold 20 dBZ). The cloud cluster's size distribution shows a large number of small cells in the morning, the increase of the cloud cluster's size in the early afternoon (between 50 and 100 km effective radius), and the typical mesoscale organization of around 220 km of effective radius at the end of the afternoon and during the night. For rain cells, which have a shorter life cycle and are related to the rainfall (not upper-level clouds, as discussed for cloud clusters), there is a large number of small rain cells at the beginning of the afternoon that grows in size through the night. In the early morning, however, there are very few small and larger rain cells; this behavior is due to the night system that occurs during the night and early morning, and this is the most important rainfall system though it is different from systems at others times where convection is very frequent. This is normalized distribution, where the cloud cover and rainfall are very different for each of these times.

Slide 7 If the size distribution is now separated by phases of a specific synoptic system, the distribution also reveals interesting features and can set up the spatial scale of the cloud organization specifically by this forcing. The threshold of 253 K clearly defines the scale of convection associated with the easterly waves. Cloud clusters are organized around a 400-km effective radius in the trough. For the convective cells merged inside, there is no typical size, but the number is different around the easterly wave phases.

Slide 8 The accumulated distribution of radar reflectivity of large deep systems, for convective and stratiform clouds, during the mature phase shows different behaviors for land and the ocean. The reflectivity profiles for land (Africa) and ocean (Atlantic Ocean) for convective clouds are also very different. These profiles were obtained from TRMM-PR (Precipitation Radar). Convection is deeper and more intense over Africa than over the Atlantic Ocean. The large concentration of ice particles in the land convective system can be clearly observed. Likewise, Stratiform rainfall is different, ocean clouds are more prevalent, producing a strong brightband signature at around 5-km altitude; however, the differences between land/ocean for this cloud type are fewer than for the convective clouds. The reason is because updrafts that are more intense over land form more intense convective cores.

Slide 9 Cloud clusters change size during their life cycles. The system is initiated and quickly grows to the mature phase, reaching its maximum size before slowly decreasing in size due to dissipation. In the dissipation phase, the warmer thresholds increase, the colder threshold decreases, and the minimum brightness temperature increases. This typical behavior corresponds to the increase of the cloud deck and reduction of the convective cells merged inside the cloud cluster. Another interesting characteristic of the cloud cluster or rain cells is the nearly linear relationship between the effective radius (a linearized measure of the area) and the lifetime duration. This behavior establishes a relationship between the space and time scale.

Slide 10 The upper left panel shows a schematic evolution of long, middle and short-lived cloud clusters. In the initiation phase, it is possible to theorize about the life duration as a function of the area expansion. Systems with strong area increases will last longer than systems with more moderate area expansion. Based on climatology, it is possible to set up the initial area expansion as a function of the life cycle duration. The area expansion is related to the mass flux, and cloud clusters with a strong mass flux will be larger and last longer than clouds with less intense dynamics. For longer time scales, this initial area expansion only matters at the beginning of the life cycle, later the cloud development is driven by the synoptic forcing.

Slide 11 The normalized area expansion can be represented by two components: the area expansion with a constant condensation that is the upper-level divergence and the area expansion at a constant upper level divergence. This term is related to the rates of cloud top increase and the condensation–evaporation process. Considering a cloud cluster that is defined at a cold threshold, the cloud top is close to the tropopause; therefore, this term can be neglected, and the normalized area expansion is equal to the upper level divergence and rate of the condensation–evaporation process. Measuring the normalized area expansion and upper-level wind divergence (satellite winds), it was possible to deduce the rate of condensation–evaporation. The figure shows schematically the relative importance of each term during the cloud cluster life cycle. In the initiation, the rate of condensation–evaporation is the dominant term, followed by the upper-level divergence that occurs just before the mature phase. During dissipation, the upper-level convergence and evaporation are responsible for the cloud cluster dissipating.

Slide 12 The nearly linear pattern of the normalized area expansion during the life cycle can help to forecast the changes in the size of a specific cloud cluster. The system initiate with a high increase in its area, reaches near zero area expansion at the mature stage, and becomes negative in the dissipation phase. With the information of the normalized area expansion, it is possible to predict the change in the cloud cluster size by using the near-linear curve.

Slide 13 These figures present the average diurnal cycle of rainfall, rain cells (defined by the 20 dBZ on CAPPI 2 km), and cloud cluster (235 K) initiation and dissipation in Central Amazonas during the TRMM-LBA experiment. In an hour, after the maximum initiation of rain cells and cloud clusters, the maximum rain rate is observed. The rain cells dissipate an hour later, but the cloud cluster has a longer lifetime of about 4 to 6 hours. The maximum normalized area expansion and minimum brightness temperature occur at the time of initiation and maximum rainfall.

Slide 14 By using NOAA satellite data co-located with a ground lightning sensor (cloud ground electrical discharge), it was possible to understand the cloud cluster processes of electrification. Cloud to ground lightning is frequent, but only if the ice particle's effective diameter is larger than 1 mm and its minimum brightness temperature is colder than ~215 K. The ice water path is nearly linearly related to the lightning activity.

Slide 15 The cloud cluster electrification processes are related to their lifetime phases. Maximum electrical activity occurs between the initiation and mature stages. The larger the cloud cluster is, the more lightning is observed.

Slide 16 Cloud electrification depends on a complex microphysical process. The hydrometeor mixed phase happens between the melting and glaciation layers. In this layer, strong updrafts bring raindrops to the bottom of the glaciation layer. This forms new ice particles as the graupel, one of the key hydrometers in the electrification process. In general, clouds first produce intracloud (IC), followed by cloud-to-ground (CG) lightning flashes.

Slide 17 Cloud cluster properties were studied using several years of ISCCP satellite images for tropical clouds. These properties were employed in radiative simulation to study the heating and cooling profiles for each part of the cloud cluster for the short and long waves radiation. This figure shows the average profile for all components of the cloud cluster. Basically, the net effect is that the middle atmosphere is heated, and the upper level is cooled. The short and longwave effect is illustrated in this figure, where the details for each cloud cluster component can be viewed via the cited reference in this slide. However, the net radiative effect for the land and ocean at 1200 LST can be seen in the table. A cloud cluster has a net cooling effect of around 35 to 64 $W.m^{-2}$. The same table shows the contribution to each part of the cloud cluster.

Slide 18 Take-aways messages

References

Futyan JM, Del Genio AD. (2007) Deep convective system evolution over Africa and the tropical Atlantic. *Journal of Climate* **20**:5041–60.

Machado LAT, Rossow WB. (1993) Structural characteristics and radiative properties of tropical cloud clusters. *Monthly Weather Review* **121**:3234–60.

Machado LAT, Desbois M, Duvel JP. (1992) Structural characteristics of deep convective systems over tropical Africa and the Atlantic Ocean. *Monthly Weather Review* **120**:392–406.

Machado LAT, Duvel JP, Desbois M. (1993) Diurnal variations and modulation by easterly waves of the size distribution of convective cloud clusters over West Africa and the Atlantic Ocean. *Monthly Weather Review* **121**:37–49.

Machado LAT, Laurent H, Lima AA. (2002) Diurnal march of the convection observed during TRMM-WETAMC/LBA. *Journal of Geophysical Research* **107**(D20):8064.

Martins RCG, Machado LAT, Costa AA. (2010) Characterization of the microphysics of precipitation over Amazon region using radar and disdrometer data. *Atmospheric Research* **96**(2–3):388–94.

Mattos EV, Machado LAT. (2011) Cloud-to-ground lightning and mesoscale convective systems. *Atmospheric Research* **99**:377–90.

Mattos EV, Machado LAT, Williams ER, *et al*. (2017) Electrification life cycle of incipient thunderstorms. *Journal of Geophysical Research: Atmospheres* **122**:4670–97.

Nakazawa T. (1988) Tropical super clusters within intraseasonal variations over the western. *Journal of the Meteorological Society of Japan* **66**:777–86.

Lecture 5

The Role of Deep Convection on the Initiation of African Easterly Waves

Ademe Mekonnen

Applied Science and Technology Program & Department of Physics
North Carolina A&T State University (NCAT)
Greensboro, NC, USA

Ademe Mekonnen is an Associate Professor of Atmospheric Sciences at the Applied Science & Technology Program and Department of Physics at North Carolina A&T State University (NCAT). His main research focus is on tropical convection and wave activity and their influence on daily rainfall and tropical cyclogenesis.

Introduction

William Rossow has played a leading role in planetary and earth sciences research. He mentored several postdocs and graduated students in various new areas of research. This contributor is one of the beneficiaries of Bill's mentorship. Bill planned and led the application of weather state (WS) data from the International Satellite Cloud Climatology Project (ISCCP) to investigate the interaction between deep convection and atmospheric wave disturbances (e.g., [Mekonnen and Rossow, 2011, 2018; Tromeur and Rossow, 2010]). Bill's work on African easterly waves (AEWs) contributed to our understanding of the wave initiation and variability problem.

The AEWs are synoptic-scale westward-moving disturbances that have a significant influence on the boreal summer daily rainfall of North Africa. It is long known that AEWs also have an important impact on the genesis of Atlantic tropical cyclones. About half of the tropical disturbances (TDs) and the great majority of Atlantic Hurricanes are formed in association with AEW vortices. However, while about 50–60 waves form and cross the West African coast annually, only a few of them trigger TDs. We still do not fully understand why some of the AEWs are instrumental in TD-genesis and hurricanes and some are not.

AEWs are among the most studied weather systems. Despite long years of research dating back to the 1940s and 1950s, the initiation and variability of AEWs remain an important area of research. Several theories have been proposed over years to describe the AEWs initiation.

Among the early studies, Carlson (1969) suggested that easterly wave disturbances are initiated in East Africa as a result of enhanced convection associated with the highlands there (Slide 5). Later, Burpee (1972) suggested that waves form in association with an unstable mid-level African easterly jet (AEJ) from equally important barotropic and baroclinic processes. Subsequent studies (e.g., [Albignat and Reed, 1980; Norquist *et al.*, 1977]) also suggested the AEJ as the cause of AEW initiation. In contrast, more recent studies argued that the AEJ is only marginally unstable (Hall *et al.*, 2006; Thorncroft *et al.*, 2008), and that its presence is not a necessary condition for wave initiation (e.g., [Hsieh and Cook, 2005]). These studies argue that AEWs are formed in association with deep convection over East Africa (Berry and Thorncroft, 2005; Mekonnen *et al.*, 2006; Kiladis *et al.*, 2006). But the process of how smaller-scale convection initiates larger-scale waves is not clear. Using WS data from the ISCCP, Mekonnen and Rossow (2011, 2018) argue that stronger latent heating and negative radiative cooling over eastern Africa are associated with larger and better-organized convective clouds compared with weaker and scattered convective clouds. They observed that scattered and less well-organized convection east of the East African Mountains triggers a chain of events that leads to a well-organized and larger-scale deep convection over the highlands, suggesting a transition of scattered deep convection into well-organized mesoscale convective system (MCS) type. This transition of scattered deep convection into MCSs leads to the emergence of wind perturbations westwards. Mekonnen and Rossow concluded that such a transition is an important pathway for AEW initiation.

The ISCCP provides various satellite observed cloud products, including a joint histogram distribution of cloud top pressure (CTP) and optical thickness (τ) from July 1983 to the present. Pattern analysis of CTP and τ for the global tropics within 35°S–35°N identifies eight different characteristics of cloud regimes, three of which represent deep convection and five of them suppressed convection. The three deep convection types are categorized as WS1, WS2 and WS3 and represent three different types of cloud systems. WS1 represents MCSs that cover a wide area and that live longer than six hours. WS2 represents the thick anvil structure of deep clouds. WS1 and WS2 exist together. Scattered and less organized and small-scale convection (e.g., isolated Cb) is represented by the WS3 category. Several studies used these cloud regimes to investigate the interaction between atmospheric waves and deep convection over various regions of the globe. However, optical thickness information is available during sunlit hours and the CTP-τ dataset is not suitable for diurnal cycle studies. Mekonnen and Rossow (2018; MR18 for short) used an infrared (IR) weather states (IR-WS) dataset that was produced by Tan *et al.* (2013). CTP retrievals in the ISCCP data have IR information available at all times of the day. Tan *et al.* (2013) developed an algorithm that considers IR-only CTP-histograms and determined IR-CTP histograms using pattern analysis. The cloud regimes in IR-WS pattern analysis also identify three types of deep convection and five suppressed convection with similar frequencies over the global tropics within 35°S–35°N. Therefore, IR-WSs are WSs extended for nighttime. So, WS1 are IR-WS1, WS2 are IR-WS2, and WS3 are IR-WS3.

MR18 used IR-WSs and reanalysis datasets to investigate the interaction between convective systems and AEWs. They studied two potential pathways of AEW initiation; namely, (1) IR-WS3→AEW→IR-WS1 (small-scale scattered deep convection develops into wave perturbation and waves organize MCS-type deep convection), and (2) IR-WS3→IR-WS1→AEW

(WS3 develops into WS1 and AEWs initiated as a result of large-scale deep convection). Below, we summarize the main findings of MR18.

They suggested that increasing intradiurnal activity and atmospheric instability (associated with increasing positive equivalent potential temperature anomaly) and positive specific humidity anomalies precede the development of well-organized and large-scale convection over the East African highlands (Slides 5, 6 and 8). They also noted that atmospheric instability favors a higher frequency of scattered, isolated-type convection (i.e., IR-WS3 types) to the east of the highlands. But increased low-level shear and a dynamical environment favor the organization of MCS-type growth over the high terrain and to the west of the mountains. Thermodynamic processes suggest that the dominant process is IR-WS3→ IR-WS1→AEW.

Dynamic parameters were also examined in (Mekonnen and Rossow, 2018). Peak relative vorticity precedes maximum convection over East African highlands (Slides 9 and 10). Slightly before the IR-WS1 peaks over the mountains and to the west of the highlands, low-level moist westerlies, low to mid-level increasing wind shear, and positive vorticity increase between central Sudan and Ethiopia (Slides 8 and 9). The AEW signatures over East Africa are weak during the time of convective transitioning. The large-scale and local environment, including a moderate increase in shear, enables the scattered deep convection (IR-WS3) to merge and form larger and organized convective systems (IR-WS1). IR-WS1 further develops to the west of the highlands. Also, increased vorticity in association with increased moderate wind shear creates an environment favorable for the development of deep and well-organized convection (i.e., IR-WS1). Based on this evidence, Mekonnen and Rossow concluded that the dominant pathway is IR-WS3→IR-WS1→AEW.

In summary, upper tropospheric mean flow, including upper-tropospheric wave disturbances, enhances the IR-WS3 types over the Arabian Sea and southern Arabia and steers these clouds westward. As they move westward and interact with the mountains, they encounter additional lifting by the high terrain that aids the development of more vigorous and well-organized larger-scale clouds. The low-level vertical wind shear maximizes over the high terrain, which enhances the further growth of MCS-type activity.

We note that the above summary highlights that well-organized and larger-scale deep convection is the main mechanism for AEW initiation over East Africa. However, this finding does not discount the role of the mid-tropospheric easterly jet environment that supports AEWs over West Africa. We suggest that AEWs first form in East Africa as a result of deep synoptic-scale convection but strengthen as they propagate across the continent in relation to the mid-level jet environment.

Acknowledgment: This research is supported by NSF AGS Award 1461911.

References

Albignat JP, Reed RJ. (1980) The origin of African wave disturbances during Phase III of GATE. *Monthly Weather Review* **108**:1827–39.

Berry GJ, Thorncroft CD. (2005) Case study of an intense African easterly wave. *Monthly Weather Review* **133**:752–66.

Burpee RW. (1972) The origin and structure of easterly waves in the lower troposphere of North Africa. *Journal of the Atmospheric Sciences* **29**:77–90.

Carlson TN. (1969) Synoptic histories of three African disturbances that developed into Atlantic hurricanes. *Monthly Weather Review* **97**:256–75.

Hall NMJ, Kiladis G, Thorncroft C. (2006) Three dimensional structure and dynamics of African easterly waves. Part II: Dynamical modes. *Journal of the Atmospheric Sciences* **63**:2231–45.

Hsieh J-S, Cook KH. (2005) Generation of African easterly wave disturbances: Relationship to the African easterly jet. *Monthly Weather Review* **133**:1311–27.

Kiladis GN, Thorncroft CD, Hall NG. (2006) Three-dimensional structure and dynamics of African easterly waves. Part I: Observations. *Journal of the Atmospheric Sciences* **63**:2212–30.

Mekonnen A, Rossow WB. (2011) The Interaction between deep convection and easterly waves over Tropical North Africa: Convective transitions and mechanisms. *Monthly Weather Review* **146**:1945–61.

Mekonnen A, Rossow WB. (2018) The Interaction between deep convection and easterly waves over Tropical North Africa: A Weather State perspective. *Journal of Climate* **24**:4276–94.

Norquist CD, Recker EE, Reed RJ. (1977) The energetics of African wave disturbances as observed during Phase III of GATE. *Monthly Weather Review* **105**:334–42.

Tan J, Jakob C, and Lane TP. (2013) On the Identification of the large-scale properties of tropical convection using cloud regimes. *Journal of Climate* **26**:6618–32.

Thorncroft C, Hodges K. (2001) African Easterly wave variability and its relationship to Atlantic Tropical cyclone activity. *Journal of Climate* **14**:1166–79.

Thorncroft CD, Hall NMJ, Kiladis GN. (2008) Three-dimensional structure and dynamics of African easterly waves. Part III: genesis. *Journal of the Atmospheric Sciences* **65**:3596–607.

Tromeur, E., and W. B. Rossow, 2010: Interaction of tropical deep convection with the large-scale circulation in MJO. *Journal of Climate* 23, 1837–53.

Slide 1

The Role of Deep Convection on the Initiation of African Easterly Waves (AEWs)

Ademe Mekonnen
Department of Physics & Applied Science and Technology Program
North Carolina A&T State University (NCAT)
Greensboro, NC

Slide 2

AEWs

- Key synoptic scale waves dominant during the June–September rain season over Africa (λ ~3,000–4,000 km, τ ~3–5 days, phase speed ~7–8° per day).
- Play a critical role on daily rainfall over Africa and tropical cyclogenesis over the Atlantic (sometimes over eastern Pacific).
- Where do they form and how do they maintain?
    - Two different but complementary views of AEW initiation (pathways!)
        1. AEWs form in association with an unstable mid-tropospheric jet over West Africa (dominant view until recently).
        2. AEWs form in association with large-scale deep convection over East Africa and strengthen due to the mid-level jet environment.

Slide 3

Critics on AEW Pathway #1:

- The mid-level jet, a.k.a. African easterly jet (AEJ), is about 40–50 degrees long and can have about 2 waves — too short (~3,000 km east-west)
- Recent studies (mainly idealized models) showed that the AEJ is nearly stable to small perturbations and the question is:

What would account for their initiation?

Hypothesis:

- AEWs are initiated in association with large-scale deep convection associated with elevated topography in East Africa.
  - ✓(this was first suggested by Carlson in the late 1960s)

Slide 4

Data Sources

We use infrared weather State (IR-WS) from ISCCP and ECMWF Interim reanalysis (ERA-I) products. Different types of deep convection are represented by the IR-WS and thermodynamic and dynamic variables are derived from the ERA-I dataset. NOAA-CPC African Rainfall Climatology V2 is also used.

❖The ISCCP provides various satellite-observed cloud products, including a joint histogram distribution of cloud top pressure (CTP) and infrared (IR) radiation from July 1983 to 2009.

❖Pattern analysis of CTP and IR information for the global tropics within 35°S–35°N identifies eight different characteristics of cloud regimes, three of which represent deep convection and five of them suppressed convection. The three deep convection types are categorized as IR-WS1, IR-WS2 and IR-WS3, and represent three different types of cloud systems (Tan *et al.*, 2013; Tan and Jakob, 2013).

- IR-WS1 identify large-scale, well-organized convection associated with MCSs.
- IR-WS2 identify large-scale thick anvil cloud systems
- IR-WS3 identify scattered, less well-organized convection (e.g., isolated Cb and cumulus-congestus)

Slide 5

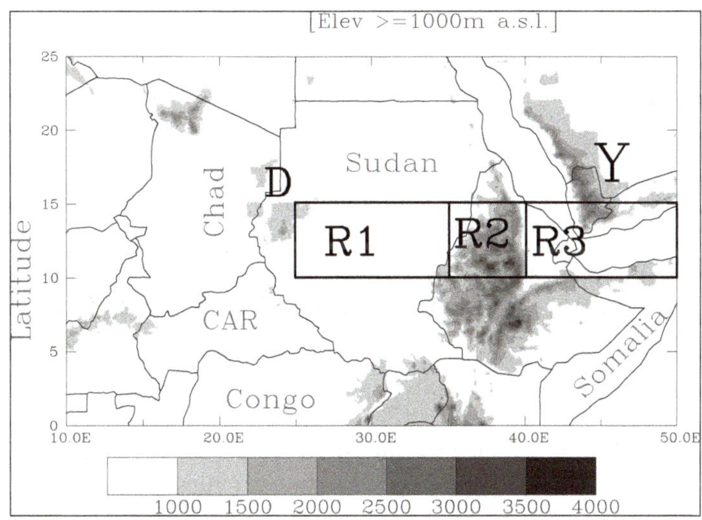

Elevation Map of the Study Area

Slide 6

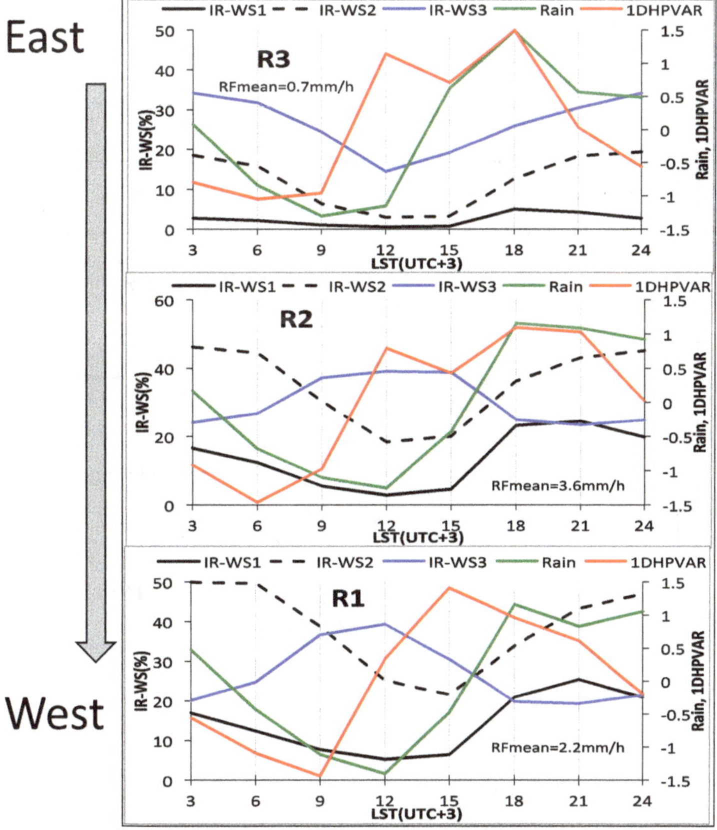

Climatology of diurnal cycle

Slide 7

Pathways to AEW Initiation:

Two potential pathways of AEW initiation are investigated:

- IR-WS3→AEW→IR-WS1 (small-scale scattered deep convection develops into wave perturbation and waves organize MCS-type deep convection)

- IR-WS3→IR-WS1→AEW (IR-WS3 develops into IR-WS1 and AEWs initiated as a result of large-scale deep convection).

Slide 8

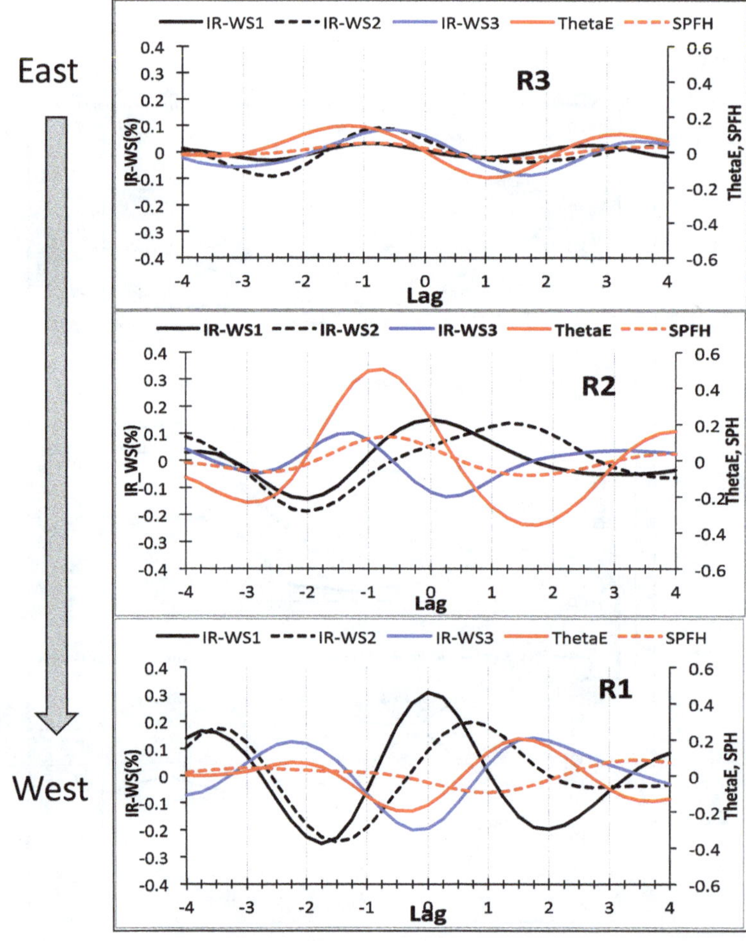

Regression composite anomalies

Slide 9

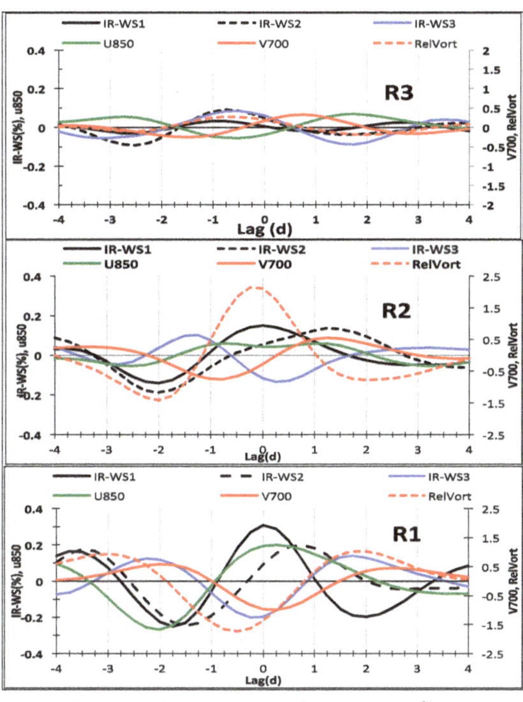

Regression composite anomalies

Slide 10

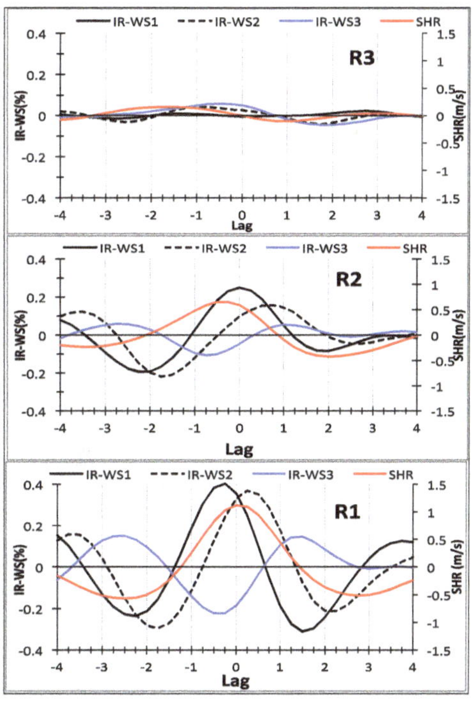

Regression composite anomalies

Slide 11

Summary ...

- Atmospheric instability favors higher frequency of a scattered, isolated convection (i.e., IR-WS3 types) to the east of the highlands.
- Increased low-level shear and dynamical environment favors organization of MCS-type growth over the high terrain and to the west of the mountains.

  ⟹ Thermodynamic processes suggest IR-WS3→IR-WS1→AEW.
- Peak relative vorticity precedes maximum convection over East African highlands
- Low-level moist westerlies, low to mid-level increasing wind shear, and positive vorticity increase to the west of highlands leading to further increase in IR-WS1

  ⟹ Dynamic measures also suggest IR-WS3→IR-WS1→AEW

Therefore, AEWs are initiated in association with convective transition from scattered small-scale convection to well-organized perturbations.

Slide Captions

Slide 1 Title page

Slide 2 Background — African Easterly Waves (AEWs): importance of AEWs, wave initiation and maintenance.

Slide 3 AEW pathways — Hypothesis.

Slide 4 Date sources: ISCCP, reanalysis and NOAA-CPC rainfall.

Slide 5 Central and East Africa (elevation ≥ 1,000 m are shaded). Boxes R1 (10°N–15°N, 25°E–35°E), R2 (10°N–15°N, 35°E–40°E; Ethiopian highlands), and R3 (10°N–15°N, 40°E–50°E) denote the study areas where the transition in weather states appear to occur (Mekonnen and Rossow, 2011; see text for details). The major highlands in the region are the Darfur Mountains (denoted by D; just north-west of R1), the Ethiopian highlands (R2), and the Yemen highlands (denoted by Y; north of R3).

Slide 6 Mean diurnal evolution of different convective cloud regimes (IR-WS 1–3), the total rainfall, and intradiurnal variance anomalies (1DHPVAR; K^2) during JAS over R1, R2 and R3. The vertical axis on the left shows the percentage frequency of the total cloudiness. The vertical axis on the right shows rainfall and intradiurnal variance anomalies standardized with their respective standard deviations. The rainfall totals are based on a 1°×1° horizontal grid. The abscissa is in local standard time (UTC+3). The mean daily rainfall and mean diurnal variances are subtracted from each to form anomalies. Daily rainfall averages are indicated for each region.

Slide 7 What is the dominant pathway for AEW initiation? Two main pathways are investigated.

Slide 8 The regression composite anomalies of equivalent potential temperature (θ_E in K; red line), specific humidity (q; g/Kg; red-dashed line), and IR-WSs (%) over East African regions. The ordinate on the right is for IR-WS, and on the left is for θ_E and q. Regressions are based on a 2–10-day 700 hPa meridional wind at 12.5°N, 32.5°E.

This suggests IR-WS3→IR-WS1→AEWs.

Slide 9 Regression composite anomalies of IR-WS1, IR-WS2, IR-WS3, zonal (U850) and meridional (V700) winds (ms^{-1}) and relative vorticity (ζ; s^{-1}; scaled by 10^{-6}) at 700 hPa. The ordinate on the right is for IR-WS and U850, and on the left is for meridional wind and ζ. Regression composites as in Slide 8.

Slide 10 Regression anomalies of IR-WS and a 600–850 vertical wind shear (SHR) based on a 2–10-day filtered 600–850 hPa wind shear at 12.5°N, 32.5°E. The ordinate on the left shows the IR-WS anomalies. The ordinate on the right shows shear anomalies. Lag times are every 0.25-day (6-hr).

∴ This suggests IR-WS3→IR-WS1→AEWs.

Slide 11 Summary and major findings.

References

Mekonnen A, Rossow WB. (2011) The Interaction between deep convection and easterly waves over Tropical North Africa: Convective transitions and mechanisms. *Monthly Weather Review* **146**:1945–61.

Tan J, Jakob C. (2013) A three-hourly dataset of the state of tropical convection based on cloud regimes. *Geophysical Research Letters* **40**:1415–19.

Tan J, Jakob C, and Lane TP. (2013) On the Identification of the large-scale properties of tropical convection using cloud regimes. *Journal of Climate* **26**:6618–32.

LECTURE 6

Examining Global Precipitation with Mesoscale Atmospheric Systems Represented by Cloud Regimes

Jackson Tan

Associate Research Scientist, University of Maryland Baltimore County, Baltimore, MD, USA
NASA Goddard Space Flight Center, Greenbelt, MD, USA

Jackson Tan is an Associate Research Scientist with University of Maryland Baltimore County at NASA Goddard Space Flight Center. He currently works on the Integrated Multi-satellitE Retrievals for GPM (IMERG) precipitation algorithm and cloud regimes defined from passive satellite observations. His research interests include the retrieval of precipitation from space, characterizing global precipitation patterns, and the relationship between clouds, convection, and precipitation.

Introduction to Cloud Regimes

One of the products from the International Satellite Cloud Climatology Project (ISCCP) is a dataset of joint-histograms summarizing the cloud field over a large area. Over a 3-hour period during the day, a network of geostationary and polar-orbiting satellites retrieve cloud properties and aggregate them into histograms for each grid box measuring a few hundred kilometers in length. This joint-histogram sorts each pixel according to how high the cloud top is (cloud top pressure) and the optical extent of the cloud vertically (optical thickness), providing a summarized view of the cloud field. They describe the populations of clouds within a particular area over a particular time period but without the detailed spatial distributions of the clouds. Nevertheless, despite the reduction of information, the joint-histograms can be extremely varied, so a method to categorize these joint-histograms statistically is needed.

One leading way to classify these joint-histograms using their patterns is to apply a clustering algorithm. The *k*-means clustering algorithm is a simple, unsupervised machine learning technique that seeks to separate observations into similar classifications based on their patterns. Given a large set of points (in this case, joint-histograms) and a small set of centroids, the algorithm will iteratively assign the points to the closest centroids and update the centroids based on the average of its associated points, stopping only when the centroids no longer change. With a set of centroids, each joint-histogram is then assigned to the closest centroid one final time and given the membership of that cluster. Therefore, instead of the diverse range of joint-histograms, *k*-means reduces them into a manageable set of discrete states, each with a representative joint-histogram. These clusters effectively represent recurring, archetypal cloud fields in the atmosphere and are hence called Cloud Regimes (CRs)

(Jakob and Tselioudis, 2003). They are also referred to as "Weather States" due to the fact that they can represent different weather conditions (Rossow *et al.*, 2005). However, we avoid this term so as to be explicit about its origins and acknowledge potential imperfections in the correspondence between clouds and weather. CRs have also been derived from other datasets such as the Moderate Resolution Imaging Spectroradiometer (MODIS), which also produces ISCCP-like joint-histograms (Oreopoulos *et al.*, 2014; 2016).

One disadvantage of the *k*-means algorithm is the need to specify the number of clusters beforehand. While there are other machine learning algorithms such as self-organizing maps that avoid this problem (McDonald *et al.*, 2016), a pseudo-objective set of criteria has also been devised for the *k*-means algorithm to provide guidelines in determining this number (Rossow *et al.*, 2005). Regardless of the approach, the value of CRs lies in their ability to allow us to better understand the weather and climate system. The usefulness of a set of CRs is limited either when they cannot delineate systems with vastly different properties or when there are multiple CRs representing similar systems. Hence, regardless of the path taken to derive them, the ultimate choice for the number of clusters should be the correspondence of CRs to actual atmospheric systems to be studied.

For example, one of the prominent CRs that consistently appears in various CR datasets — regardless of differences in region, clustering technique, or cloud dataset — is that which represents organized convection. Convection arises from an unstable atmosphere, such as due to solar heating of the surface or converging surface winds, and results in a rising mass of air that is typically moist and forms clouds. If the atmosphere is sufficiently unstable, these clouds can be deep (i.e., have a vertical extent starting from about 2 km above the surface and reaching the top of the troposphere) and may produce precipitation. Under favorable conditions, these isolated deep convective clouds may organize themselves into larger, coherent systems such as mesoscale convective systems (e.g., [Moncrieff, 2010]). These systems of organized convection have a distinctive signature of cloud fields with substantial deep convective clouds accompanied by stratiform anvils and other convective clouds. This cloud signature corresponds to one of the CRs; in other words, this CR can identify organized convection and thus serve as a proxy to further our understanding of their properties (Tan *et al.*, 2013).

The CRs can also represent a convective environment in which the convection is not as well organized. The centroid describes a mix of stratiform anvils and cirrus clouds. Such cloud fields may depict a scene in which isolated "popcorn" convection occurs, leading to individual thunderstorms scattered over a large area, but the conditions are not favorable for them to be organized into a larger system. Other CRs can represent suppressed conditions such as fair weather states and fields of stratocumuli (Tan *et al.*, 2013). The range of atmospheric systems that can be represented by the CRs makes it a useful tool for studying various atmospheric states.

Precipitation Properties of Different Atmospheric Systems

By compositing precipitation data to CRs (i.e., matching precipitation data to CR occurrence at the same location and time window), we can collect precipitation values of the CRs and analyze their distributions and properties. Compositing is a simple but effective tool to understand the aggregated behavior of different phenomena. For example, we can examine the mean or median to get a sense of how much precipitation is associated with each CR, or we can compute its 90th percentile to assess its potential to produce extreme precipitation. Such information is useful to assess whether climate models can reproduce the observed precipitation statistics associated with similar cloud fields. As opposed to a simple climatology of precipitation (e.g., maps of precipitation), this approach provides a more in-depth picture of the atmosphere, providing details of the processes related to precipitation. A further advantage of compositing is that it generally gives larger sample sizes in the statistics, thus mitigating possible outliers and reducing random errors in the observed data, which can be relatively high for satellite-based data, especially for precipitation.

Compositing precipitation data to CRs reveal that most of them are associated with very little precipitation at the global scale (Lee *et al.*, 2013; Tan *et al.*, 2013). These CRs may produce some very isolated and scattered precipitation somewhere in the grid — e.g., a localized drizzle — but are negligible over the length scales of several hundreds of kilometers that we are concerned about. Among the CRs with appreciable precipitation, they have distinct distributions of precipitation — a strong indication that they represent different atmospheric systems (Jakob and Schumacher, 2008; Lee *et al.*, 2013; Tan *et al.*, 2013). These CRs also reveal a diversity of fine-scale precipitation patterns within the area covered by the CR (Tan and Oreopoulos, 2019).

One CR stands out in particular: organized convection. Its precipitation values and chance of precipitation are unusually high (Tan *et al.*, 2013; Rossow *et al.*, 2013), such that it produces nearly half the rainfall in the tropics despite occurring less than 10% of the time (Lee *et al.*, 2013). Within the system, it is precipitating in about half the area it occupies on average. Interestingly, the contrast between this 50% precipitation fraction and the near-100% cloud cover reflects the fact that most clouds are not associated with precipitation (Tan and Oreopoulos, 2019).

In the tropics, the other precipitating CRs, representing convection that is less organized, have weaker precipitation rates. In fact, despite each being more frequent than organized convection, the collective contribution to the total tropical precipitation is less (Lee *et al.*, 2013; Tan *et al.*, 2013). This illustrates the skewed nature of how different atmospheric systems contribute to the total precipitation in the tropics: a small number of systems is responsible for most of the precipitation. The fact that the same dataset can separate between organized and less organized convection enables studies relating to the precipitation behavior due to the transition between these two states in different regions of the world (Mekonnen and Rossow, 2018; Worku *et al.*, 2019).

Extending beyond the tropics to include the mid-latitudes (60°N/S), organized convection is still associated with the most intense precipitation, with the highest contribution and chance of precipitation (Tan and Oreopoulos, 2019). However, other systems emerge in importance. Summer extratropical storms are also associated with intense precipitation, though not to the degree of organized convection. These storms have the highest precipitation rates and fractions after organized convection. On the other hand, winter extratropical storms produce much lower precipitation despite the similarity in cloud patterns and annual geographical distribution, demonstrating the importance of meteorological conditions in affecting the CR properties (Leinonen *et al.*, 2016).

Changes in Tropical Precipitation due to Organized Convection

Using satellite data to study changes can be tricky due to potential artifacts arising from factors such as orbital drifts, satellite replacement (and associated improvements in instrument capabilities), and changes in calibration. Nevertheless, it is possible with data that have been quality controlled to ensure data homogeneity. Indeed, the ISCCP CR representing organized convection is less vulnerable to satellite viewing artifacts and has been cross-checked with satellite records to rule out known artifacts. As such, it can be used to diagnose changes in the occurrence of organized convection in the last 25 years (Tselioudis *et al.*, 2010). Furthermore, this increase is not globally homogeneous: organized convection has become more frequent in some regions but declined in others. Given the intense precipitation associated with organized convection, what does this mean for how precipitation has changed and will change in the future?

By compositing precipitation with the CR representing organized convection and then analyzing how these precipitation values have changed over time, we can identify the contribution of organized convection to changes in precipitation (Tan *et al.*, 2015). Furthermore, we can find out whether the changes are due to organized convection becoming more or less frequent or each organized convective system producing more or less precipitation. As it turns out, it is the former that dominates precipitation changes in the tropics; the increased occurrence of organized convection is primarily responsible for driving an increase in tropical precipitation. Indeed, geographically, the match between the observed change in precipitation and the change in the occurrence of organized convection is striking. This increase overwhelms the slight decrease in the average precipitation per event. In other words, organized convection is producing slightly less precipitation on average, but it is becoming much more often. The increased precipitation associated with more frequent organized convection also dominated over the changes associated with other CRs. Therefore, the change in the occurrence of organized convection alone can explain most of the precipitation change in the tropics.

However, the close relationship between the change in the occurrence of organized convection and the change in tropical precipitation has an implication for projections of future precipitation. Many climate models used to project future changes do not have an explicit representation of organized convection. Because of the coarse resolution of climate models (~50–100 km) compared to individual convective cells (~2 km), climate models do not simulate individual clouds, but instead, represent their statistical properties through an ensemble of unorganized buoyant plumes. This lack of a representation of the organization of convection introduces some uncertainty into these future projections. This is a longstanding problem, but one that has attracted significant attention recently, so further progress is likely in future generations of climate models.

The degree to which climate models fail to reproduce the precipitation of organized convection can be assessed. By implementing tools called "satellite simulators" in their codes, climate models can produce joint-histograms of clouds like those from the ISCCP, thus allowing CRs to be defined in the models. With the model CRs, we can then evaluate the model performance in terms of the precipitation of these atmospheric systems (Tan *et al.*, 2018). Consistent with our understanding of climate models, most models underestimate the precipitation associated with each occurrence of organized convection. However, some models compensate for this by simulating more organized convection. Therefore, if models struggle to produce the main source of precipitation recently, projections of precipitation likely contain some inaccuracies and must be interpreted cautiously with its associated uncertainties in mind. However, with the continual improvement of climate models, both in its representation of convection and in its resolution, engenders hope that this important aspect of the climate system can be projected with greater accuracy.

References

Jakob C, Schumacher C. (2008) Precipitation and latent heating characteristics of the major tropical western pacific cloud regimes. *Journal of Climate* **21**:4348–64. doi:10.1175/2008JCLI2122.1

Jakob C, Tselioudis G. (2003) Objective identification of cloud regimes in the Tropical Western Pacific. *Geophysical Research Letters* **30**:2082. doi:10.1029/2003GL018367

Lee D, Oreopoulos L, Huffman GJ, *et al.* (2013) The precipitation characteristics of ISCCP tropical weather states. *Journal of Climate* **26**:772–88. doi:10.1175/JCLI-D-11-00718.1

Leinonen J, Lebsock MD, Oreopoulos L, Cho N. (2016) Interregional differences in MODIS-derived cloud regimes. *Journal of Geophysical Research: Atmospheres* **121**:11648–65. doi:10.1002/2016JD025193

McDonald AJ, Cassano JJ, Jolly B, *et al.* (2016) An automated satellite cloud classification scheme using self-organizing maps: Alternative ISCCP weather states. *Journal of Geophysical Research: Atmospheres* **12**:13009–30. doi:10.1002/2016JD025199

Mekonnen A, Rossow WB. (2018) The Interaction between deep convection and easterly wave activity over Africa: Convective transitions and mechanisms. *Monthly Weather Review* **146**:1945–61. https://doi.org/10.1175/MWR-D-17-0217.1

Moncrieff MW. (2010) The multiscale organization of moist convection and the intersection of weather and climate. In *Climate Dynamics: Why Does Climate Vary?* Eds. Sun D-Z, Bryan F. American Geophysical Union, Washington, pp. 3–26. doi:10.1029/2008GM000838

Oreopoulos L, Cho N, Lee D, Kato S. (2016) Radiative effects of global MODIS cloud regimes. *Journal of Geophysical Research: Atmospheres* **121**:2299–317. doi:10.1002/2015JD024502

Oreopoulos L, Cho N, Lee D, *et al.* (2014) An examination of the nature of global MODIS cloud regimes. *Journal of Geophysical Research: Atmospheres* **119**:8362–83. doi:10.1002/2013JD021409

Rossow WB, Mekonnen A, Pearl C, Goncalves W. (2013) Tropical Precipitation Extremes. *Journal of Climate* **26**:1457–66. doi:10.1175/JCLI-D-11-00725.1

Rossow WB, Tselioudis G, Polak A, Jakob C. (2005) Tropical climate described as a distribution of weather states indicated by distinct mesoscale cloud property mixtures. *Geophysical Research Letters* **32**:L21812. doi:10.1029/2005GL024584

Tan J, Jakob C, Lane TP. (2013) On the Identification of the large-scale properties of tropical convection using cloud regimes. *Journal of Climate* **26**:6618–32. doi:10.1175/JCLI-D-12-00624.1

Tan J, Oreopoulos L. (2019) Subgrid precipitation properties of mesoscale atmospheric systems represented by MODIS cloud regimes. *Journal of Climate* **32**:1797–812. https://doi.org/10.1175/JCLI-D-18-0570.1

Tan J, Jakob C, Rossow WB, Tselioudis G. (2015) Increases in tropical rainfall driven by changes in frequency of organized deep convection. *Nature* **519**:451–54. doi:10.1038/nature14339

Tan J, Oreopoulos L, Jakob C, Jin D. (2018) Evaluating rainfall errors in global climate models through cloud regimes. *Climate Dynamics* **50**:3301–14. doi:10.1007/s00382-017-3806-7

Tselioudis G, Tromeur E, Rossow WB, Zerefos CS. (2010) Decadal changes in tropical convection suggest effects on stratospheric water vapor. *Geophysical Research Letters* **37**:L14806. doi:10.1029/2010GL044092

Worku LY, Mekonnen A, Schreck CJ. (2019) Diurnal cycle of rainfall and convection over the Maritime Continent using TRMM and ISCCP. *International Journal of Climatology* **39**:5191–200. https://doi.org/10.1002/joc.6121

Slide 1

Examining Global Precipitation with Mesoscale Atmospheric Systems Represented by Cloud Regimes

Jackson Tan

University of Maryland Baltimore County / NASA Goddard Space Flight Center

Slide 2

ISCCP Joint-Histograms Summarizes Cloud Distributions

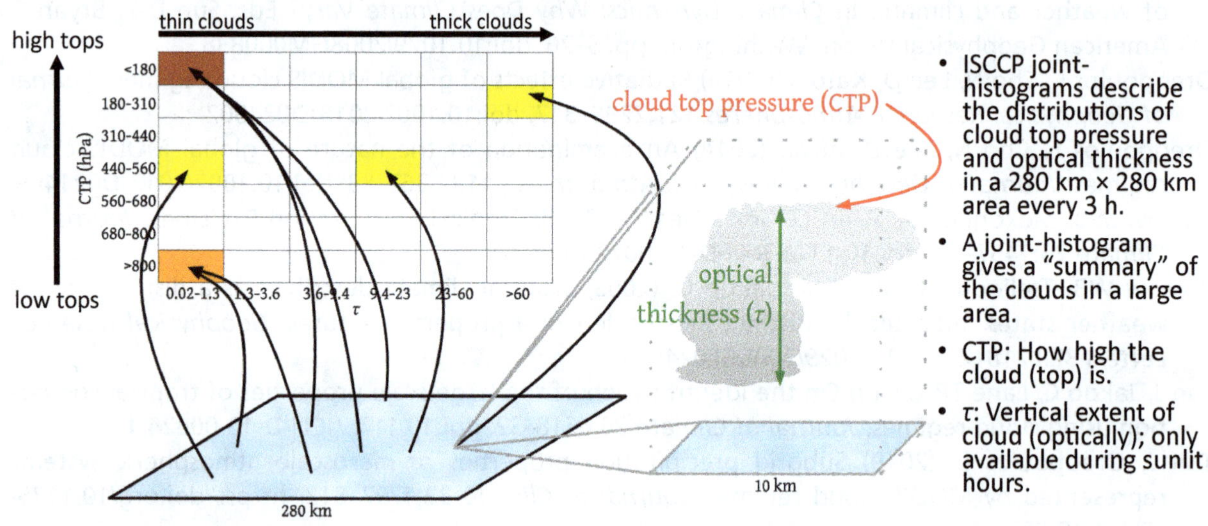

- ISCCP joint-histograms describe the distributions of cloud top pressure and optical thickness in a 280 km × 280 km area every 3 h.
- A joint-histogram gives a "summary" of the clouds in a large area.
- CTP: How high the cloud (top) is.
- τ: Vertical extent of cloud (optically); only available during sunlit hours.

Slide 3

How to Organize the Joint-Histograms?

Despite the reduction in information, the joint-histograms are still highly varied.

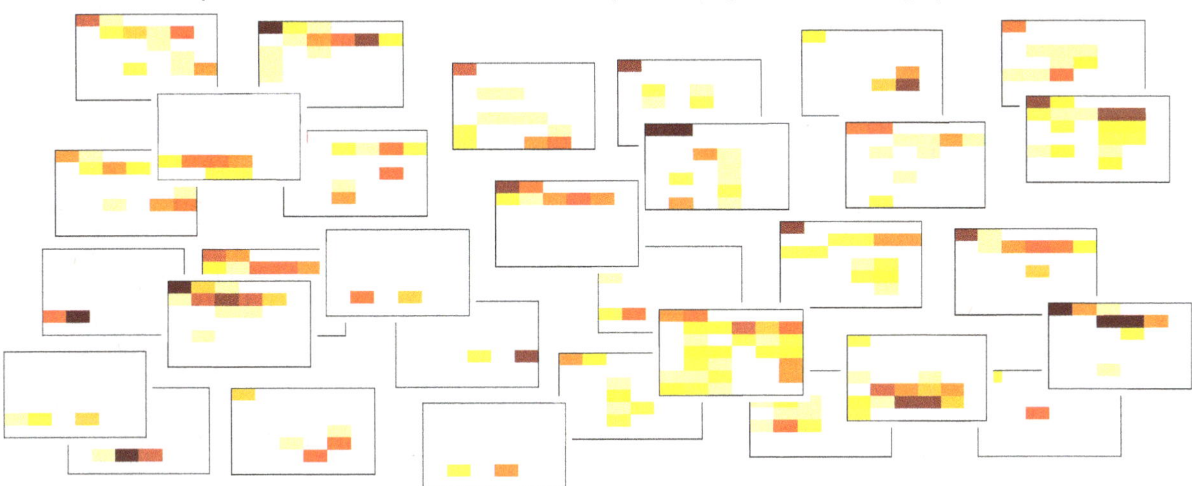

Slide 4

Solution: Find Pattern Clusters → Cloud Regimes

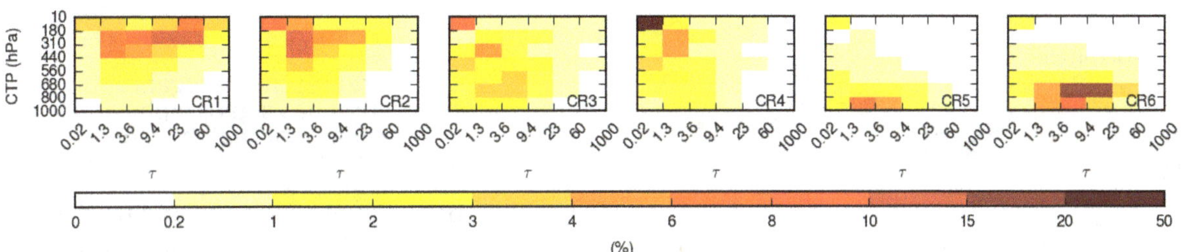

- Use *k*-means clustering algorithm to identify "clusters" of similar patterns in the joint-histograms; figure shows the mean pattern (centroids) of each cluster in the tropics.
- Clusters represent recurring, archetypal cloud fields and are hence called Cloud Regimes (CRs).
- CRs are useful because they correspond to different mesoscale atmospheric systems. Because of this, they are also sometimes called Weather States (WSs).

Slide 5

Cloud Regimes Represent Atmospheric Systems

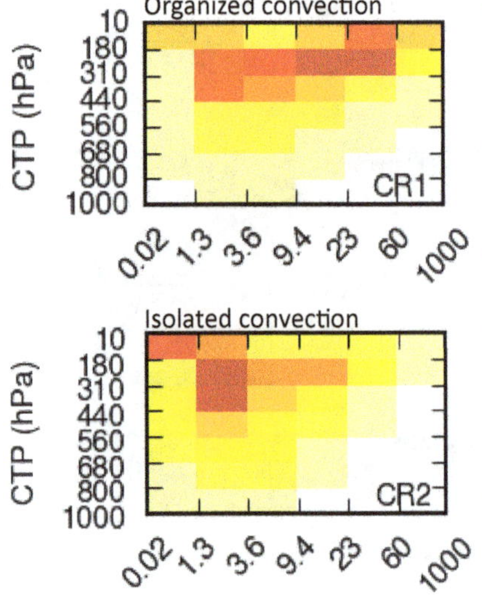

- *Organized convection or mesoscale convective systems*. With deep convective clouds (thick clouds with high tops) and stratiform anvils (clouds with high tops and medium thickness), CR1 represents large and organized storm systems.
- *Isolated (or less organized) convection*. With thinner clouds, CR2 has some deep convective clouds but more stratiform anvils and cirrus (thin clouds with high tops).
- Other CRs can represent systems such as extratropical storms, stratocumuli fields or "fair weather" states.
- The usefulness of the CRs comes from the fact that they can represent actual atmospheric systems that are multifaceted and complex. The nomenclature of the CRs (e.g., CR1, CR2) may vary from study to study, but it is what they represent physically that is important.
- CRs can be derived from other instruments such as MODIS or even other clustering methods. The fact that major CRs are so robust is an indication that it represents actual atmospheric systems.

Slide 6

Organized Convection Reflects Rainfall Pattern

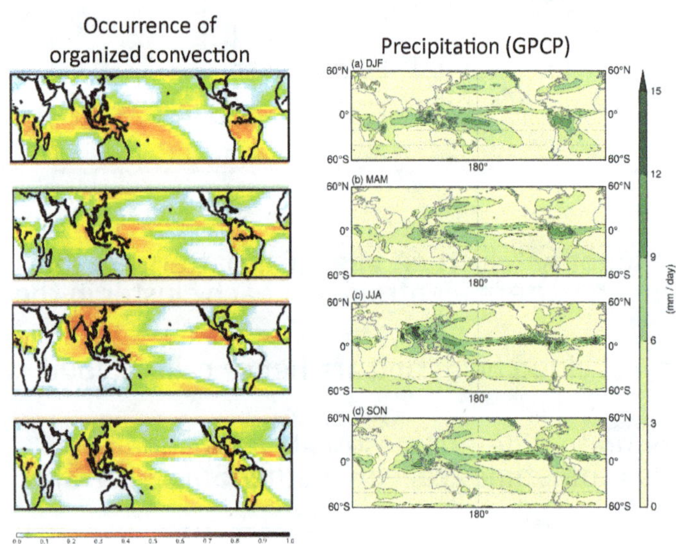

- The CR representing organized convection has a geographical distribution that matches the seasonal "march" of the rainfall. This suggests a strong connection between precipitation and the CRs.
- While this is not surprising because clouds and precipitation are intimately linked processes, the fact that precipitation is primarily associated with just one CR is suggestive of its disproportionate impact.
- The focus of these slides is to explore the connection between precipitation and recurring atmospheric phenomena as represented by the CRs.

Slide 7

Organized Convection Produces Most Tropical Rainfall

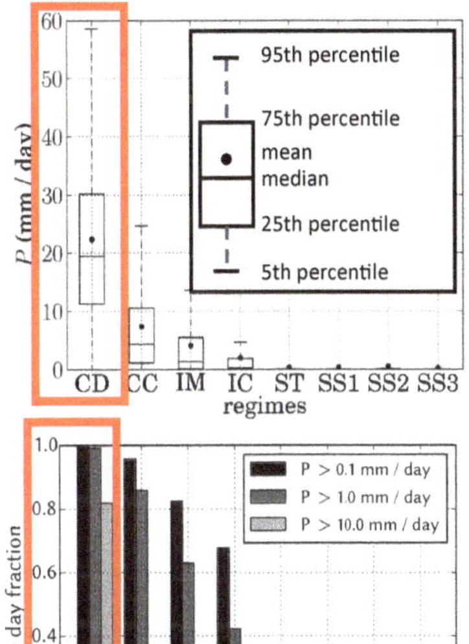

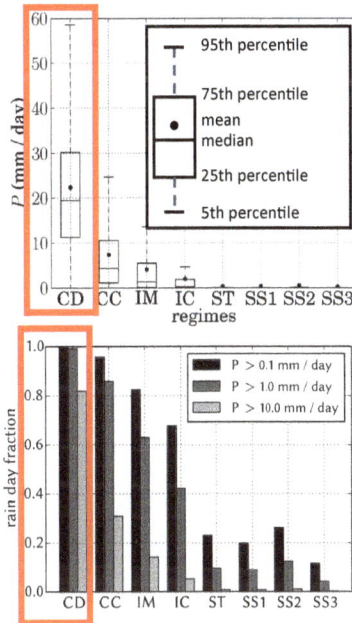

- Comparing CR occurrences with satellite-observed precipitation reveals that each CR has a distinct distribution of precipitation rates. In fact, precipitation is generally associated with only a few CRs; other CRs are generally dry. This concurs with daily experience: there are many cloud types, but only a few produce precipitation.
- In particular, one CR, associated with organized convection (marked by red boxes here and in the next slide), stands out in its precipitation profile: unusually high median rate of 20 mm/day, near 100% chance of any precipitation (>0.1 mm/day), and more than 80% chance of heavy precipitation (>10 mm/day).
- In fact, despite occurring less than 10% of the time (RFO), it produces nearly half the rainfall in the tropics (contribution); see figure in next slide.
- Physically, this tells us that the phenomenon of organized convection is the dominant mean by which precipitation is produced in the tropics. It illustrates how "unequal" precipitation patterns are: most of the precipitation amount is produced by a minority of atmospheric states.

Slide 8

Precipitation is Controlled Primarily by a Few CRs

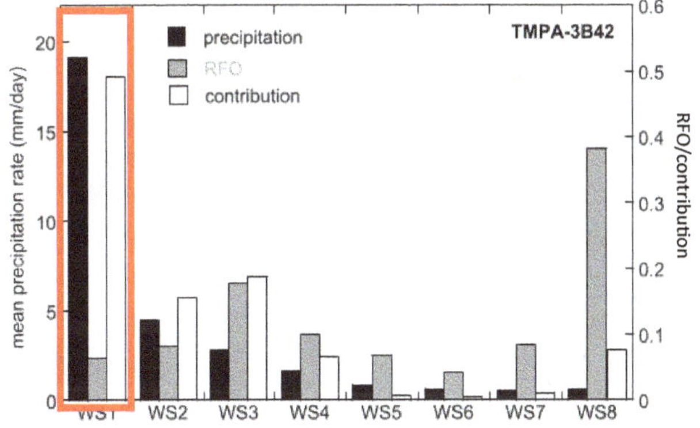

- There are other convectively-weaker CRs than is associated with lower precipitation rates, but most CRs, such as the fair weather "WS8" or "ST", are dry. For example, "WS8" occurs nearly 40% of the time but produces less than 10% of the rainfall. These "dry" CRs are important for other purposes, such as radiative effect, but they often can be ignored when looking at precipitation.
- Two other CRs with nonnegligible precipitation is "CC"/"WS2" and "IM"/"WS3". Representing isolated convective regimes, the rainfall associated with these two CRs are usually isolated thunderstorms.

Slide 9

Tropical Rainfall Change Driven by Organized Convection

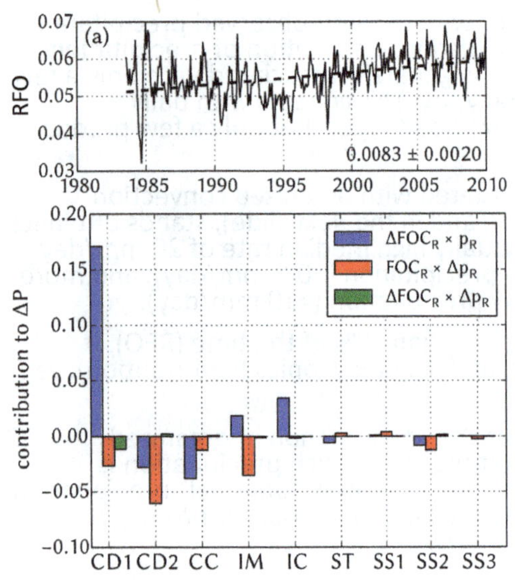

- If most precipitation is produced by organized convection, what happens when organized convection changes?
- Across the entire tropics, the occurrence of the organized convection CR has been increasing over the last two to three decades.
- This is the largest contribution to the change in rainfall over the same period (blue bar of "CD1"), overcoming the decreased contribution per event (red bar of "CD1" and "CD2").

Slide 10

Tropical Rainfall Change Driven by Organized Convection

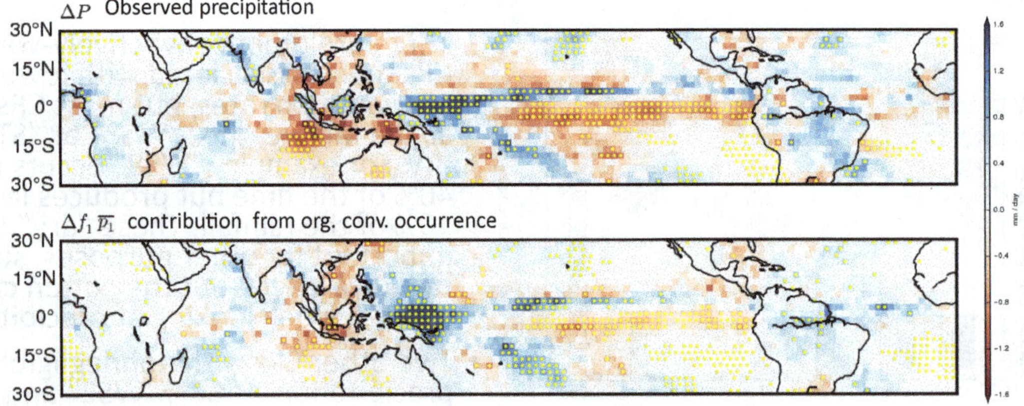

Indeed, where more organized convection occurs, the precipitation increases. Likewise for decrease in precipitation. It is the dominant contribution compared to other factors. On the global scale, organized convection largely controls how tropical precipitation has changed and will change in the future. This has implications on the accuracy of future projections of precipitation because climate models currently struggle in properly representing such organized convective systems.

Slide 11

Compensating Precipitation Errors in Climate Models

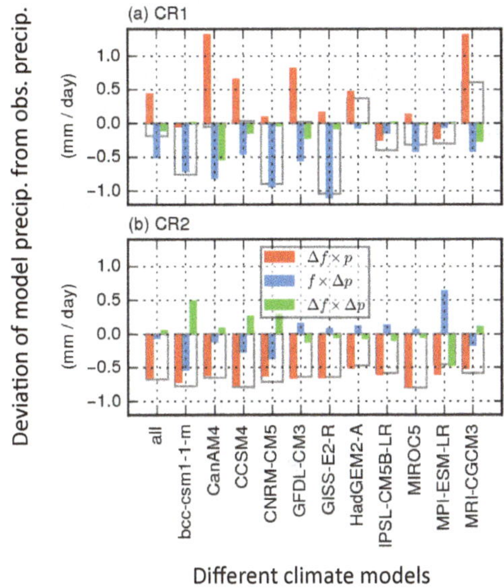

- Climate models underestimate the intense rainfall associated with organized convection (CR1) in the tropics (blue bars). This is a known longstanding problem with the current generation of models; significant efforts are ongoing to resolve this issue.
- Some models compensate for this deficiency by producing more organized convection (red bars), resulting in some compensating errors to the total precipitation.
- Models also underestimate how often isolated convection (CR2) occurs, resulting in a persistent under-contribution to the total tropical precipitation.

Slide 12

Summer Extratropical Storms in Mid-Latitudes

Tropics and mid-latitudes (60°N/S)

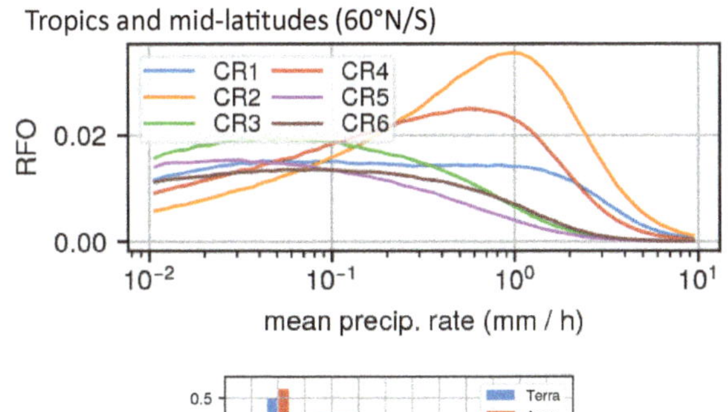

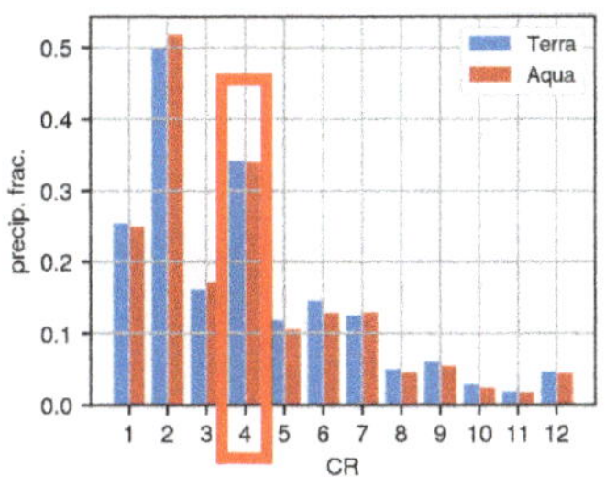

- Beyond the tropics, other systems come into play, though organized convection still dominates.
- Extratropical storms during summer (CR4) are associated with significant precipitation, second to that of organized convection (CR2).
- It also has a relatively high fraction of the grid that is precipitating (about 35% on average).
- Note: Contrast this with winter extratropical storms (CR5) — both the precipitation rates and fractions are much lower.

Slide 13

Summary

- Cloud regimes are classifications of cloud fields observed by satellites. They represent — and hence can serve as proxies for — atmospheric systems globally.
- Because of the intimate link between clouds and precipitation, CRs have distinct precipitation properties. Precipitation is limited to only a handful of CRs. Most of the rainfall in the tropics is associated with organized convection. Isolated convection produces less rainfall.
- Since organized convection has been increasing over the last two to three decades, this means that the contribution to total tropical rainfall from organized convection has been increasing. This increase is more than the slight decline in the average rainfall of each occurrence of organized convection. The contributions from other CRs are small.
- As the current generation of climate models does not represent organized convection, an evaluation of organized convective rainfall in models confirmed their reduced contribution, which is in some models compensated by organized convection occurring more frequently.
- In the mid-latitudes, summer extratropical storms are associated with significant precipitation, just below that of organized convection.
- These results illustrate how cloud regimes can effectively and objectively identify different systems in the global atmosphere and advance our understanding of its properties, including precipitation.

Slide Captions

Slide 1

Slide 2 One of the products from the International Satellite Cloud Climatology Project (ISCCP) is a dataset of joint-histograms summarizing the cloud field over a large area. Over a three-hour period during the daytime, a network of geostationary and polar-orbiting satellites observes the cloud properties over different grid cells around the world. These grid cells are much larger than individual clouds, with areas of about 280 km × 280 km (or 100 km × 100 km in the new "H" version). A joint-histogram is produced for each grid cell every three hours, a sampling of the population of different clouds in its area. These joint-histograms sort each observation of the cloud according to how high the cloud top is (cloud top pressure) and the optical extent of the cloud vertically (optical thickness), providing a simplified view of the cloud field.

Deep convective clouds (or cumulonimbus clouds) have high cloud tops and a great vertical extent; they are sorted into the top-right corners of the joint-histogram. Cirrus clouds, being high but thin, go to the top left. Puffy cumulus clouds that appear on a dry, sunny day have low tops and low optical thickness; they are binned to the bottom left. Cumulus congestus clouds, often the halfway point between the low cumulus and the towering cumulonimbus, have a moderate thickness in the middle level of the atmosphere; these populate the center portion of the joint-histogram. Therefore, these joint-histograms describe the populations of clouds within a particular area over a particular time period.

Slide 3 The joint-histogram summarizes the cloud fields without the detailed spatial locations of the clouds within the grid cell. Nevertheless, despite the reduction of information, the joint-histograms can be extremely varied. It is not uncommon to have multiple distinct cloud types — e.g., cumulonimbus, cirrus, cumulus — all coexisting at the same time, just as one might see on a humid summer afternoon. Furthermore, at 280 km × 280 km, there are 6,596 grid cells globally. (At 100 km × 100 km, there are 41,252 grid cells globally.) So while we can look at individual joint-histograms and interpret the cloud fields accordingly, it is laborious to do so globally and over a long period of time. A method to categorize these joint-histograms statistically is needed.

Slide 4 The solution to the problem of organizing the varied joint-histograms lies in cluster analysis. Cluster analysis seeks to identify groups in a dataset such that members in the same group are more similar — based on some specified metric — to each other than to members in other groups. While there are many clustering algorithms with wide-ranging complexity, one of the simplest — and yet effective for our purpose — is the *k*-means clustering algorithm, an unsupervised machine learning technique that seeks to identify recurring patterns. Given a large set of points and a small set of centroids, the algorithm will iteratively assign the points to the centroids and update the centroids based on the average of its associated points, typically stopping only when the centroids no longer change.

Applied to a collection of joint-histograms, the algorithm, despite its simplicity, can identify clusters of joint-histograms. The figure shows the centroids (the average joint-histograms of all members in a cluster) that emerge when *k*-means is applied to the joint-histograms in the tropics. Many of these centroids also appear when *k*-means is applied to joint-histograms in other parts of the globe, illustrating the robustness of this approach in identifying clusters of cloud fields. One disadvantage of the *k*-means algorithm is the need to specify beforehand the number of clusters; however, a pseudo-objective set of criteria has been devised to provide guidelines in determining this number. At the same time, from a scientific perspective, whether the clusters are statistically "well chosen" (however this is defined) is less important than whether they are physically useful and representative of the real world. Indeed, multiple studies have demonstrated that these clusters effectively represent recurring, archetypal cloud fields in the atmosphere, and even more, have used them to advance our understanding of the climate system. Hence, these clusters are endowed with the label of "Cloud Regimes" (CRs), though occasionally — due to the fact that they represent different weather conditions — they are also referred to as Weather States. In this presentation, "CR" is used.

Slide 5 This slide selects two of the six CRs in the tropics for further discussion to exemplify the rich information content that is captured by the CRs and — more importantly — their connection to a physical system. One CR, which is so prominent that it appears in nearly all CR datasets from different regions of the world, represents organized convection. Based on the centroid, this CR represents cloud fields that have deep convective clouds (thick clouds with high tops) and stratiform anvils (slightly thinner clouds with high tops). This combination of clouds occurs in large, complex, organized storm systems in the real world, such as mesoscale convective systems and even hurricanes. The other example is also a CR that describes a convective environment, but the convection is less organized. The centroid describes a mix of stratiform anvils and cirrus clouds (thin clouds with high tops). Such cloud fields may depict a scene in which isolated "popcorn" convection occurs, leading to small, individual thunderstorms scattered over a large area, but the conditions are not favorable for them to combine and develop into a larger system. One can imagine scattered thunderstorms on a hot summer afternoon — bringing rain to one neighborhood and leaving the next suburb dry — as an excellent example of this isolated convection CR. Other CRs may represent different precipitating cloud fields, such as extratropical storms or congestus and nimbostratus fields; they can also represent suppressed (nonprecipitating) conditions such as fields of stratocumuli or scenes of "fair weather".

The value of CRs comes from the fact that they represent actual systems in the atmosphere, multifaceted and complex systems that often defy a simple way of characterizing them. Many studies have used different classification techniques, such as sorting a grid cell by mid-tropospheric vertical velocity, but CRs turn out to be an effective and objective way of identifying different atmospheric systems. In fact, CRs have been derived using observations from other instruments such as the Moderate Resolution Imaging Spectroradiometer (MODIS) as well as other clustering techniques like self-organizing maps, but quite often they result in the same key CRs. This implies that these major CRs are robust and represent actual atmospheric systems.

Slide 6 As we would expect based on our interpretation of its cloud structure, the organized convection CR is associated with lots of rainfall. On a seasonal basis, the geographical distribution of the organized convection matches that of precipitation (using the Global Precipitation Climatology Project version 2.3 dataset). This is compelling evidence that organized convection is a crucial knob controlling the rainfall in the tropical atmosphere. While this is not surprising given the intimate link between clouds and precipitation, it is the strong connection with only one CR — and very much less with other CRs — that is suggestive of an "unequal" effect of various phenomena on global patterns of precipitation. It is the focus of this presentation to advance our understanding of precipitation by sorting them based on recurring atmospheric phenomena as identified by CRs.

Slide 7 By matching (or compositing) precipitation data to CR occurrence in space and time, we can collect the precipitation values associated with each CR and analyze its distribution and properties. This allows us to examine, for example, the mean precipitation rate associated with organized convection. As opposed to a simple climatology of precipitation (e.g., maps of precipitation), this approach paints a deeper picture of the atmosphere, providing the processes that are giving rise to precipitation. The aggregation of precipitation values also helps reduce random errors in the observed data, which can be somewhat high for satellite-based data, especially of precipitation.

The two figures show simply the precipitation properties of various CRs, which are by themselves already very informative. To begin, most of the CRs are associated with very little precipitation at the global scale. There may be some precipitation — e.g., scattered drizzle within an area of 280 km in length — but they are negligible in the larger picture. Among the CRs with appreciable precipitation, they have distinct distributions in their associated precipitation. One CR, in particular, stands out: organized convection (emphasized with red boxes). Its precipitation values and the chance of rain are exceedingly high, with a median rain rate of nearly 20 mm/day and a greater than 80% chance of heavy precipitation (>10 mm/day). This is remarkable considering that these precipitation rates are averaged over a large 280 km × 280 km area; it is common to have intense rainfall over a small area of several kilometers, but it is rare to have similar rates over an area thousands of times larger. Other CRs, even those representing different states of isolated convection, have much lower precipitation rates. This illustrates the skewed contribution of precipitation systems: a small number of systems are responsible for most of the precipitation in the world.

Slide 8 The skewed contribution of precipitation in the previous slide is clearly on display in this figure here. Contribution depends not only on how much rain is associated with each CR but also on how frequently the CR occurs. Organized convection (WS1), despite occurring less than 10% of the time, produces nearly half the rainfall in the tropics by virtue of its extraordinarily high precipitation. In fact, it is precipitating in about half the area it occupies, much higher than other CRs. (Interestingly, its cloud cover is nearly 100%, which is a reflection that most clouds are not associated with precipitation.) The other two isolated convective regimes (WS2 and WS3) are more frequent, but they have lower contributions because of their lower precipitation rates; in fact, collectively, their contributions are still lower than that of organized convection. The fair weather regime (WS8) is the most frequent CR, occurring about 40% of the time, but it has a very low contribution because it is in general not precipitating. Such regimes may be important for other purposes such as studying the global cloud radiative effects, but for purposes of precipitation, they can typically be ignored.

Slide 9 If just one CR has such a disproportionately large influence over rainfall, this means that it is the primary factor controlling how rainfall changes. Using carefully quality-controlled data with checks to ensure satellite homogeneity, several studies on changes in organized convection have found an increase in its occurrence over the last 25 years embedded in a highly variable record (top figure). Furthermore, this increase is not globally homogeneous: organized convection is more frequent in some regions while less common in other regions. To trace this change in organized convection to precipitation, a decomposition of the contribution to how precipitation has changed in the tropics (bottom figure) shows the dominance of the increased occurrence of organized convection in driving an increase in rainfall. This increase overwhelms the slight decrease in the average rainfall per event. In other words, organized convection is producing slightly less rain on average, but it is becoming much more often and overwhelming the decreased rain. The increased rainfall associated with more frequent organized convection also dominated over the changes associated with other CRs.

Slide 10 Indeed, looking at the geographical distributions shown here in the two figures, the match between the observed change in precipitation and the change in the occurrence of organized convection is striking. The contribution to the change in precipitation from how frequently organized convection occurs is remarkably close to how precipitation has been observed to change. Other factors, such as the average rainfall of each occurrence of organized convection or even the sum of contributions from other convective regimes, pale in comparison. Therefore, the change in the occurrence of organized convection alone can explain most of the rainfall change in the tropics. However, this has a severe implication. Climate models used to project future changes do not have an explicit representation of organized convection. This is a longstanding problem — one that has so far been resistant to attempts at resolving it, though it has attracted significant attention recently and progress will likely be made in the coming generations of climate models. Until then, climate models will likely struggle to reproduce the rainfall associated with organized convection.

Slide 11 The previous slide suggested through the limitations in how climate models represent organized convection that they will be deficient in reproducing precipitation associated with organized convection. As climate models produce the joint-histograms that form the basis of CRs, it is possible to define CRs in the models and thus evaluate their performance in terms of their precipitation of these atmospheric systems. This figure breaks down the contribution of the organized convection CR and the isolated convection CR to the error in total tropical precipitation for each model.

We can separate the contribution due to the error in how frequent it occurs (red bars) and the error in reproducing the precipitation of that system. As hypothesized, most models underestimate the rainfall associated with each occurrence of organized convection (blue bars of CR1). However, some models compensate for this by producing more organized convection (red bars of CR1). For one of the isolated convective regimes, the models consistently underestimate its occurrence (red bars of CR2), leading to a persistent underestimation of its contribution to the total tropical precipitation. Therefore, precipitation in climate models contains considerable uncertainties. Given that climate model projections are used to guide societal and policy planning, this means that future projections of precipitation change need to be interpreted cautiously with its associated uncertainty in mind. However, the continual improvement of climate models, both in their representation of convection and their resolution, will help reduce these uncertainties.

Slide 12 The analysis thus far has been focused on the tropics. This is primarily because precipitation is in general most intense and frequent in the tropics. However, there are atmospheric systems in higher latitudes that also result in significant precipitation. Extending beyond the tropics to include the mid-latitudes (60°N/S), organized convection (CR2) remains the system associated with the most intense precipitation rates (top figure) and large area coverage of precipitation (bottom figure). However, extratropical storms during summer (CR4) now play a significant role. These storms have the highest precipitation rates and fractions after organized convection. Interestingly, in contrast, winter extratropical storms (CR5) are associated with much lower precipitation despite their similarity in cloud pattern and annual geographical distribution to summer extratropical storms. This demonstrates that a similarity in cloud pattern does not necessarily lead to a similarity in precipitation distribution, despite the intimate connection between clouds and precipitation. It also demonstrates the power of CRs in distinguishing similar systems and thus their ability to represent atmospheric systems globally.

Slide 13 Summary

Lecture 7

Satellite-derived Water Vapor Transport — Use For Diagnosing the Tropical Circulation Change

B. J. Sohn

School of Earth and Environmental Sciences
Seoul National University, Seoul, S. Korea

B. J. Sohn is a Professor of Earth and Environmental Sciences at Seoul National University, Korea. He teaches Radiative Transfer and Satellite Remote Sensing. His research focuses on satellite remote sensing in general, and its applications to understand weather and climate phenomena, including radiation budget, rain mechanism, and Arctic climate change.

Introduction

The main point of the global water cycle should be water transfer from one water reservoir to another; the hydrosphere containing the water consists of five reservoirs (so-called sub-climate systems: ocean, cryosphere, land, atmosphere, and biosphere). With direct and diffuse solar absorption by the surface, water evaporates in a form of vapor from the hydrosphere, and water in the vapor phase is transported from one source region to a sink region by atmospheric circulation (Slide 2). In association with atmospheric circulation, low-level convergence or mechanical forcing such as induced upward motion carries the water vapor upward as well. As the water vapor moves upward, because of decreasing pressure and temperature with height, it condenses into liquid/ice water, forming cloud while releasing the latent heat acquired at the source region. Large amounts of latent heat is released in this step, and it often shapes the circulation pattern, in particular, over the tropics.

On the other hand, clouds, simultaneously formed with the condensation, are capable of altering the radiation balance through shortwave cloud albedo and longwave greenhouse effects, whose net effect is, in general, cooling due to the stronger cloud albedo effect. Cloud particles grow into precipitation size particles and fall out as a form of precipitation. Some precipitation falls as snow and can be cumulated and hardened as ice caps and glaciers. Other precipitation falls over the ocean and land and returns to the vapor source region over time, continuing the water cycle (Slide 2).

Among components involved in the above water cycle, water vapor transport and its associated convergence/divergence shape spatial distributions of heating through the latent heat release. Because the differential heating induced by latent heating distribution is the main force driving the atmospheric circulation, in particular over the tropics, improved knowledge on water vapor transport is important for better understanding our weather/climate system. One example of demonstrating the link of convergence or divergence of water vapor transport and associated latent release with circulation change may be associated with the El Niño and Southern Oscillation (ENSO), which shows extreme interannual variations of rainfall as well as evaporation within the east-west circulation cell connecting the western Pacific to the eastern Pacific (Sohn *et al.*, 2013).

Water vapor transport has been diagnosed from conventional radiosonde observations or reanalysis products. However, the picture of water vapor transport is still incomplete due in large part to the lack of observations of atmospheric hydrological data, in particular over the oceans where conventional observations are sparse. Although conventional radiosonde measurements are very helpful in diagnosing regional water vapor transport features and the associated water budget, persistent uncertainty exists in measuring water vapor transport for a particular climate event occurring globally on an interannual basis. Long-term statistics of the global water vapor transport were documented based on conventional measurements of dynamic and hydrological variables (e.g., [Peixoto and Oort, 1983]), but only in a climatological sense with very coarse spatial resolution. Thus, if spatially and temporally homogeneous satellite measurements of water vapor transport are available, results will lead to a better understanding of water vapor transport in the water cycle and the role of water vapor in climate. We here offer an indirect calculation method, which uses satellite-borne data for inferring water vapor transport on a 2-D basis.

Methodology

The horizontal total water vapor transport (**Q**) crossing the boundary between two adjacent atmospheric columns can be given as follows:

$$Q=\frac{1}{g}\int_0^{p_s} q\,\mathbf{V}\,dp, \qquad (1)$$

where g is the acceleration of gravity, q is specific humidity, **V** is the horizontal wind vector, and p_s is the surface pressure. Thus, **Q** is a transport vector. Here, calculation methods are introduced, depending on different data sets used.

Direct Flux Integration Using Analysis Data

Although reanalysis data of atmospheric variables are not observations, associated results often give a reference for the comparison of satellite-derived results. We here refer to the method using reanalysis data as an analysis method. When vertical profiles of specific humidity (q) and horizontal wind fields (**V**) are available from datasets like ERA-Interim and NCEP, Eq. (1) can be directly integrated with the given profile data. For example, ERA-Interim reanalysis has q and **V** at 23 levels from 1,000 to 1 hPa, and it is straightforward to calculate the water vapor transport **Q** (Slide 3).

Satellite Method 1

Since satellite measurements of the V profile are not available so far, the water vapor transport may be indirectly calculated using the concept of water budget (Sohn *et al.*, 2004). For a given atmospheric column, the water balance is achieved as follows:

$$\frac{\partial W}{\partial t}+\nabla\cdot \mathrm{Q}=E-P, \tag{2}$$

$$W=\frac{1}{g}\int_0^{p_s} q\,dp, \tag{3}$$

where W is the total precipitable water, and E and P are evaporation and precipitation, respectively. Since **Q** is a vector, it can be separated into two components: rotational ($\mathbf{Q}_R$) and divergent ($\mathbf{Q}_D$) components. If the water vapor transport potential function (Φ) is introduced, we effectively remove the rotational component ($\mathbf{Q}_R$), i.e.,

$$\nabla\cdot \mathrm{Q}_D=E-P-\frac{\partial W}{\partial t}=-\nabla^2\Phi, \tag{4}$$

$$\mathrm{Q}_D=-\nabla\Phi. \tag{5}$$

Equations (4) and (5) are solved for Φ with inputs of evaporation, precipitation, and total precipitable water (Slide 3). Time variation of precipitable water is counted for estimating the total vapor storage at the given time and given column. Those input data are available from passive microwave measurements from space such as the Special Sensors for Microwave Imager (SSM/I) onboard the Defense Meteorological Satellite Program (DMSP) satellite and the Tropical Rainfall Measuring Mission (TRMM) Microwave Imager (TMI). Thus, such satellite-derived water budget data offer an opportunity of studying the atmospheric water budget, including water vapor transport, although the accuracy of space-borne water budget data may still be subject to questions (Slide 4).

Solving Eq. (4) for the water vapor transport potential function (Φ) requires known boundary conditions for a given domain. To avoid the difficulty in specifying boundary conditions, Eq. (4) can be solved over the global domain using spherical harmonics in which boundary conditions vanish at both poles. Since microwave-based water budget data are normally available over the ocean, for solving water vapor transports in a global domain, land areas void of microwave-based precipitation and evaporation can be filled with reanalysis products. Thus, the method is not truly based on satellite measurements only. In the future, if sounder products of both q and **V** profiles are available over land, the satellite-only method can truly be used for estimating the water vapor transport. However, in this study, influences of filling the land void of SSM/I-based E-P data were studied by filling it with NCEP values. It was indicated that the adjustments made only over the ocean data, while keeping

NCEP-based land data, brought in a weaker SSM/I water vapor transport from the ocean to the land (Park and Sohn, 2007).

Satellite Method 2

Another satellite-based method of calculating the water vapor transport was introduced by Liu and Tang (2005). In their study, an equivalent velocity (V_e) was defined in such a way that $V_e = Q/W$. Then V_e was predicted from surface wind (V_s) measured by the scatterometers. In order to obtain the relationship between V_e and V_s, the Artificial Neural Network approach was taken by applying blended radiosonde/NCEP data collocated with ocean surface wind vectors from QuikSCAT. A 5-day running average was applied to each parameter to remove highly fluctuating short-term variations. Used inputs are: (1) ocean surface wind speed, (2) direction, (3) latitude, (4)–(5) sine and cosine of longitude, and (6)–(7) sine and cosine of time.

In this chapter, we present water vapor transport obtained by solving the potential function from the use of Special Sensor Microwave Imager (SSM/I) derived water budget data in Hamburg Ocean Atmosphere Parameters and Fluxes from Satellite Data (HOAPS) (Anderson *et al.*, 2010), see Slide 5.

Use of the Water Vapor Transport for Studying the Change in Tropical Circulation

In response to the global warming trend caused by the increase of anthropogenic greenhouse gas emissions, the tropical circulation has been undergoing significant changes and is expected to be continuously changing. One of the debates is associated with large-scale thermally direct circulation over the Indo-Pacific oceans. The long-term trend of the Walker circulation over the tropical Pacific is weakening over the course of the 20th century because of a reduced east-west SST gradient across the Pacific Ocean basin (Vecchi *et al.*, 2006). It is further argued that the circulation, in response to CO_2 increase, will be weakened because of thermodynamic and hydrological constraints under the warmed earth-atmosphere system (Held and Soden, 2006). It was contended that the much smaller rainfall increase in surface temperature increase (2.0% per degree) in comparison to water vapor increase (7.5% per degree) would require a slower circulation. If such a slowdown of the circulation is accepted, we should be able to see facets of circulation changes from satellite measurements. We can examine the circulation changes using the aforementioned satellite-derived water vapor transport.

In order to use the water vapor transport as an index to examine whether circulation has indeed been weakened we define an effective wind and use it as a circulation index. If we take the divergent component Q_D and insert it into Eq. (1), then:

$$V_E = Q_D/TPW, \tag{6}$$

where V_E can be referred to as an effective wind for water vapor transport (Slide 7). It was proved that the effective wind represents the speed at which the water vapor is carried in the lower returning branch of the circulation cell, generally at around an 850-hPa level.

When the Walker circulation cell is considered, the surface-returning branch takes a path from the eastern Pacific sinking region to the western Pacific ascending region, carrying the water vapor to the ascending convection region. Thus, the direction and magnitude of V_E can be considered to be a circulation index (Slide 8). In fact, satellite-based water budget data is converted into a dynamical quantity, which can give us information on the circulation and its trend (Slide 9).

Focusing on the tropical circulation, circulation indices for the north-south Hadley circulation and east-west Walker circulation can be defined using V_E. For example, instead of using the surface pressure difference between the eastern and western Pacific as a Walker circulation index, the circulation index, obtained by taking the average of V_E over the tropical area connecting ascending and descending epicenters, can be used for monitoring the Walker circulation intensity and temporal variation. From long-term trend analysis of satellite-based V_E, the Walker circulation has intensified over the last three decades (Sohn and Park, 2010), in contrast to the general view of weakening in climate simulations such as the Coupled Model Intercomparison Project (CMIP), see Slide 10.

Satellite-derived water vapor transports were also used in conjunction with SST, ISCCP cloud data (Rossow *et al.*, 2005), NOAA HIRS channel brightness temperature, and sea level pressure data to test the hypothesis that the Pacific Walker circulation has intensified in a 30-year period (1979–2008). Temporal and spatial variations of those variables, recognized from independent datasets, occur in tandem with each other, strongly supporting the intensified Walker circulation over the tropical Pacific Ocean (Sohn *et al.*, 2013). Since the SST trend was attributed to more frequent occurrences of central Pacific-type El Niño in recent decades, it is suggested that the decadal variation of El Niño caused the intensified Walker circulation over the past 30 years. Climate models, however, show that model results deviate greatly from the observed intensified Walker circulation. These studies, initiated from water vapor transport based on satellite-based water budget data, led to the conclusion that the uncertainties in current climate models may be due to the natural variability dominating the forced signal over the tropical Pacific during the last three decades in the 20th-century climate scenario runs by CMIP5 CGCMs (Slides 11–14).

References

Anderson A, Fennig K, Klepp C, *et al.* (2010) The Hamburg ocean atmosphere parameters and fluxes from satellite data — HOAPS-3. *Earth System Science Data* **2**:215–34. doi:10.5194/essd-2-215-2010.

Held IM, Soden B. (2006) Robust responses of the hydrological cycle to global warming. *Journal of Climate* **19**:5686–99.

Liu WT, Tang W. (2005) Estimating moisture transport over oceans using space-based observations. *Journal of Geophysical Research* **110**:D10101. doi:10.1029/2004JD005300

Park S-C, Sohn BJ, Wang B. (2007) Satellite assessment of divergent water vapor transport from NCEP, ERA40, and JRA25 reanalyses over the Asian summer monsoon region. *Journal of the Meteorological Society of Japan* **85**:615–32.

Peixoto JP, Oort AH. (1983) The atmospheric branch of the hydrological cycle and climate. In *Variation in the Global Water Budget.* Eds. Street-Perrott A, Beran M, Ratcliffe R. Reidel, pp. 5–65.

Rossow WB, Tselioudis G, Polak A, Jakob C. (2005) Tropical climate described as a distribution of weather states indicated by distinct mesoscale cloud property mixtures. *Geophysical Research Letters* **32**:L21812. doi:10.1029/2005GL024584

Sohn BJ, Park S-C. (2010) Strengthened tropical circulations in past three decades inferred from water vapor transport. *Journal of Geophysical Research* **115**:D15112. doi:10.1029/2009JD013713

Sohn BJ, Smith EA, Robertson FR, Park S-C. (2004) Derived over-ocean water vapor transports from satellite-retrieved E-P datasets. *Journal of Climate* **17**:1352–65.

Sohn BJ, Yeh S-W, Schmetz J, Song H-J. (2013) Observational evidences of Walker circulation changes over the last 30 years contrasting with GCM results. *Climate Dynamics* **40**:1721–32.

Vecchi GA, Soden BJ, Wittenberg AT, *et al.* (2006) Weakening of tropical Pacific atmospheric circulation due to anthropogenic forcing. *Nature* **441**:73–6.

Slide 1

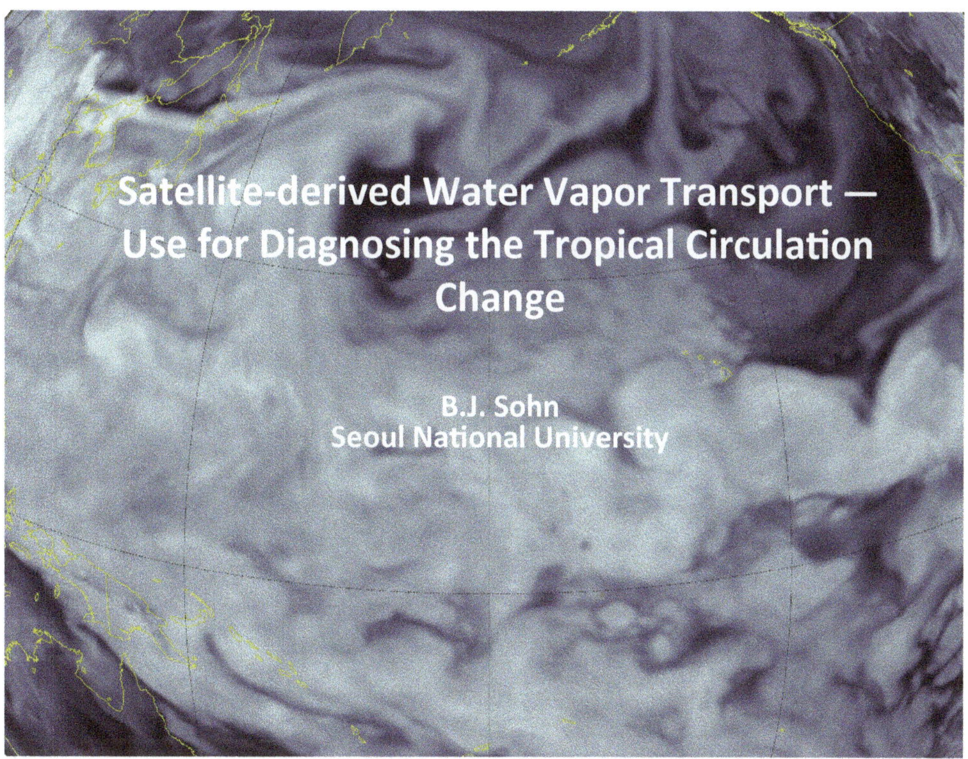

Slide 2

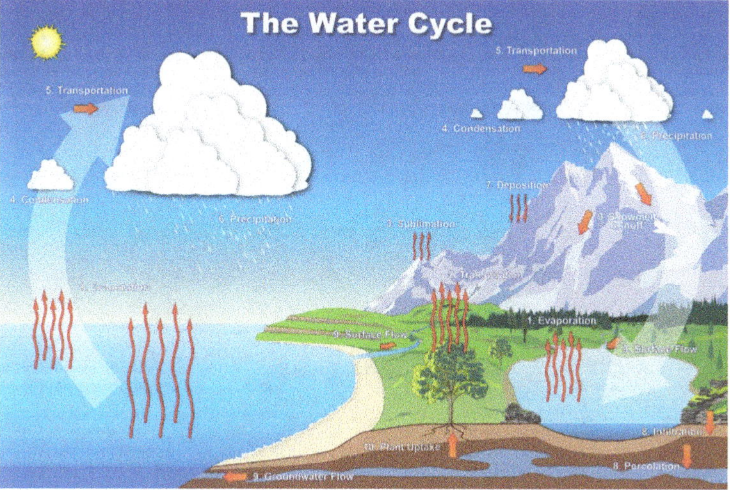

Image courtesy of NOAA

The global water cycle transfers water from one water reservoir to another; the hydrosphere consists of the ocean, cryosphere, land, atmosphere, and biosphere. Through the atmospheric circulation, evaporated water vapor is transported from one source region to a sink region, effectively carrying latent heat. Because the latent heat release is, in turn, the main force driving the atmospheric circulation, improved knowledge of water vapor transport is important for the better understanding of our climate system.

Slide 3

Water Vapor Transport (WVT) Calculation

M1: Use of water budget data from satellite

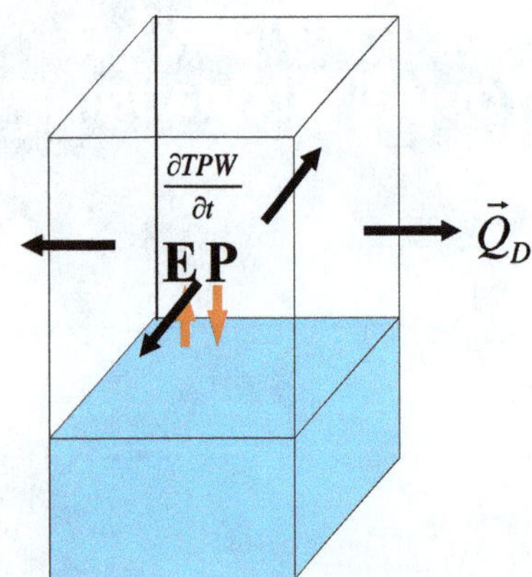

$$\frac{\partial(\boldsymbol{TPW})}{\partial \boldsymbol{t}} + \nabla \bullet \boldsymbol{Q} = \boldsymbol{E} - \boldsymbol{P}$$

E-P = Evaporation - Precipitation

$Q = Q_R + Q_D$

Q_R : Rotational component

Q_D : Divergent components

$$\nabla \bullet \mathrm{Q}_D = E - P - \frac{\partial TPW}{\partial t} = -\nabla^2 \Phi$$

$$\mathbf{Q}_D = -\nabla \Phi$$

M2: Direct integration for WVT with wind (V) and humidity (q) profiles

$$\mathrm{Q} = -\frac{1}{g}\int_{p_0}^{0} q\vec{\mathrm{V}}\, dp$$

Slide 4

Evaporation & Precipitation Climatology (1987-2005)

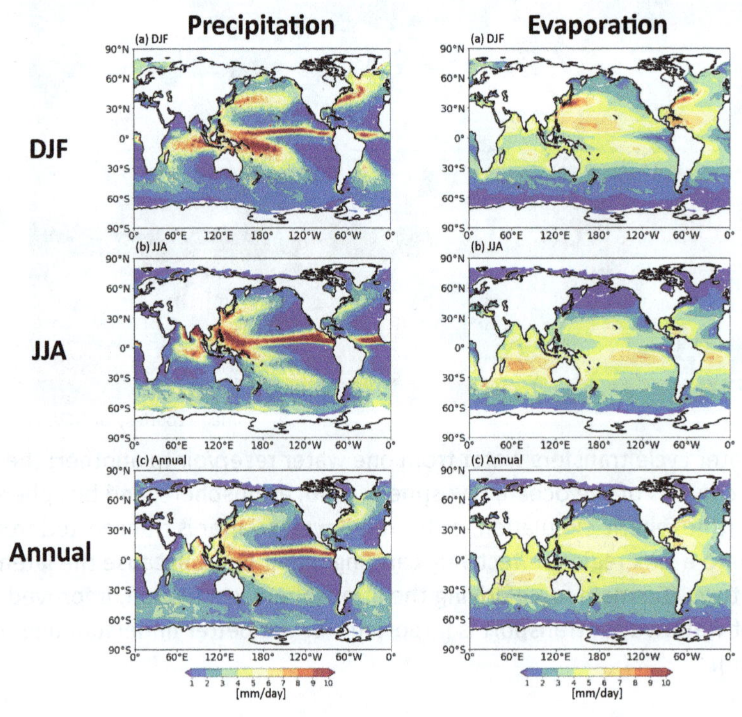

Slide 5

Divergent Water Vapor Transport (from HOAPS)

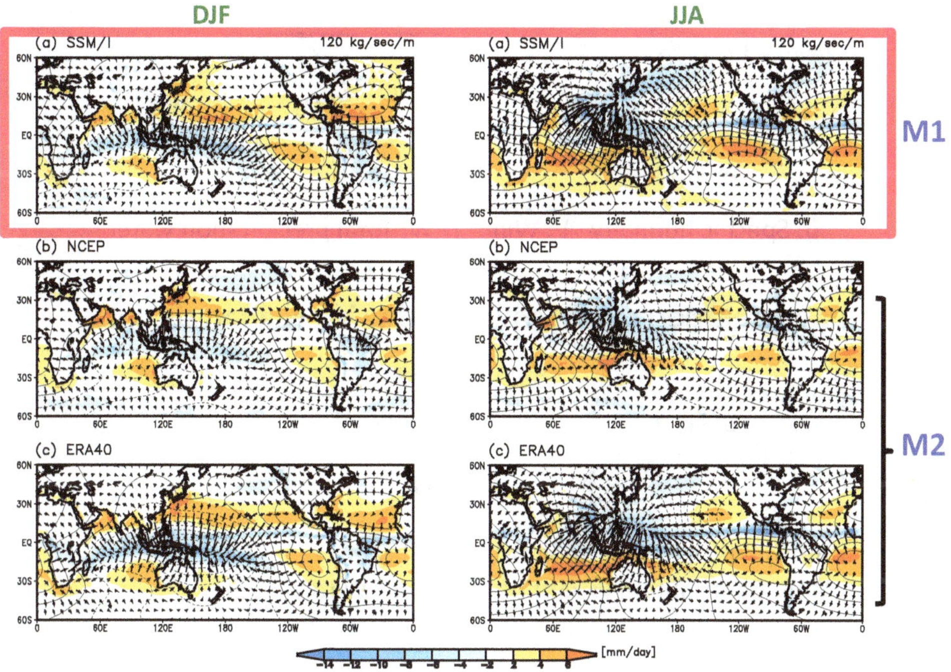

Slide 6

Validation using RAOB observations

$$divQ = -\frac{1}{g}\int_{Ps}^{0}\oint(q\vec{v})\bullet dl\,dp = E - P$$

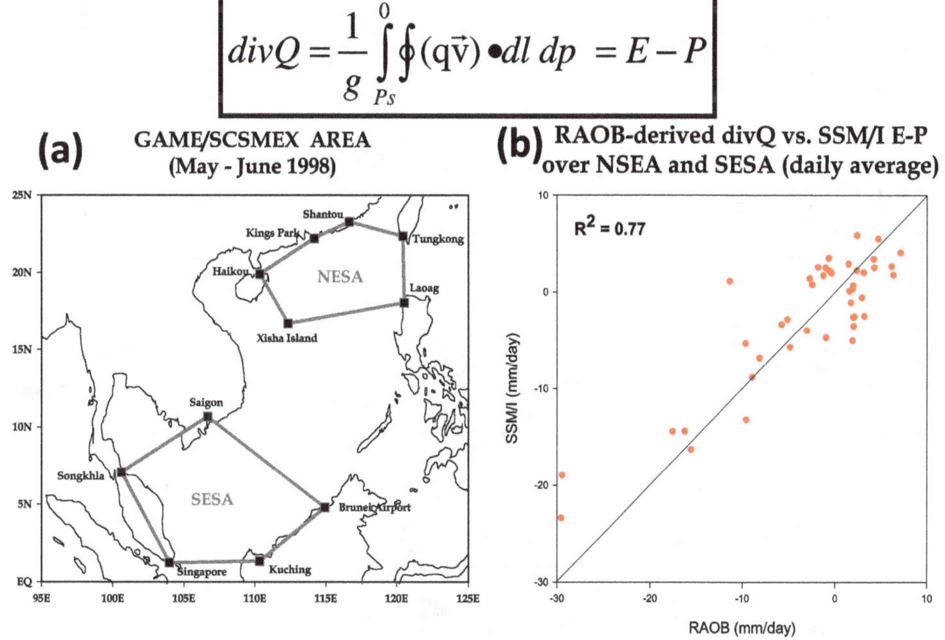

Park and Sohn, 2007, *Journal of Meteorological Society of Japan*

Slide 7

Effective Wind for Water Vapor Transport (V_E)

$$\vec{Q} = -\frac{1}{g}\int_{Ps}^{0} q\,\vec{V}\,dp = -\vec{V}_E \cdot \frac{1}{g}\int_{Ps}^{0} q\,dp = \vec{V}_E \cdot TPW$$

$$\vec{V}_E = \frac{\vec{Q}}{TPW}$$

q: Specific humidity *V*: Wind *TPW*: Total precipitation water

Interpretation of effective wind using reanalysis data

$$\vec{Q}_D = -\frac{1}{g}\int_{Ps}^{0} q\,\vec{V}_D\,dp = TPW\sum_{i=1}^{N}\frac{PW(i)}{TPW}\vec{V}_D(i)$$

$$V_E = \vec{Q}\;/\;TPW = \sum_{i=1}^{N} W(i)\vec{V}_D(i)$$

PW(i): Precipitable water at the i-th layer
W(i): Contribution weight of i-th layer PW to TPW

Slide 8

Definition of Hadley and Walker Circulation Indices

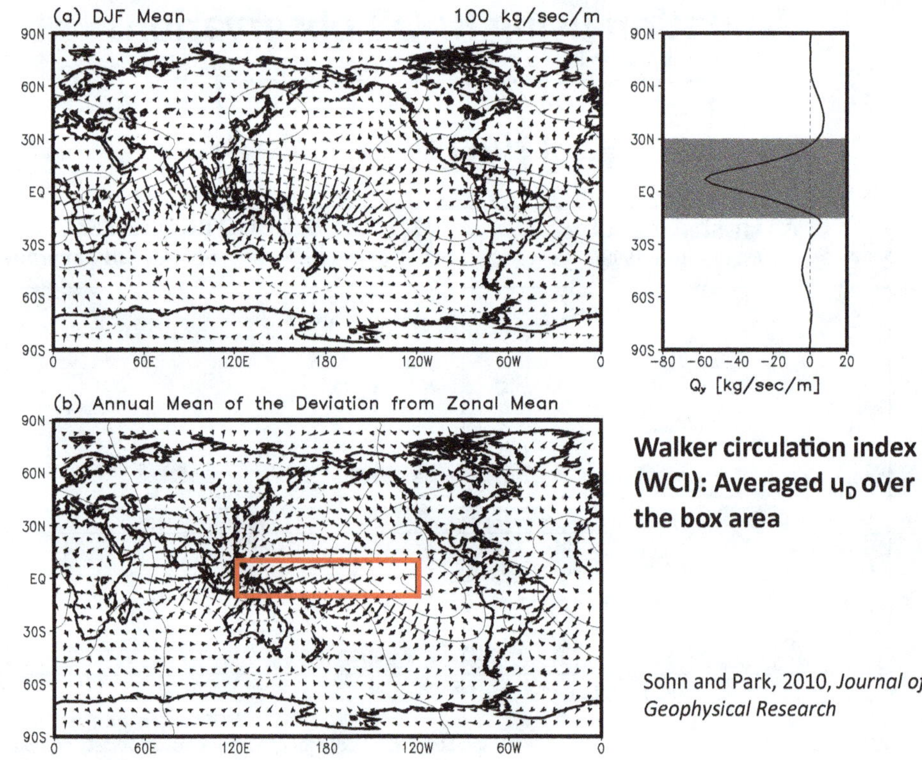

Walker circulation index (WCI): Averaged u_D over the box area

Sohn and Park, 2010, *Journal of Geophysical Research*

Slide 9

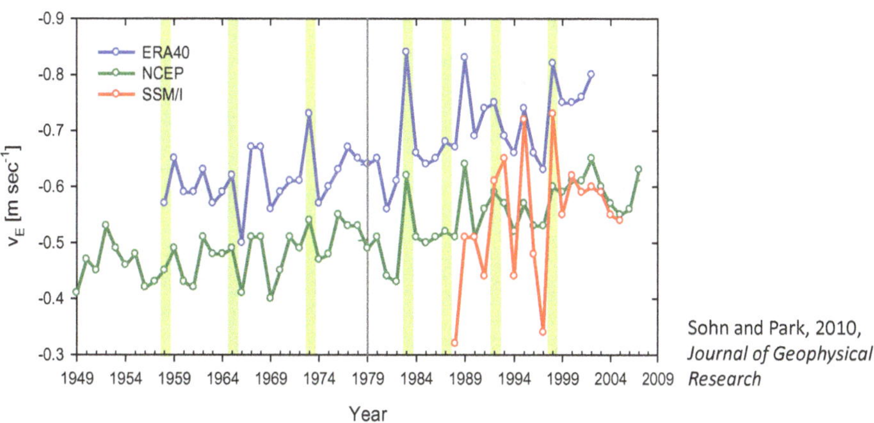

The intensity of the Hadley circulation has increased in both ERA40 and NCEP reanalyses. The CSI from SSM/I over the 18-year-period also shows an increasing trend, with larger fluctuations than were shown with reanalysis results. It is of interest to note that the strength of the Hadley circulation has generally intensified during the El Niño years.

Slide 10

Walker Circulation trend

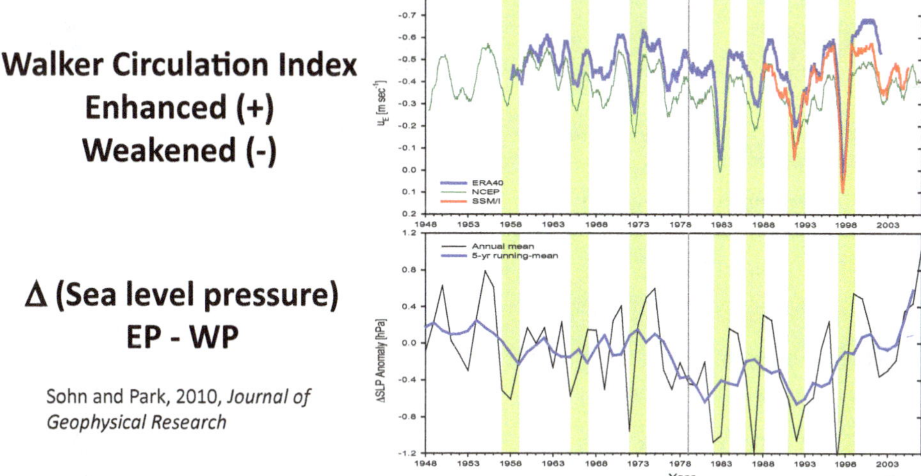

Walker circulation index (top). Sea level pressure (SLP) difference between the east Pacific and west Pacific (160°W–80°W, 5°N–5°S; 80°E–160°E, 5°N–5°S). The increasing trend shown in the CSI for the Walker circulation (strengthened Walker circulation) is consistent with expected results from changes in SLP differences, suggesting that the Walker circulation has been strengthened over the past three decades.

Slide 11

Hypothesis: Intensified Walker Circulation

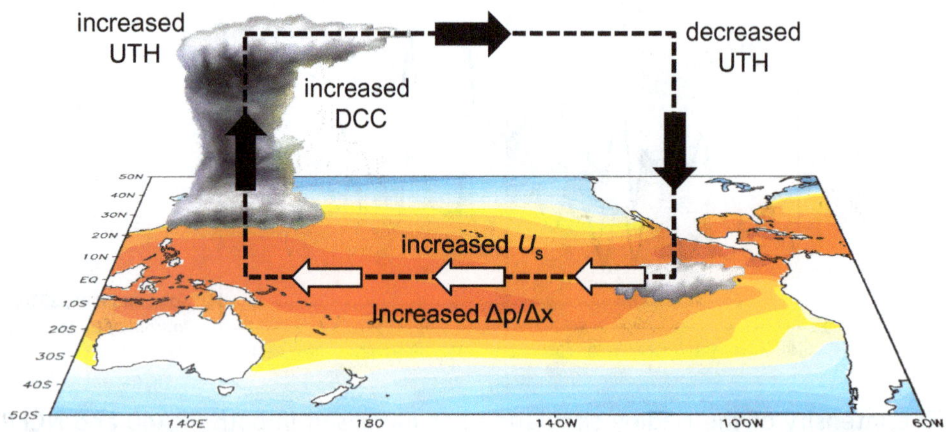

Enhanced Walker circulation: Enhanced deep convective activities over the western Pacific, moistening of the upper troposphere over tropical convective regions, increased upper tropospheric humidity (UTH) over the western Pacific, increased SLP over the eastern Pacific descending region, stronger east-west SLP gradient in conjunction with the decreased SLP over the western Pacific ascending region.

Slide 12

Datasets and 30-year (1979–2008) Climatology

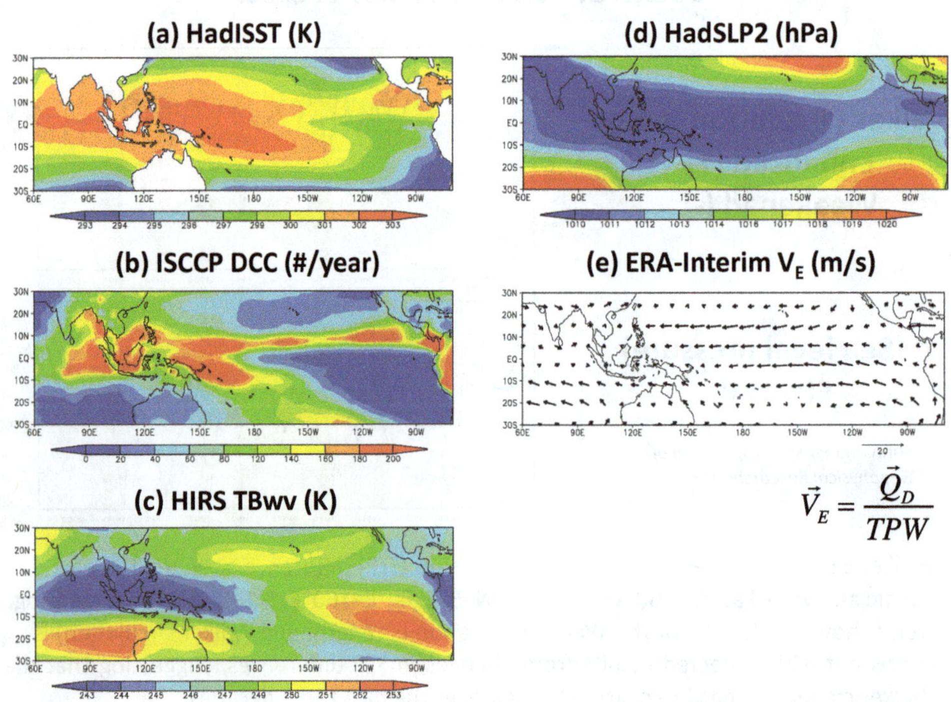

Slide 13

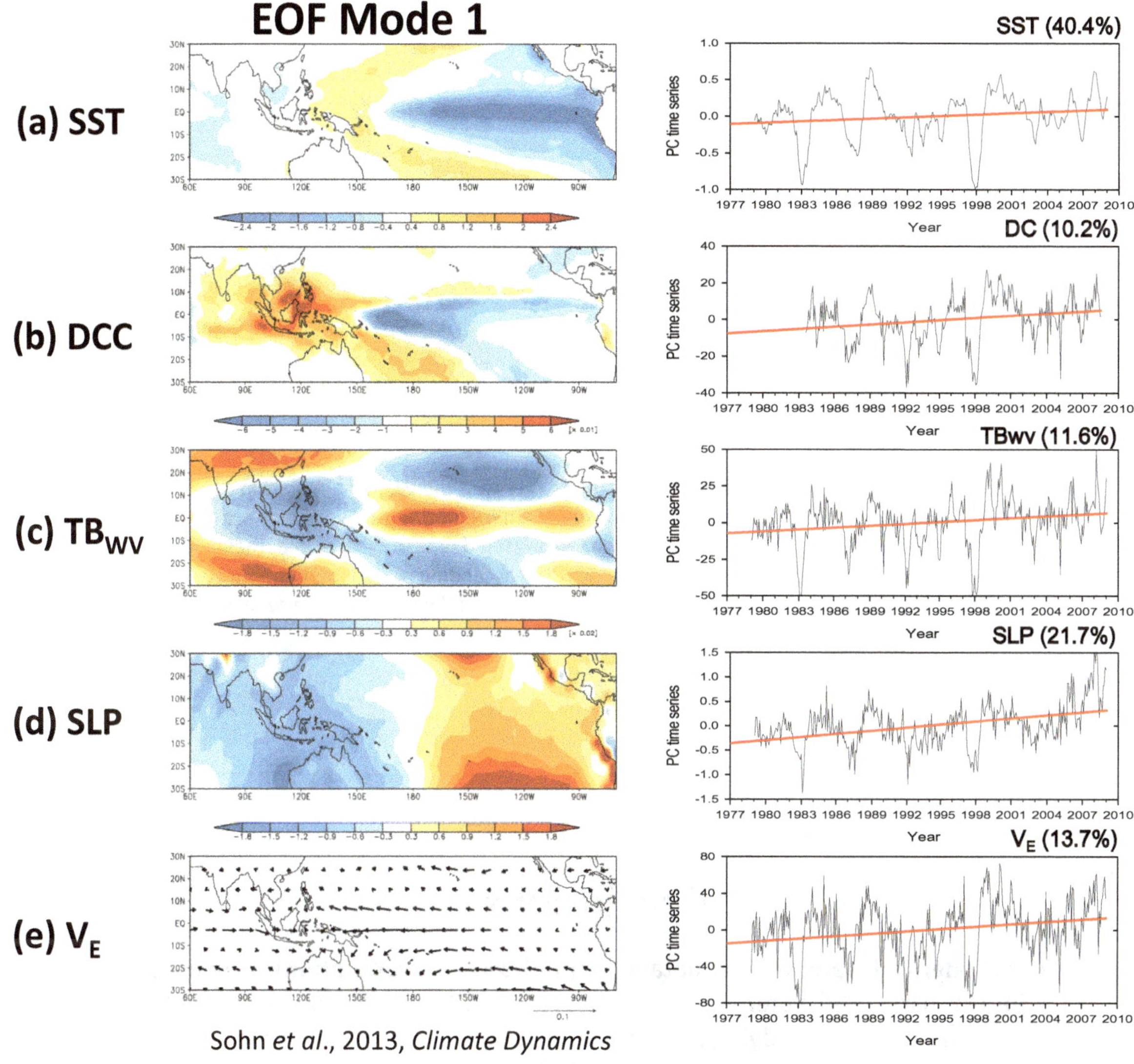
EOF Mode 1
(a) SST
(b) DCC
(c) TB_{WV}
(d) SLP
(e) V_E
SST (40.4%)
DC (10.2%)
TBwv (11.6%)
SLP (21.7%)
V_E (13.7%)
PC time series
Year
Sohn *et al.*, 2013, *Climate Dynamics*

Slide 14

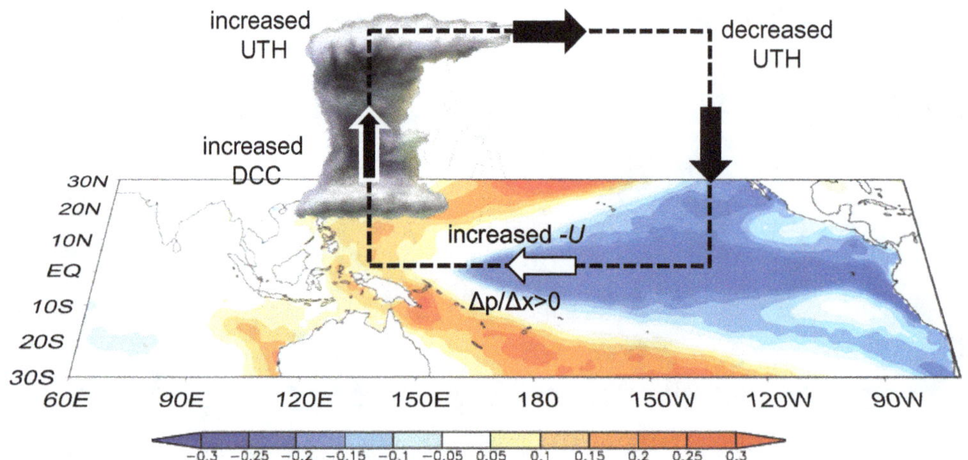

Thirty years of cumulated satellite data demonstrate that the Walker circulation has intensified over the past three decades. Variations in meteorological variables are coherent with the intensified circulation, i.e., increased deep convective activities and associated upper tropospheric moistening over the ascending region, and enhanced easterly wind in the boundary layer over the entire tropical Pacific basin in response to the increased east-west SLP gradient.

Slide 15

Conclusions

- **SSMI-derived water budget data (precipitation and evaporation) were used for estimating the water vapor transport (WVP). Then, effective wind (Ve) was estimated for studying the tropical circulation change using the relationship of V_E = WVP/TPW, where TPW is the total precipitable water.**
- **In contrast to the climate model prediction of the weakening of the Walker circulation trend, the effective winds indicate that the Walker circulation has been strengthened for the recent three decades from 1979.**
- **EOF composite analysis of independent observation data over the recent 30 years (1979–2008) all show coherent behaviors of an intensified Walker circulation.**

  WP: Warmer SSTs, enhanced convection, wetter UT, lower SLP
  EP: Cooler SSTs, decreased UTH, higher SLP

  Then ΔSLP/Δt > 0, and thus the easterly became strengthened within BL, as shown in WVT.
- **These are found to be due to "more occurrences of Central Pacific El Niños".**

Slide Captions

Slide 1

Slide 2 Under direct and indirect solar absorption by the surface, water evaporates in a form of water vapor from the ocean, and water in the vapor phase is transported from a source region to a sink region through the atmospheric circulation, effectively transporting heat as well.

Slide 3 Method 1 (M1) — satellite water budget method

For a given atmospheric column, the water balance is achieved between evaporation, precipitation, and water vapor storage. By separating the total transport **Q** into its rotational ($\mathbf{Q}_R$) and divergent ($\mathbf{Q}_D$) components and introducing the water vapor transport potential function (Φ), Φ can be solved over the whole global domain, using spherical harmonics to avoid boundary value problems. Inputs for this model can be from passive microwave measurements of total water vapor, precipitation, and evaporation. Once Φ is solved, the associated divergent component of water vapor transport ($\mathbf{Q}_D$) can be obtained by taking the gradient of potential function.

Method 2 (M2) — direct integration method

Total water vapor transport at the given atmospheric column can be directly obtained by vertically integrating water flux from the surface to the TOA if vertical profiles of specific humidity (q) and horizontal wind fields (**V**) are available. Inputs may be from reanalysis products.

Slide 4 Eighteen year mean winter (DJF) and summer (JJA) evaporation and precipitation were derived from SSM/I-based HOAPS data. Prominent features found in the precipitation map (left) include the local maxima over the Intertropical Convergence Zone (ITCZ) north of the equator from the western Pacific Maritime Continent to the eastern Pacific, over the equatorial Atlantic, and over the South Pacific Convergence Zone (SPCZ) stretching from New Guinea. Seasonal difference between winter and summer is evident, as noted in the northern hemispheric monsoon regions such as in Southeast Asia and East Asia during the summer.

Mean evaporations are given in the right-hand side panels (top: DJF, middle: JJA, bottom: Annual mean). Maximum evaporation rates larger than 6 mm/day are found over the North Pacific off East Asia and over the North Atlantic off North America because of the frequent dry and cold-air advection from the continents eastward onto the warm oceans during the DJF period.

Evaporation rates larger than 6 mm/day are also found in the subtropical trade wind zones over the North Pacific and Atlantic Oceans. It is of interest to note that minimum evaporation areas over the Tropics are generally found to be located with precipitation maximum areas. Zonally extended patterns are noticeable in the Southern Hemispheric oceans. Strong seasonal dependence is noted.

Slide 5 The resultant divergent components of water vapor transport ($\mathbf{Q}_D$) from the satellite method (M1) and NCEP and ERA40 reanalysis are presented, along with the potential function (contour lines), for DJF (left panels) and JJA (right panels). Because the spatial gradient of the potential function gives rise to water vapor flux whose divergence must be equal to $[E - P]$, the overall pattern of the potential function is analogous to the water vapor divergence field but has a much smoother shape because of the Laplacian operator. Arrows, representing transport direction and magnitude, nicely connect between the water vapor source region and sink region.

During the DJF, in the Pacific, the ridges of the potential function extend from the eastern Pacific to the East China Sea, and to the southeast Pacific, in contrast to the trough from Madagascar to the east of New Guinea. Such distributions of ridges and troughs establish a great southward transport band extending from the equatorial Indian Ocean to the southeast Pacific through the Maritime Continent and the central Pacific. In the Atlantic, a high potential area centered over the east of the West Indies and a local high in the south mid-Atlantic form a relatively strong vector transport across the equator.

During the summer (JJA), divergent water vapor transport from the SSM/I is characterized by cross-equatorial transports from the Southern Hemisphere, which are well consistent with large-scale circulation features known in association with the Asian summer monsoon; i.e., (1) the cross-equatorial flow associated with monsoon circulation over the Indian Ocean, (2) the southeast flow associated with the anticyclonic circulation over the western Pacific, and (3) the cross-equatorial flow north of New Guinea. These transports result in a strong water vapor convergence over the Bay of Bengal, the tropical western Pacific, and the East Asian monsoon region.

Results obtained from reanalysis products (using method M2) are given in (b) and (c). In the case of NCEP (b), the magnitudes of the potential function and moisture transports are substantially smaller than those of SSM/I over the entire oceanic area, indicating weak moisture convergence and divergence in the NCEP. The magnitude of moisture transports of ERA40 (c) is larger than those of NCEP but substantially smaller than suggested from SSM/I, although the general patterns of flux convergence are similar to SSM/I.

Slide 6 Satellite-estimated $(E - P - \frac{\partial W}{\partial t})$ is compared against $\nabla \cdot \mathbf{Q}$ from radiosonde observations, in order to assess the accuracy of the satellite approach of estimating water vapor transport. Flux divergence from radiosonde observations is based on the line integral of the water vapor flux into the domain of interest. In principle, Green's theorem is applied over the domain, yielding the area mean moisture flux divergence over the domain:

$$\nabla \cdot \mathrm{Q} = \frac{1}{A}\int_{1000hPa}^{200hPa}\left(\oint q\mathrm{V}_n \cdot dl\right)\frac{dp}{g} = E - P - \frac{\partial W}{\partial t},$$

where $\mathbf{V}_n$ is the wind component normal to the boundary segment $d\mathbf{l}$, and A is the domain area.

Two oceanic areas (NESA, SESA) surrounded by radiosonde observation sites were chosen for the calculation. Radiosonde observations were made at each site, with 6-hourly soundings during the May–June 1998 period as a part of the Global Energy and Water Cycle Experiment (GEWEX) Asian Monsoon Experiment (GAME), i.e., GAME/SCSMEX (South China Sea Monsoon Experiment). On a daily basis, SSM/I-derived flux divergence is in good agreement with that suggested from radiosonde observations, showing a correlation coefficient of 0.88, implying that the satellite results are comparable to radiosonde-derived values.

Slide 7 Water vapor transport varies with the combination of water vapor and wind speed. Thus both of these factors contribute to any trends in water vapor flux, as exemplified by the fact that water vapor flux can increase under a static velocity field when the amount of water vapor increases. Thus, to understand trends in circulation strength using water vapor flux, the concept of "effective wind (V_E) for the water vapor transport" is defined by scaling Q_D with TPW.

In order to understand what factors control most of the changes in effective wind, time series of ERA40 precipitable water (PW), weighting (W = PW/TPW), divergent wind ($\mathbf{V}_D$), and contribution to the effective wind (WV_D) were examined, at lower tropospheric layers. Results indicate that the effective wind represents the speed at which the water vapor is carried in the lower returning branch of the circulation cell. Because of this reasoning, the effective wind can be used as an index delineating a measure of circulation intensity.

Slide 8 The 18-year DJF (December–January–February) means of water vapor transport potential (Φ) and the divergent component of water vapor transport ($\mathbf{Q}_D$) are given in the top-left panel. The zonal mean distribution of the DJF mean transport, shown in the top-right panel, reflects a facet of the zonal mean circulation that is particularly associated with the Hadley circulation. During the winter, the Hadley circulation is thought to be a large-scale mass overturning, with rising air in the ITCZ in the summer hemisphere and descending air in the subtropics in the winter hemisphere. The formed cell manifests a strong cross-equatorial current near the surface. In the zonal mean, zero fluxes are located at ~30°N and ~10°S, and the maximum zonal flux at ~10°N, indicating that the return flow of the Hadley circulation carries water vapor from the 30°N–10°N region to the 10°N–10°S region. Because stronger Hadley circulation requires more water vapor transport associated with the stronger return flow in the surface boundary, the magnitude of water vapor flux (averaged over the 30°N–10°S area: shaded area of the zonal mean flux distribution) can be used as an index of the Hadley circulation intensity.

A similar water vapor flux can be used as an index of examining intensity changes in large-scale east-west circulation across the equatorial Pacific, which is known as the Walker circulation. This is based on taking the annual mean deviation of water vapor transport from the zonal mean. The Walker circulation is assumed to be independent of the Hadley circulation. By connecting the climatologically known ascending branch over the Western Pacific to the descending branch of the Walker circulation, we expect a westward return flow that would convey water vapor onto the ascending region through the boundary layer of the return branch from the east Pacific to the west Pacific. As shown in the bottom figure, the flux average over the lower-level returning branch is used as an index of the Walker circulation intensity.

Slide 9 The north-west component of the effective wind, averaged for the DJF period over the 30°N–15°S latitudinal band, is used as a DJF Hadley circulation strength index (CSI). An SSM/I-derived CSI is given together with CSIs from ERA40 and NCEP reanalyses. Shaded bars represent the El Niño years. It is evident that the intensity of the Hadley circulation has increased throughout the available data periods. However, there seems to be a discontinuity in the trend around 1979; the trend in the post-1979 period appears to be larger than it was pre-1979 for both ERA40 and NCEP. The CSI from SSM/I over the 18-year-period also shows an increasing trend, with larger fluctuations than were shown with reanalysis results. It is of interest to note that the strength of the Hadley circulation has generally intensified during the El Niño years.

This positive trend implies that the moist ascending region gets wetter while the dry subsidence region gets drier because the stronger effective wind means that more water vapor is transported from the subtropics to the ITCZ. Over the subtropical subsidence region, the increased sinking motion within the stronger Hadley circulation cell induces drier adiabatic warming, which should be balanced by increased radiative cooling from the drier upper troposphere. By contrast, the stronger sinking motion should induce a shallower boundary layer due to the stronger inversion above the boundary layer, resulting in reduced TPW.

Slide 10 In this slide, the Walker circulation index is defined to be the effective wind averaged over the area outlined in Slide 7 because the effective wind reflects the mass flow associated with water vapor transport in the lower branch of the Walker circulation cell. In the upper figure, the negative value represents an enhanced Walker circulation, and the positive value weakened the Walker circulation. Again, the shaded bars represent El Niño years. Both ERA40 and NCEP reanalyses capture the weaker Walker circulation during El Niño years, in which 1983 and 1998 show a near breakdown or a reversed pattern of the east-west circulation. In the period from 1988 to 2005, for which SSM/I data is available, the magnitude of the SSM/I-derived circulation index falls between the values from both reanalyses, and the interannual variations are very similar to those found in the reanalysis. All trends within the SSM/I-available period are quite similar, suggesting that the Pacific Walker circulation has intensified over the past three decades.

The lower figure is the difference in sea level pressure (SLP) between the east Pacific and west Pacific, i.e., between two regions (160°W–80°W, 5°N–5°S; 80°E–160°E, 5°N–5°S). SLP anomalies are used after removing monthly long-term mean averages from the monthly data. It is clear that the increasing trend shown in the CSI for the Walker circulation is consistent with expected results from changes in SLP differences, strongly suggesting that the Walker circulation has strengthened over the last three decades.

Slide 11 We look for further observational evidences supporting the strengthening of the Walker circulation over the past three decades. When enhanced Walker circulation is taken as a working hypothesis, the following can be expected. Deep convection should increase over the western Pacific ascending region. Because deep convective clouds are the main contributors to the moistening of the upper troposphere over tropical convective regions, increased upper tropospheric humidity (UTH) can be expected in accordance with the increased convection over the western Pacific. Because of the mass overturning caused by the intensified upward motion over the ascending branch, the eastern Pacific descending region should experience increased SLP, establishing a stronger east-west SLP gradient in conjunction with the decreased SLP over the western Pacific ascending region. As a consequence, in the lower branch of the Walker cell, the easterly trade wind should increase. These hypothesized variations that might be sensitive to changes in the Walker circulation are examined using various datasets.

Slide 12 Datasets to examine the expected changes in variables are:

HadISST, HadSLP2 (1979~2008)

- HadISST: [Rayner *et al.*, 2003] — interpolated
- Resolution: 1° × 1° grid, monthly mean
- HadSLP2: [Allan and Ansell, 2006]
- Resolution: 5° × 5° grid, monthly mean

ISCCP D1 (July 1983 to June 2008)

- ISCCP "cloud regimes: DCC, Anvil, Cirrus..." defined by the *k*-means clustering algorithm for 35°N–35°S applied to cloud top pressure (CTP) – cloud optical thickness (COT) histograms [Rossow *et al.*, 2005]
- Resolution: 2.5° × 2.5° grid, monthly frequency

HIRS TB6.7 (1979–2008)

- Calibrated by Lei Shi [Shi and Bates, 2011]
- Resolution: 2.5° × 2.5° grid, monthly mean

Hadley Centre Sea Ice and Sea Surface Temperature data are used, together with the frequency of occurrence of a deep convective cloud obtained from the weather state (WS) analysis of the ISCCP cloud data. In order to relate changes in the UTH to convective activities over the warm pool region, we use the three-decade-long NOAA HIRS channel brightness temperature (TB) around the 6.7-μm water vapor absorption band. Positive (negative) TB anomalies from the mean condition represent drier (wetter) conditions than normal. The monthly mean SLP data from the Hadley Centre data version 2 are also put together with the ERA-Interim effective wind.

Slide 13 In order to examine the spatially coherent patterns of temporal variations in climate variables that are considered to be facets of the Walker circulation changes, the Empirical Orthogonal Function (EOF) analysis is conducted for the SST, deep convection (DC), HIRS 6.7-μm water vapor channel TB (hereafter referred to as TB_{WV}), sea level pressure (SLP), and water vapor transport (WVP) effective wind (V_E). A trend analysis of the principal component (PC) time series is also conducted, and results are given in this slide.

The V_E EOF first mode (e), explaining approximately 13.7% of the total variance, generally shows easterly loadings over the tropical Pacific and westerly loadings over the tropical Indian Ocean, forming a convergence loading over the western Pacific warm pool region. The PC time series is remarkably similar to that found from four other variables ((a)–(d)), suggesting that all variables expressed by EOF mode 1 are consistent with each other in the context of a strengthened Walker circulation over the last three decades.

From the analysis, it is conjectured that the time series of SST mode 1 reflects fewer occurrences of conventional Eastern Pacific El Niño in the recent decade, in which their magnitude has become weaker.

Slide 14 We set up an intensified Walker circulation in recent decades as a working hypothesis, and we expected changes in certain atmospheric variables and surface parameters in response to the enhanced Walker circulation. This slide is a summary of results obtained from 30 years of cumulated satellite data, from which the working hypothesis of the enhanced Walker circulation can be tested. It is shown that variations in meteorological or surface variables are coherent with the intensified circulation, i.e., increased deep convective activities and associated upper tropospheric moistening over the western Pacific ascending region, increased SLP over the eastern Pacific descending region in contrast to decreased SLP over the western Pacific ascending region, and enhanced easterly wind in the boundary layer over the entire tropical Pacific basin in response to the increased east-west SLP gradient. The consistency between variables, under the hypothesis of the enhanced Walker circulation, should be proof that the Walker circulation has intensified over the past three decades.

Slide 15 Conclusions

References

Allan R, Ansell T. (2006) A new globally complete monthly historical gridded mean sea level pressure dataset (HadSLP2): 1850–2004. *Journal of Climate* **19**:5816–42.

Rayner NA, Parker DE, Horton EB, Folland CK, Alexander LV, Rowell DP, Kent EC, Kaplan A. (2003) Global analyses of sea surface temperature, sea ice, and night marine air temperature since the late nineteenth century. *Journal Geophysical Research* **108**:4407. doi:10.1029/2002JD002670

Rossow WB, Tselioudis G, Polak A, Jakob C. (2005) Tropical climate described as a distribution of weather states indicated by distinct mesoscale cloud property mixtures. *Geophysical Research Letters* **32**:L21812. doi:10.1029/2005GL024584

Shi L, Bates JJ. (2011) Three decades of intersatellite-calibrated High-Resolution Infrared Radiation Sounder upper tropospheric water vapor. *Journal Geophysical Research* **116**:D04108. doi:10.1029/2010JD014847

Sohn BJ, Park S-C. (2010) Strengthened tropical circulations in past three decades inferred from water vapor transport. *Journal of Geophysical Research* **115**:D15112. doi:10.1029/2009JD013713

Sohn BJ, Yeh S-W, Schmetz J, Song H-J. (2013) Observational evidences of Walker circulation changes over the last 30 years contrasting with GCM results. *Climate Dynamics* **40**:1721–32.

Section 2

Clouds and Radiation

LECTURE 8

Understanding Global Cloud Radiative Impacts

Lazaros Oreopoulos
NASA Goddard Space Flight Center, Greenbelt, MD, USA

Lazaros Oreopoulos is a Senior Research Scientist at NASA Goddard Space Flight Center. He examines the radiative role of clouds in the planet's climate, the global signatures of their interaction with aerosols, and their climatological representation in Global Climate Models.

Introduction

The influence of clouds on the radiation budget is one of the most important facets of the Earth's climate. Satellite observations in the last 35 years have helped us understand the bulk radiative effects of different classes of clouds. Bill Rossow has spearheaded efforts in that direction, demonstrating the value of the International Satellite Cloud Climatology Project (ISCCP) cloud retrievals as a foundation for the full reconstruction of the Earth's Radiation Budget components.

The main aim of this document is to provide an overview of the *global* radiative impacts of clouds at *annual* time scales. While geographical and seasonal aspects receive some attention, it is cloud radiative effect decomposition by cloud category (or "class") that is the main focal point. Furthermore, this document addresses the large-scale radiative behavior of clouds only for the *current climate*, with no further discussion of how this behavior may change in the future and what its climatic impact may be (refer to [Del Genio, 2018] for an excellent overview of this topic).

Cloud classes can be defined in many ways, and are often referred to with terms such as "cloud type" and "cloud regime". Some of the definitions are more meaningful at the level of individual instrument footprints ("pixels" for passive or "rays" for active satellite observations) while others for larger areas (e.g., "mesoscale"). This distinction will become apparent below.

Global Annual Cloud Radiative Influence

The most common framework for analyzing and discussing cloud radiative impacts separates atmospheric radiation into solar or shortwave (SW) radiation originating from the Sun and thermal infrared (IR) or longwave (LW) radiation emitted by the Earth (atmosphere and surface). A cloudless Earth is characterized by radiative warming of the surface (due to absorption of SW and LW radiation) and radiative cooling of the atmosphere (due to LW emission exceeding LW and SW absorption). *Our goal here is to examine how clouds affect this fundamental distinctive behavior.* We have conveniently defined the Cloud Radiative Effect (CRE) as a quantity to measure the impact of clouds on radiative fluxes (irradiances) over an area. CRE is defined as the difference between all-sky (including clouds) and cloud-less-sky (clear-sky) fluxes. The common convention to ascribe a negative CRE when we want to convey a cooling effect of clouds compels us to define CRE in terms of net, downward minus upward, fluxes. Thus:

$$CRE_{LW,SW} = F^{all}_{LW,SW} - F^{clr}_{LW,SW}, \tag{1}$$

where $F = F^{\downarrow} - F^{\uparrow}$.

This definition applies to both the top of atmosphere (TOA) and the surface (SFC). The radiative impact of clouds on the atmospheric column as a whole (ATM) is obtained by subtracting the SFC from the TOA CRE:

$$CRE^{ATM} = CRE^{TOA} - CRE^{SFC}. \tag{2}$$

Finally, the LW and SW CREs are often combined to obtain the overall ("total", often called "net") radiative effect of clouds:

$$CRE_{TOT} = CRE_{LW} + CRE_{SW}. \tag{3}$$

Satellite and other observations, assisted by radiative transfer modeling, indicate that the radiative effect of clouds on the current climate (i.e., on a global annual basis) can be summarized as follows:

- Clouds cool the planet (climate), i.e., $CRE^{TOA}_{TOT} < 0$.
- Clouds cool the Earth's surface, i.e., $CRE^{SFC}_{TOT} < 0$.
- Clouds have a near-neutral radiative effect on the Earth's atmosphere, i.e., $CRE^{ATM}_{TOT} \approx 0$.

The above statements can be further elaborated as follows:

- The cooling effect of clouds on the climate results from SW cooling ($CRE^{TOA}_{SW} < 0$) exceeding LW warming ($CRE^{TOA}_{LW} > 0$), i.e., $|CRE^{TOA}_{SW}| > |CRE^{TOA}_{LW}|$.
- The cooling effect of clouds on the surface results from SW cooling ($CRE^{SFC}_{SW} < 0$) exceeding LW warming ($CRE^{SFC}_{LW} > 0$), i.e., $|CRE^{SFC}_{SW}| > |CRE^{SFC}_{LW}|$.
- The near-neutral radiative effect of clouds on the atmosphere results from a small overall LW effect $CRE^{ATM}_{LW} \approx 0$ being combined with a small overall SW effect $CRE^{ATM}_{SW} \approx 0$; however, as we will see below, the range and distribution of LW atmospheric radiative effect values are much broader than that of the SW.

Again, the above summary of cloud radiative effects pertains to global annual conditions. As we will see below, deviations for different cloud classes are numerous and important. The overall impact depends on the mean radiative effect of the individual cloud classes and how often they occur. Deviations also arise because of seasonal effects, mainly because of the great hemispheric seasonal contrasts in $F_{SW}^{TOA,\downarrow}$.

We will now explore the radiative behavior of clouds in more detail through a closer examination of how they alter the radiative behavior of a cloudless atmosphere in both the LW and SW part of the electromagnetic (EM) spectrum (cf. Slide 2).

Radiative Effects of Clouds in the LW

The temperature of the troposphere generally decreases with height. Keeping this in mind is essential for understanding the LW radiative effects of clouds. Also, clouds increase atmospheric opacity, which means that they add extinction optical depth to the atmosphere (i.e., increase the atmosphere's ability to block LW radiation). Because the scattering effect of clouds can largely be neglected in the LW, the extinction effect of clouds mostly comes in the form of absorption. A completely opaque cloud (not allowing any LW radiation to be transmitted) absorbs all incoming radiation and is often called a "black" cloud. Many clouds are black or nearly black, but many are also transmissive of LW radiation. Like any object in thermodynamic equilibrium, clouds also emit EM energy, which per our conventions belongs to the LW part of the spectrum (energy emitted by the Earth's surface and atmosphere), the bulk of which resides between 5 and 50 μm (at terrestrial temperatures EM energy outside this range is considered negligible). So, what happens when a cloud is inserted into the Earth's atmosphere? The opacity of the atmosphere increases. More LW is absorbed, and more is emitted (upward and downward) for a fixed temperature. A surface observer in a cloudless atmosphere senses the cumulative effect of the atmosphere's radiation emitted from all levels that are partially absorbed as they traverse the atmospheric column. The higher (colder temperatures) the radiation originates from, the fewer chances it has to survive its downward journey. If one inverts $F_{LW}^{SFC,\downarrow}$ using the Stefan–Boltzman law, an *effective temperature of emission* T_{eff}^{SFC} can be calculated, corresponding to the temperature of a black-body atmosphere emitting towards the surface on average the same amount as the actual atmosphere. Using a standard atmospheric profile, this temperature can be converted to an effective emission height. Clouds increase $F_{LW}^{SFC,\downarrow}$, and therefore reduce the *effective emission height* H_{eff}^{SFC} of LW radiation towards the surface. In other words, with a cloud present, it is as if the LW radiation comes from lower in the atmosphere. This effect depends not only on the additional opacity added by the cloud but also on the location of the cloud (mostly its base for a relatively opaque cloud). If the cloud base is above the H_{eff}^{SFC} of the clear atmosphere (i.e., *high* cloud), the increase in $F_{LW}^{SFC,\downarrow}$ will be smaller because the LW radiation has greater chances of being absorbed and re-emitted by the atmosphere below the cloud. The increase in opacity brought by a *low* cloud has a greater effect because the cloud base temperature is likely higher than T_{eff}^{SFC} and its base is therefore below H_{eff}^{SFC}, so increases to $F_{LW}^{SFC,\downarrow}$ are more dramatic. In summary, the lower and more opaque the cloud, the greater its impact on $F_{LW}^{SFC,\downarrow}$, the greater the contrast between clear and cloudy downward fluxes, and thus the greater the value of CRE_{LW}^{SFC}. At the same time, upward fluxes at the SFC between clear and cloudy conditions are virtually the same on average.

The LW effect at TOA can be understood with similar arguments. Where clouds occur, the additional atmospheric opacity blocks a portion of the LW radiation emitted upward by the lower (warmer) parts of the atmosphere, and more of the upward emission now appears to be coming from colder cloud tops, i.e., $F_{LW}^{TOA,\uparrow}$ decreases, which can also be viewed as a decrease in T_{eff}^{TOA}, or equivalently a raising of the effective emission height H_{eff}^{TOA} compared to a cloudless atmosphere. The higher the cloud top, i.e., the higher it is above the H_{eff}^{TOA} of the *clear* atmosphere, the more dramatic the decrease in T_{eff}^{TOA} and $F_{LW}^{TOA,\uparrow}$, and the larger the contrast between clear and cloudy fluxes, thus the greater CRE_{LW}^{TOA}. Low clouds have small CRE_{LW}^{TOA} because despite the increase in atmospheric opacity, their cloud tops remain below H_{eff}^{TOA}, resulting in only a small reduction in T_{eff}^{TOA} since the radiation emitted by the cloud has a greater chance of being absorbed and re-emitted upward by the colder atmosphere above.

Given this picture of LW radiation propagation in the atmosphere, high and low clouds exhibit a rather distinct behavior: low clouds have a small instantaneous radiative impact at TOA and a large one at the SFC; high clouds have a large instantaneous radiative impact at TOA and a small one at the SFC. The impact on the atmosphere itself depends on the difference between the TOA and SFC boundaries and can be thought of in terms of the distinct net flux divergence for the two classes of clouds for the same clear-sky conditions. For high clouds, the net input from the surface is almost the same as it is for clear skies, with the upward surface flux only weakly counteracted by a downward LW, which is barely affected by high clouds; at the same time, substantially less LW flux escapes to space, so the net flux divergence is smaller than in clear conditions and therefore LW energy retention increases. For low clouds, the net energy input at the surface is smaller since upward emission is strongly counteracted by enhanced downward emission; at the same time, the LW flux escaping to space is not much reduced compared to clear skies, yielding a net flux divergence that is larger than that of clear skies, i.e., more efficient cooling. Simply put, for high clouds, almost the same LW energy as clear skies is supplied from below while less escapes, while for low clouds, far less LW energy compared to clear skies is supplied from below while almost the same LW energy as clear skies escapes to space. Reduced cooling (warming compared to clear skies) happens in the former case and enhanced cooling in the latter case. In terms of Eq. (2), $CRE_{LW}^{ATM} > 0$ for high clouds because $CRE_{LW}^{TOA} > CRE_{LW}^{SFC}$, while $CRE_{LW}^{ATM} < 0$ for low clouds because $CRE_{LW}^{TOA} < CRE_{LW}^{SFC}$.

Radiative Effects of Clouds in the SW

For the SW, interpretations of cloud radiative effects are much simpler — increased opacity (extinction) compared to clear skies manifests mostly as increased scattering, reduced transmission, and to a lesser extent, enhanced absorption. This yields enhanced reflectance to space $F_{SW}^{TOA,\uparrow}$ compared to clear skies, ($CRE_{SW}^{TOA} < 0$) and reduced transmitted flux towards the surface $F_{SW}^{SFC,\downarrow}$ ($CRE_{SW}^{SFC} < 0$) far exceeding any differences in surface upward fluxes. Column atmospheric absorption increases ($CRE_{SW}^{ATM} > 0$). While this may at first glance seem unavoidable given that clouds add absorptive liquid droplets and ice crystals to the atmosphere, the atmospheric absorption increase is actually not universal since the additional absorption by the cloud particles is counteracted by reduced water vapor absorption below the cloud due to reduced SW transmission. On a global basis, the particle absorption exceeds the reduced water vapor absorption slightly, so that $CRE_{SW}^{ATM} \approx$ *positive, but small*.

Additional Perspectives of the Global Cloud Radiative Influence

In the following, we will be attaching specific numbers to the radiative effects of clouds. Specifically, we will examine how much brighter the planet becomes because of clouds, how much dimmer clouds make the surface, how much colder clouds make the Earth appear from space, and how much warmer they make the atmosphere appear from the perspective of a surface observer. These numbers are obtained from the Clouds and the Earth's Radiant Energy System (CERES) broadband satellite radiometer, specifically the Energy Balanced and Filled (EBAF) Ed4.0 dataset (Loeb *et al.*, 2018).

Increase in planetary albedo ($F_{SW}^{TOA,\uparrow}/F_{SW}^{TOA,\downarrow}$)*:* Planetary albedo increases from 0.16 for a cloudless planet to 0.29 for a cloudy planet, an increase of 81%. This can be thought also as a decrease in planetary absorption (i.e., cooling), expressed in energy units as CRE_{SW}^{TOA} = (340-99) – (340-53) = -46 Wm^{-2}. (cf. Slide 3).

Decrease in solar radiation reaching (transmitted to) the surface ($F_{SW}^{SFC,\downarrow}/F_{SW}^{TOA,\downarrow}$): From 0.72 for a cloudless planet to 0.55 for a cloudy planet, a decrease of 23%. This can be thought also as a decrease in surface absorption (i.e., cooling), expressed in energy units as CRE_{SW}^{SFC} = (187 – 23) – (244 – 30) = -50 Wm^{-2}. (cf. Slide 4).

The enhanced greenhouse effect due to clouds can be viewed as either a colder planet from space (TOA) or a warmer planet from the surface (SFC).

From the TOA perspective: A "greenhouse factor" can be expressed as the ratio $F_{LW}^{TOA,\uparrow}/F_{LW}^{SFC,\uparrow}$ that can be thought of as the effective transmission of the atmosphere, which is different for clear (0.67 = 268/398) versus cloudy (0.60 = 240/398) conditions, i.e., ~ 11% less thermal IR radiation escapes to space with clouds present. In energy units CRE_{LW}^{TOA} = (0-240) – (0-268) = +28 Wm^{-2}. In terms of effective temperature, a cloudy atmosphere makes the planet appear colder by about 7 K since for cloudy skies $T_{eff}^{TOA} \approx 255$ K, while for clear skies $T_{eff}^{TOA} \approx 262$ K. (cf. Slide 5).

From the surface perspective: The enhanced downwelling LW flux under cloudy conditions can be expressed as the ratio of $F_{LW}^{SFC,\downarrow}$ for cloudy versus clear conditions, which is ~1.1 = 345/315, i.e., an enhancement of ~10% because of clouds. In energy units CRE_{LW}^{SFC} = (345-398) – (315-398) = +30 Wm^{-2}. The cloudy atmosphere appearing warmer than the clear atmosphere to the surface observer can also be expressed in terms of T_{eff}^{SFC}, which is 279 K for a cloudy atmosphere and 273 K for a clear atmosphere, i.e., the planet (atmosphere) appears 6 K warmer from the surface with clouds present (cf. Slide 6).

The table below provides a summary of the different components of the CRE (cf. Slide 7). The combined effects of SW and LW are under the "TOT" column and convey the earlier statement that clouds on an annual global basis cool both the Earth's surface and the planet as a whole. Because the SFC cooling is about 2 Wm^{-2} greater, this means that a cloudy atmosphere retains a small amount of additional energy compared to clear skies. Because many fluxes have to be combined to obtain this number, a high uncertainty should be attached to it. According to CERES EBAF Ed4.0, it comes from the SW side, where the small enhanced

Table 1. Global annual CRE components according to CERES EBAF Ed4.0.

Perspective	LW	SW	TOT	Impact
TOA ("Planet, "Climate")	+28	−46	−18	Cooling
Surface (SFC, BOA)	+30	−50	−20	Cooling
Atmosphere (ATM)	−2	+4	+2	Slight Warming (?)

absorption by clouds exceeds the small enhanced LW cooling (we have previously indicated that these two quantities are nearly zero). Recall that a clear atmosphere cools radiatively because the net (absorption minus emission) LW cooling far exceeds SW warming (due to absorption). The table below suggests that clouds do not change this by much on average — they add some warming in the SW and some cooling in the LW, but the changes are small and of the opposite sign (according to CERES), further contributing to a smaller overall effect. We stress again that atmospheric CRE, coming from (in the case of TOT) double differences, is more uncertain than the other components, so the exact magnitude and sign (for LW and TOT) may change slightly in the future as new measurements and improved products are introduced.

CRE Decomposition by Cloud Class

We will now examine the CRE contributions of different cloud classes. As stated earlier, "cloud class" is a general "umbrella" term meant to convey different ways of categorizing the world's clouds. Here we will show results for four different classifications: Cloud Type (CT), Cloud Regime (CR), aka Weather State (WS), Cloud Vertical Structure (CVS), and Cloud (thermodynamical) Phase (CP). Because these cloud classes are related through common cloud features, consistencies are seen among the decomposition choices of the various CRE components. Appropriately equipped GCMs that can create such cloud classes can then be evaluated based on whether the details of their own CRE decomposition are consistent with observations.

Cloud Type

The earliest decomposition of CRE employs CT and traces back to (Hartmann *et al.*, 1992), H92. CTs are defined via somewhat arbitrary boundaries in a joint histogram of cloud top pressure and cloud optical thickness defined over synoptic/mesoscale domains (~100 km and above). H92 divided the histograms first introduced by the ISCCP (Rossow and Schiffer, 1991) to create five basic CTs (Slide 8) and developed a method to collocate the CTs with CREs from the Earth Radiation Budget Experiment (ERBE) (Barkstrom, 1984). Chen *et al.* (2000), C00, used a different division of the histogram that comprised nine parts, each mapped to one of the standard cloud types common in surface observer classifications. The CRE of each CT was determined by a radiative transfer (RT) algorithm calculating irradiances for clear and cloudy skies with input coming from passive imager retrievals. Stephens *et al.* (2018), S18, used active CloudSat and CALIPSO (CC) cloud retrievals in an effort to update the H92 results whereby the nine ISCCP CTs were grouped into the five H92 CTs, and the CRE was calculated again from RT calculations operating on CC cloud retrievals. Slide 8 shows a summary of CRE findings for TOA. The H92 and S18 results are shown as total CRE zonal means by CT (separately for the winter and summer seasons), while the C00 results are provided as a table where global values with SW and LW components are given separately and are weighted by the frequency of occurrence of each CT. The common feature of all results is that the total CRE at TOA is rarely positive, with only high thin clouds (cirrus) consistently producing such values, while the negative values are mainly contributed by low clouds and are stronger for the high illumination summer season.

Cloud Regime (Weather State)

Although sometimes confused with CT, the CR is a fundamentally different concept. A CR or Weather State (WS) is commonly defined by a centroid representing the mean of joint histograms that have similar co-variations of cloud extinction (expressed by cloud optical thickness) and cloud vertical location (expressed by cloud top pressure) at scales ~100 km (Rossow *et al.*, 2005). The CRs thus come from the same joint histograms used to define CTs but are in principle mixtures of CTs, one of which is usually dominant within a CR. Besides the ISCCP, CRs have been also derived from 12 years of MODIS joint histograms, and their occurrences have been collocated with CERES gridded one-degree fluxes (Doelling *et al.*, 2013) in order to composite CREs by CR. Slide 9 shows the 12 MODIS CRs centroids (Oreopoulos *et al.*, 2016) and plots the total TOA CRE versus the total SFC CRE for each CR, weighted by their global relative frequency of occurrence (RFO). All CRs have negative total CRE at both TOA and SFC, indicating that all CRs cool the planet and the surface. The CRs above the diagonal are dominated by low clouds and cool the atmosphere ($CRE_{TOT}^{ATM} < 0$), while those below diagonal are dominated by high clouds and warm the atmosphere ($CRE_{TOT}^{ATM} > 0$). CR4 and CR5 with plenty of mid-level clouds are close to the diagonal and have almost a neutral effect on the atmosphere ($CRE_{TOT}^{ATM} \approx 0$). The plot demonstrates that CRs provide a meaningful description of the diversity of CRE produced by the Earth's cloud systems.

Cloud Vertical Structure

Active observations such as those by the CALIPSO and CloudSat satellites carrying a lidar and radar, respectively, can be used to determine the boundaries of clouds in an atmospheric column, and therefore the Cloud Vertical Structure (CVS). Oreopoulos *et al.* (2017) used a combined CC product to count the frequencies of the most common CVS defined by the occurrence of either isolated High (*H*), Middle (*M*), and Low (*L*) clouds or the co-occurrence of such clouds at various configurations (two or three occurring simultaneously in an atmospheric column, either as contiguous or non-contiguous entities), as in (Tselioudis *et al.*, 2013). Another CC product based on RT calculations applied to CC cloud retrievals allows then the calculation of CRE at TOA, SFC (BOA = Bottom of the Atmosphere) and within ATM for each configuration, at both the SW and LW (blue and red bars in Slide 10, respectively). Slide 10 results account for the frequency of occurrence of each CVS class. All CVS classes except for two have negative CRE_{TOT}^{TOA} (blue bars bigger than red bars), so most CVS classes cool the planet. All CVS classes have negative CRE_{TOT}^{SFC}, and therefore radiatively cool the surface. Six CVS classes radiatively warm the atmosphere (positive difference between TOA and BOA total CRE), three cool it, and one (CVS = "*M*") is near neutral (red, blue, and no hue in the middle part of the plot). To warm the atmosphere a CVS has to include *H* clouds;

to cool it, it has to comprise *L* clouds. CVS = "*L*" has a huge CRE contribution to planetary (TOA) cooling, followed by CVS = "*H* × *M* × *L*", but a much smaller contribution to surface cooling because of large positive CRE_{LW}^{SFC}. Note that even the CVS = "*H*" (isolated high clouds) provides instantaneous radiative cooling to the surface.

Cloud Phase

The CRE breakdown by the CVS described above is broadly consistent with results from (Matus and L'Ecuyer, 2017) based on the same dataset but relying on a cloud (thermodynamical) phase classification. Four CP categories are identified, with three representing liquid, ice and mixed-phase clouds, and one where a phase cannot be assigned because multi-layer clouds of potentially different phases coexist in the atmospheric column. Slide 11 shows the TOA, SFC and ATM total CRE for each CP class, and their combination, all weighted by their frequency of occurrence. Only ice clouds warm the planet (recall that both CVS classes with positive CRE_{TOT}^{TOA} included *H* clouds) and clouds of all phases cool the surface (again, consistent with the CVS analysis), while the atmosphere is warmed by all CP categories other than liquid clouds (recall that all CVS classes with negative CRE_{TOT}^{ATM} included *L* clouds) but with the combined effect exhibiting large spatial variability. The last column is repeated in Slide 12, which also shows the SW and LW components of the atmospheric CRE. Consistent with our earlier discussion, clouds of all CP categories warm the atmosphere in the SW ($CRE_{SW}^{ATM} > 0$), while the CRE_{TOT}^{ATM} and CRE_{LW}^{ATM} patterns are very similar, indicating that the former is just the outcome of applying an offset, in the form of a uniformly positive CRE_{SW}^{ATM}, to the latter.

References

Barkstrom BR. (1984) The Earth Radiation Budget Experiment (ERBE). *Bulletin of the American Meteorological Society* **65**(11):1170–85.

Chen T, Rossow WB, Zhang YC. (2000) Radiative effects of cloud-type variations. *Journal of Climate* **13**:264–86. doi:10.1175/1520-0442(2000)013<0264:REOCTV>2.0.CO;2

Del Genio AD. (2018) The role of clouds in climate. In *Our Warming Planet: Topics in Climate Dynamics*. Eds. Rosenzweig C, Rind R, Lacis A, Manley D. Lectures in Climate Change: **Volume 1**. World Scientific Publishing, pp. 103–30. doi:10.1142/9789813148796_0005

Doelling DRN, Loeb G, Keyes DF, *et al.* (2013) Geostationary enhanced temporal interpolation for CERES flux products. *Journal of Atmospheric and Oceanic Technology* **30**:1072–90. doi:10.1175/JTECH-D-12-00136.1

Hartmann DL, Ockert-Bell ME, Michelsen ML. (1992) The effect of cloud type on Earth's energy balance: Global analysis. *Journal of Climate* **5**:1281–304.

Loeb NG, Doelling DR, Wang H, *et al.* (2018) Clouds and the Earth's Radiant Energy System (CERES) Energy Balanced and Filled (EBAF) Top-of-Atmosphere (TOA) edition-4.0 data product. *Journal of Climate* **31**:895–918. https://doi.org/10.1175/JCLI-D-17-0208.1

Matus AV, L'Ecuyer TS. (2017) The role of cloud phase in Earth's radiation budget. *Journal of Geophysical Research: Atmospheres* **122**:2559–78. doi:10.1002/2016JD025951

Oreopoulos L, Cho N, Lee D. (2017) New insights about cloud vertical structure from CloudSat and CALIPSO observations. *Journal of Geophysical Research: Atmospheres* **122**:9280–300. doi:10.1002/2017JD026629

Oreopoulos L, Cho N, Lee D, Kato S. (2016) Radiative effects of global MODIS cloud regimes. *Journal of Geophysical Research: Atmospheres* **121**:2299–317. doi:10.1002/2015JD024502

Rossow WB, Schiffer RA. (1991) ISCCP cloud data products. *Bulletin of the American Meteorological Society* **72**:2–20.

Rossow W, Tselioudis G, Polak A, Jakob C. (2005) Tropical climate described as a distribution of weather states indicated by distinct mesoscale cloud property mixtures. *Geophysical Research Letters* **32**:L21812. doi:10.1029/2005GL024584

Stephens GL, O'Brien D, Webster PJ, *et al.* (2015) The albedo of Earth. *Reviews of Geophysics* **53**: 141–63. doi:10.1002/2014RG000449

Stephens GL, Winker D, Pelon J, *et al.* (2018) CloudSat and CALIPSO within the A-Train: Ten years of actively observing the Earth system. *Bulletin of the American Meteorological Society* **99**(3): 569–81. https://doi.org/10.1175/BAMS-D-16-0324.1

Tselioudis G, Rossow WB, Zhang Y, Konsta D. (2013) Global weather states and their properties from passive and active satellite cloud retrievals *Journal of Climate* **26**:7734–46. doi:10.1175/JCLI-D-13-00024.1

Slide 1

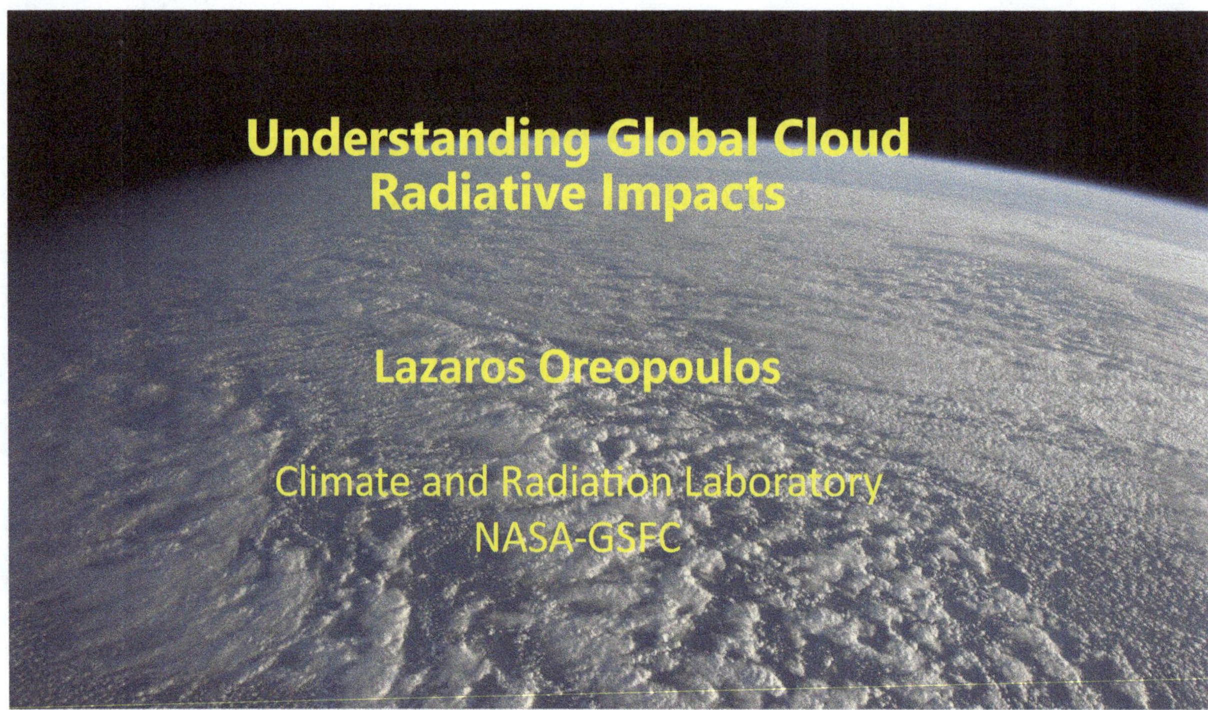

Slide 2

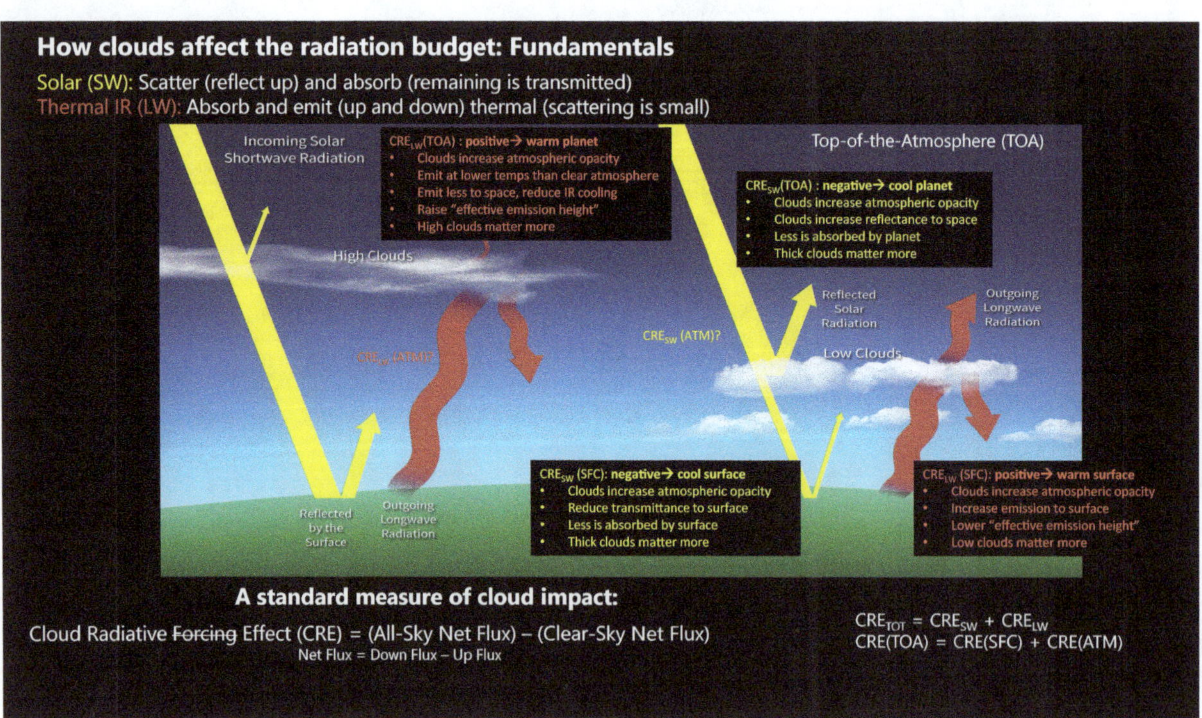

Slide 3

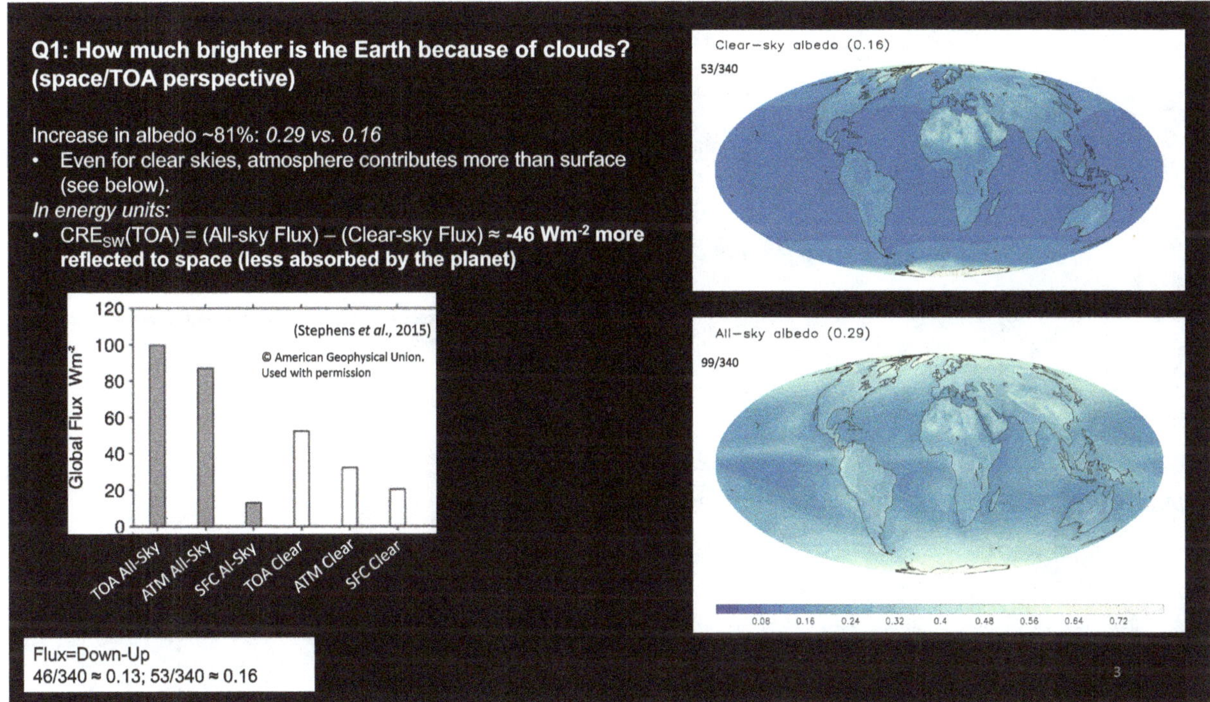

Q1: How much brighter is the Earth because of clouds? (space/TOA perspective)
Increase in albedo ~81%: 0.29 vs. 0.16
• Even for clear skies, atmosphere contributes more than surface (see below).
In energy units:
• CRE$_{SW}$(TOA) = (All-sky Flux) – (Clear-sky Flux) ≈ -46 Wm^{-2} more reflected to space (less absorbed by the planet)
(Stephens et al., 2015)
© American Geophysical Union. Used with permission
Global Flux Wm^{-2}
TOA All-Sky
ATM All-Sky
SFC All-Sky
TOA Clear
ATM Clear
SFC Clear
Flux=Down-Up
46/340 ≈ 0.13; 53/340 ≈ 0.16
Clear-sky albedo (0.16)
53/340
All-sky albedo (0.29)
99/340
3

Slide 4

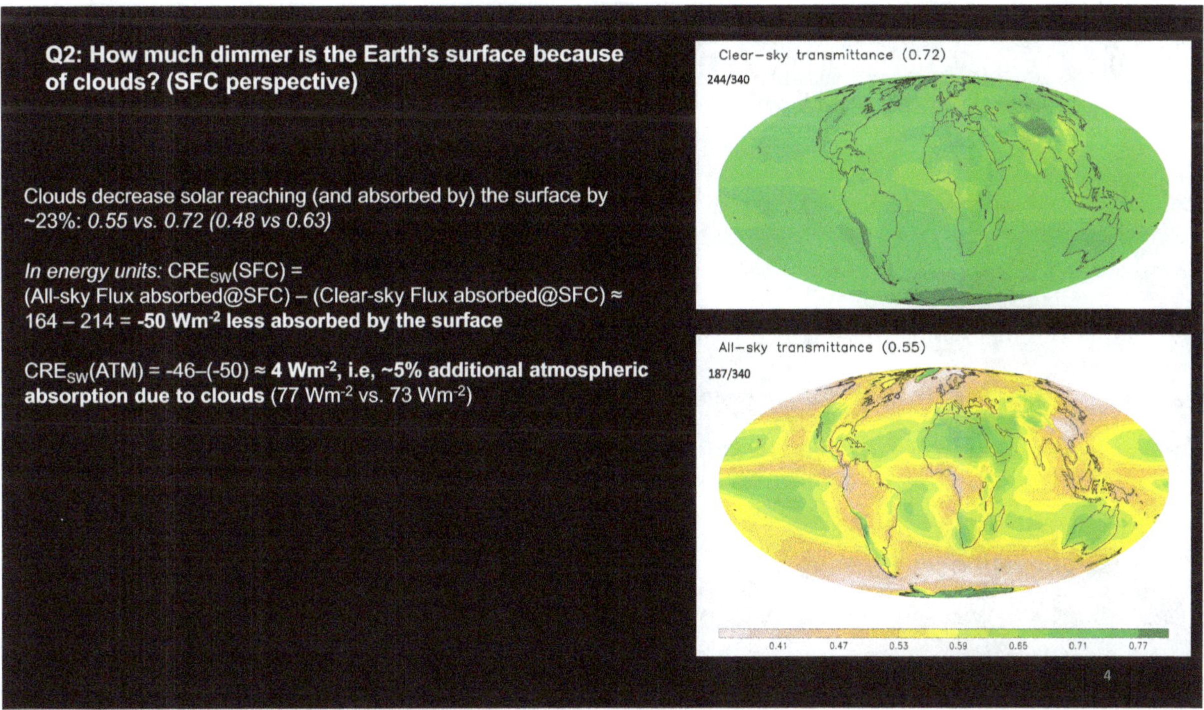

Q2: How much dimmer is the Earth's surface because of clouds? (SFC perspective)
Clouds decrease solar reaching (and absorbed by) the surface by ~23%: 0.55 vs. 0.72 (0.48 vs 0.63)
In energy units: CRE$_{SW}$(SFC) =
(All-sky Flux absorbed@SFC) – (Clear-sky Flux absorbed@SFC) ≈
164 – 214 = -50 Wm^{-2} less absorbed by the surface
CRE$_{SW}$(ATM) = -46–(-50) ≈ 4 Wm^{-2}, i.e, ~5% additional atmospheric absorption due to clouds (77 Wm^{-2} vs. 73 Wm^{-2})
Clear-sky transmittance (0.72)
244/340
All-sky transmittance (0.55)
187/340
4

Slide 7

Summary of Global Annual Cloud Radiative Effects

(Annual, according to CERES EBAF Ed4.0)

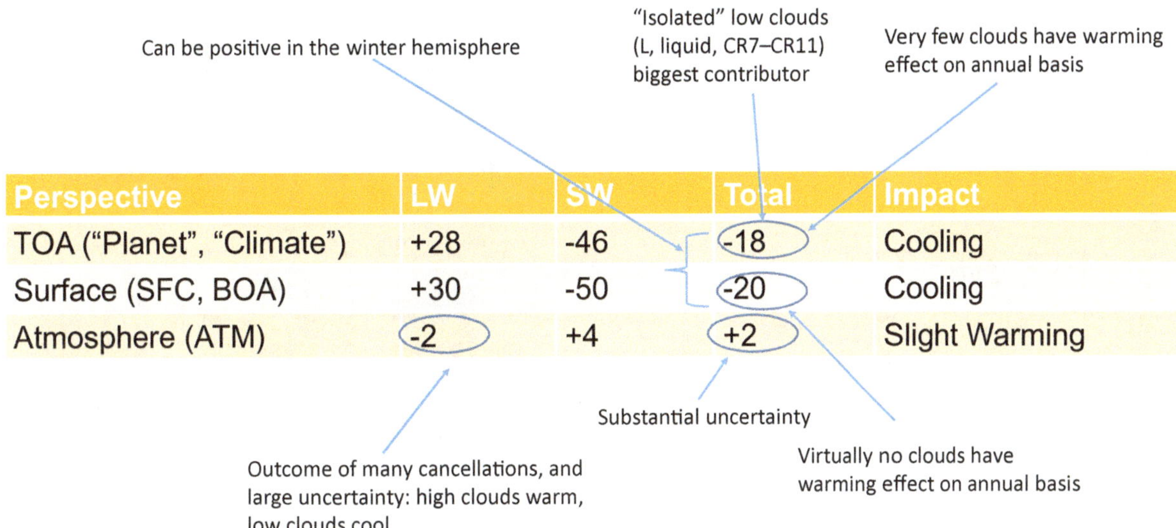

Perspective	LW	SW	Total	Impact
TOA ("Planet", "Climate")	+28	-46	-18	Cooling
Surface (SFC, BOA)	+30	-50	-20	Cooling
Atmosphere (ATM)	-2	+4	+2	Slight Warming

Slide 8

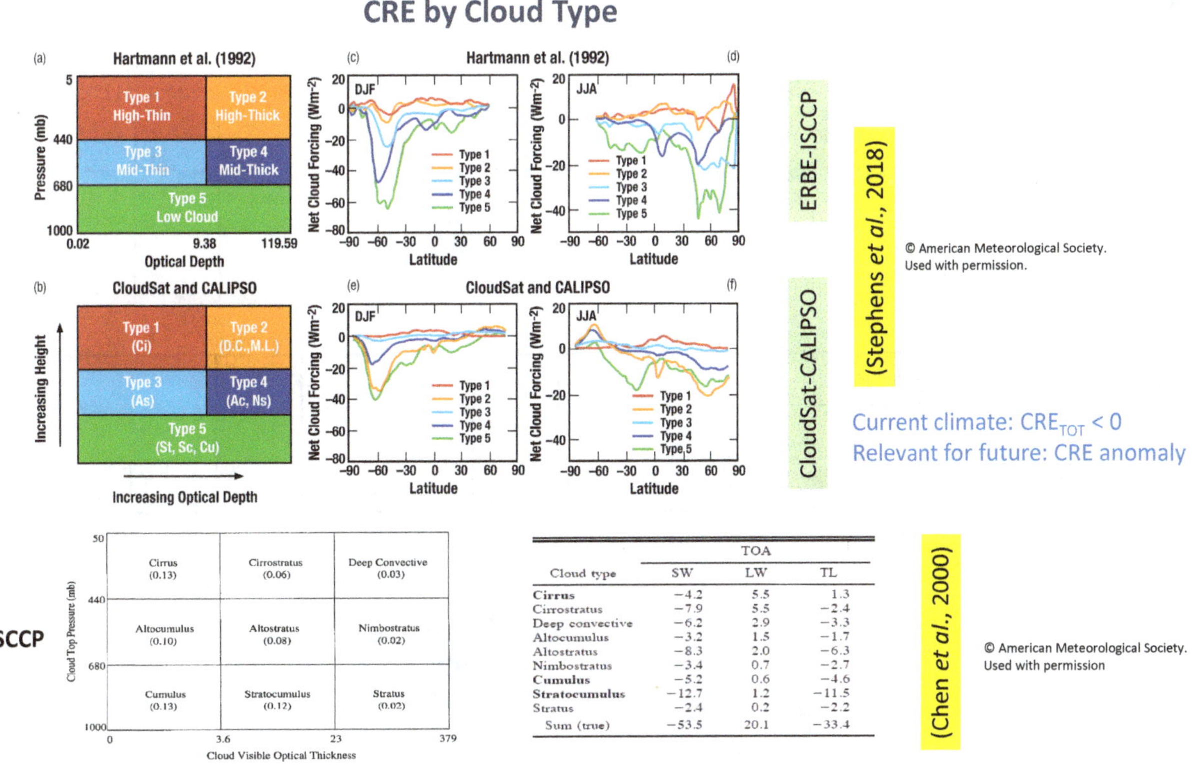

Cloud type	TOA SW	TOA LW	TOA TL
Cirrus	−4.2	5.5	1.3
Cirrostratus	−7.9	5.5	−2.4
Deep convective	−6.2	2.9	−3.3
Altocumulus	−3.2	1.5	−1.7
Altostratus	−8.3	2.0	−6.3
Nimbostratus	−3.4	0.7	−2.7
Cumulus	−5.2	0.6	−4.6
Stratocumulus	−12.7	1.2	−11.5
Stratus	−2.4	0.2	−2.2
Sum (true)	−53.5	20.1	−33.4

Slide 9

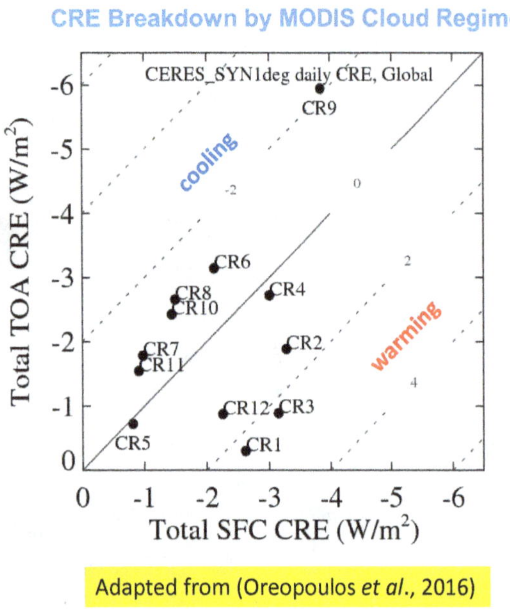

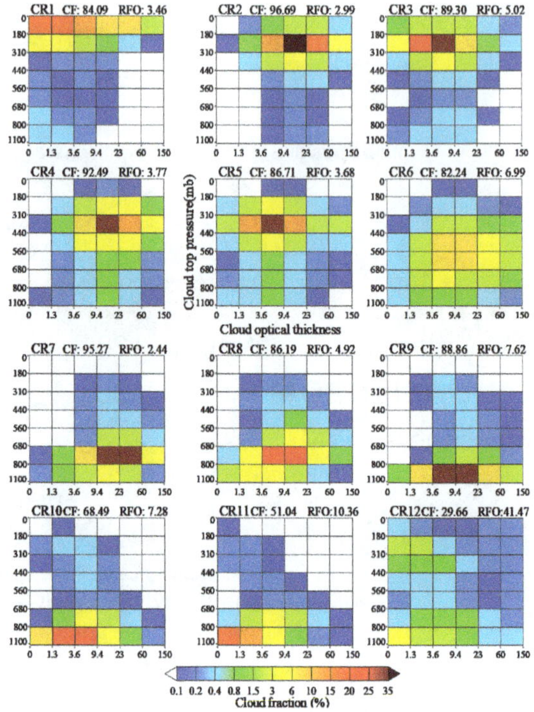

Slide 10

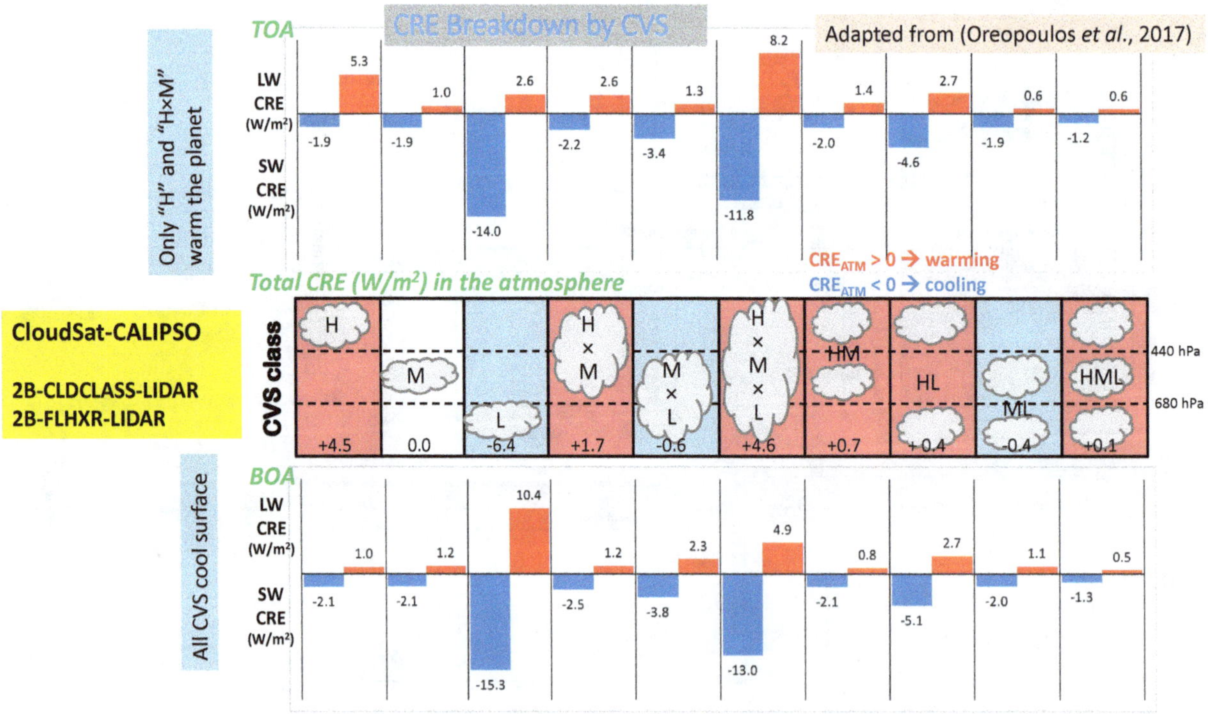

Slide 11

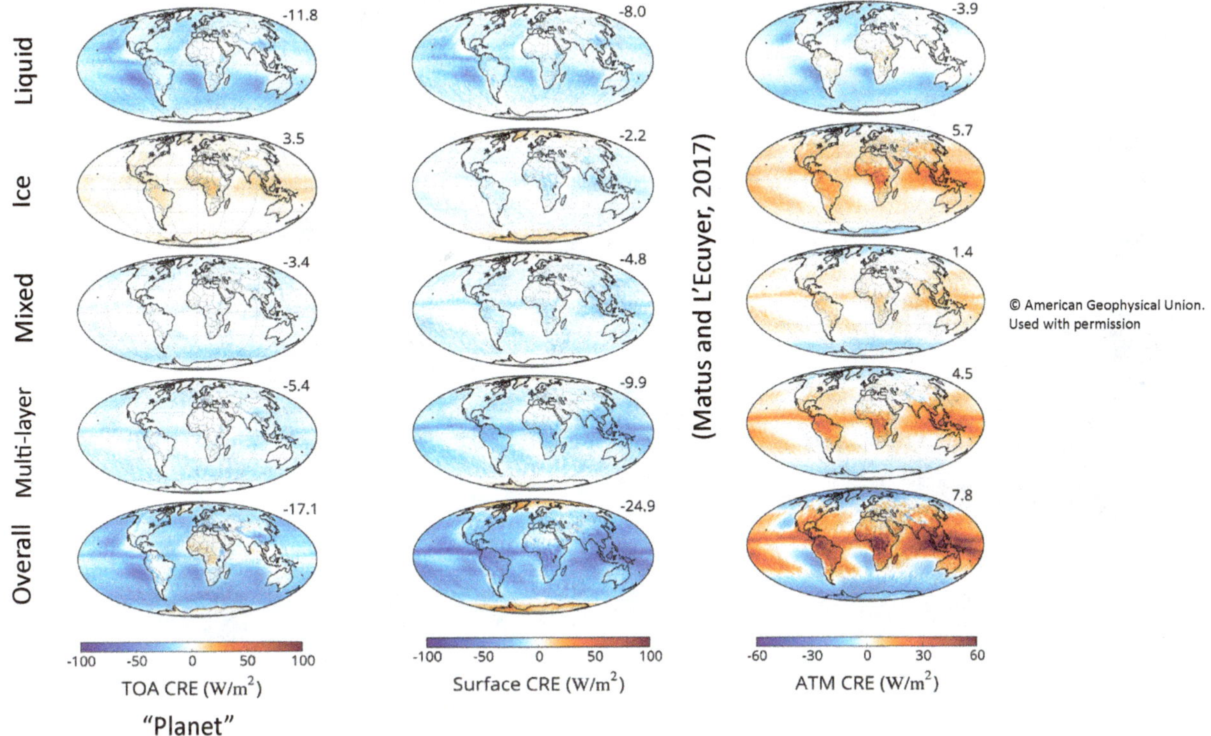

Slide 12

Breakdown of Atmospheric Cloud Radiative Effect by Phase

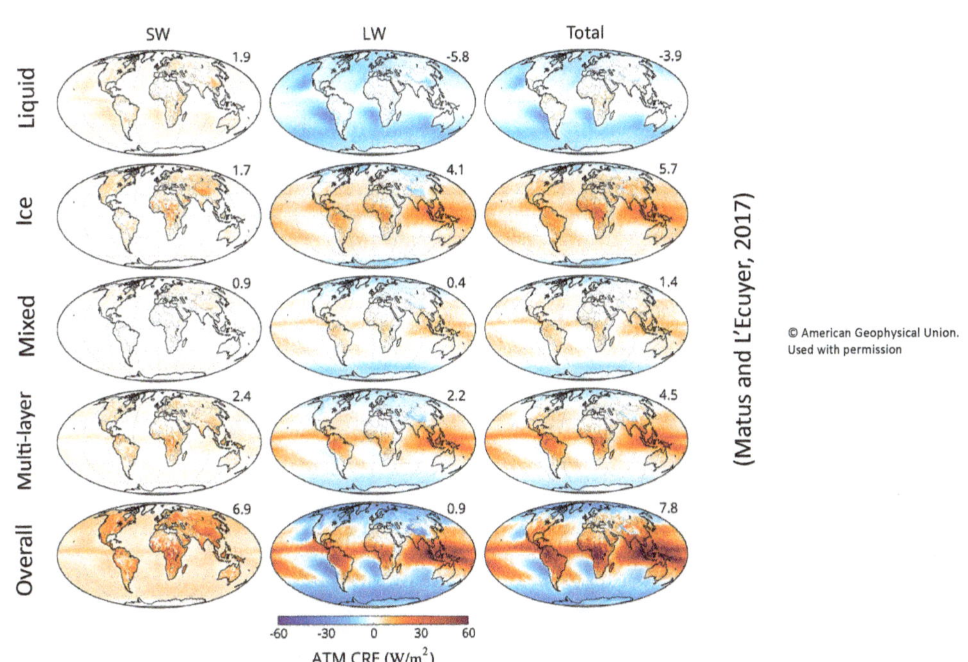

Slide 13

Summary of Global Cloud Radiative Impacts

- Overall, current clouds radiatively cool the planet and (a bit more) the surface; radiative impact on atmosphere is much smaller (slight warming, from SW, within uncertainties)
  - Atmospheric effect is the result of lots of cooling/warming cancellation in LW
- Clouds can have extensive warming TOA/SFC effect in the winter hemisphere
- Thanks to modern satellite observations and radiative transfer modeling, the radiative impact of clouds can be decomposed by cloud class:
  - Cloud classes with low/liquid clouds cool the surface and atmosphere
  - Cloud classes with high clouds still cool the surface, but warm the atmosphere
- Of climate change relevance (cloud feedback):
  - Positive feedback: Fewer clouds that cool and/or more that warm planet
  - Negative feedback: More clouds that cool and/or fewer that warm planet

13

Slide Captions

Slide 1

Slide 2 The effect of clouds on the Earth's Radiation Budget can be viewed as the story of the competition between their distinct interactions with thermal infrared (IR) longwave (LW) radiation where clouds absorb and emit (very little scattering), and the solar shortwave (SW) where clouds scatter and absorb. A standard way to measure the impact of clouds on the flow of energy over a given area is through the so-called Cloud Radiative Effect (CRE), which used to be called "Cloud Radiative Forcing", the difference between all-sky and clear-sky radiative fluxes (irradiances). The difference is properly defined with net = down – up fluxes, when the clear-sky flux is subtracted from all-sky flux so that positive values indicate warming and negative values indicate cooling *regardless of the perspective* (planet [TOA], surface [SFC/BOA], atmosphere [ATM]). The total CRE is the sum of the SW and LW CRE components. From space, we see what is commonly called the top of atmosphere (TOA) CRE. The TOA CRE gives the combined effect of clouds on the surface and atmosphere; in other words, the impact of clouds on the "planet" or "climate". In the SW and at TOA, clouds have a negative SW CRE: They reflect more than clear skies, so less is absorbed by the planet, and we state that clouds have a "cooling effect" on the planet. The thicker and more extensive the cloud is, the greater the cooling. Now at the SFC, clouds also produce a negative CRE, with the cooling stemming from the reduction of energy being available to be absorbed by the SFC. What happens in the atmosphere is not obvious since it depends on the relative magnitudes of TOA and SFC cooling (but it turns out to be positive since clouds enhance atmospheric absorption relative to clear skies). In the LW, CRE at TOA is positive, indicating warming largely because the atmospheric temperature drops with height. Clouds increase the atmospheric opacity in the IR, and because they emit to space at lower temperatures than the clear atmosphere (in other words they "raise the upward effective emission height"), we say that they help the planet "trap more radiation", i.e., they induce a stronger greenhouse effect. The higher and colder the cloud is, the more reduced the emission to space; thus, the stronger the trapping and the planetary warming. The increase in IR opacity also means greater emission toward the surface ("lowering the downward effective emission height"), which is a warming effect (positive LW CRE) at the surface as well. The effect on the surface is greater the lower the cloud (base) is. Again, the effect is on the atmosphere is not obvious as it depends on the relative magnitudes of TOA and SFC LW CRE (it turns out to be a slight cooling but within the uncertainty of the measurements).

Slide 3 Here we address the SW and the TOA perspective. The fundamental question is: How much more reflective is the Earth because of clouds? Satellite observations (CERES) indicate that without clouds, the Earth would reflect about 16% (i.e., clear-sky planetary albedo = 0.16) of incoming solar energy and exhibit a large contrast between the ice-free ocean (less reflective) and continents/sea-ice (more reflective). The atmosphere contributes more to the clear-sky reflectance than the surface, according to an analysis by Stephens *et al.* (2015). Adding clouds increases the albedo to 29% (i.e., current planetary albedo = 0.29). The albedo increase contributed by clouds is almost universal (except over highly reflective surfaces) but is more prominent over the ocean. The relative increase in albedo due to clouds is about 81% and in energy units, namely SW CRE, about 46 Wm^{-2} (negative) on an annual basis. The additional reflection due to clouds has a cooling effect on the planet since it reduces solar absorption.

Slide 4 Moving on to the surface, one can look at how much difference clouds make in the amount of solar radiation surviving its propagation through the atmosphere and reaching the surface (transmittance), or perhaps more importantly, in the fraction of solar radiation absorbed by the surface. The top right panel shows that the spatial variations of clear-sky transmittance are quite substantial and depend mainly on water vapor (and ozone) amounts and the angle of illumination. When clouds are added, the transmittance map changes dramatically (lower right panel). The absolute change is as big or bigger than albedo, but the relative changes are smaller because clear-sky transmittance and clear-sky surface absorption are much larger than clear-sky albedo. Ultimately, clouds decrease the solar radiation reaching (and being absorbed by) the surface by 23%. In energy units, the cooling effect at the surface in terms of SW CRE is 50 Wm^{-2} (negative), according to CERES EBAF Ed4.0, indicating less surface absorption. The difference between TOA and SFC SW CRE is 4 Wm^{-2}, i.e., the atmosphere absorbs an additional 4 Wm^{-2} with clouds present, or 5% more in relative terms. This result is the outcome of some compensation between additional absorption by cloud particles and reduced water vapor absorption since clouds partially shield the water vapor-rich lower levels of the troposphere from incoming solar radiation. We should note here that surface radiative fluxes are not directly observed but are obtained by performing a significant amount of modeling while the observed TOA fluxes are used as a constraint.

Slide 5 We now examine cloud LW effects, namely, cloud "greenhouse effects". Again, either a space (TOA) or surface perspective can be adopted. Starting with the TOA perspective, one way to measure the greenhouse (GH) effect is the GH factor, namely, the ratio of LW emitted to space to the LW emitted upward from the surface. This provides an LW emission "reduction factor", which can also be thought of as a conversion factor that scales surface emission to planetary emission. Without clouds, the global value is 0.67, i.e., 67% of the surface emission escapes to space. There is a fair amount of spatial variability in the GH factor with smaller values in the more humid tropics (upper right panel). When clouds are added, the global value reduces to 0.60, and the regional contrasts become stronger (lower right panel). In terms of this measure, clouds then increase the GH effect by 11% ($0.60/0.67 \approx 0.89$). But we really do not have to compare ratios. We can directly compare TOA upward LW fluxes between clear and all-sky conditions by either taking a ratio or a difference (i.e., CRE). The cloud impact either as a ratio of fluxes or GH factors is the same because the upward flux at the surface is virtually the same for clear and cloudy conditions. In energy units, the warming effect in the LW expressed as CRE is 28 Wm^{-2} (positive, less emitted to space). Yet another way to measure the GH effect from space is the planetary effective temperature, i.e., the temperature perceived from space of a blackbody planet, obtained from the Steffan–Boltzmann law. With clouds, the Earth appears about 7 K colder (255 K versus 262 K).

Slide 6 From the surface perspective, the cloud GH effect can be measured as either the ratio or difference between downward LW fluxes under clear and cloudy conditions. The ratio indicates about 10% more downwelling LW due to clouds, which as a difference, represents an additional 30 Wm^{-2}. The maps look rather similar in terms of variability between clear and cloudy conditions. The surface increase is larger (within uncertainties) than the TOA decrease by only 2 Wm^{-2} globally, with the difference representing a reduction in LW absorption (i.e., a cooling) due to clouds. The global -2 Wm^{-2} value is the net result of large spatial variations from various cloud classes, as we will see later. Blackbody effective temperatures can be used again as a measure of the cloud GH effect on the surface. In this case, what is being measured is how much warmer, on average, a cloudy atmosphere appears to a surface observer compared to a clear atmosphere. When inverted through the Steffan–Boltzmann law, the downwelling fluxes from the CERES product indicate a 6 K warming (279 K for cloudy versus 273 K for clear).

Slide 7 This table provides a summary of the global annual CREs, with additional interpretive information coming from Slides 8 to 12. The SW cooling effect exceeds the LW warming effect at both TOA and SFC. Because the SFC cooling is greater than the TOA cooling, clouds have a warming effect on the atmosphere. Although atmospheric warming originates from the SW, the variability is much larger in the LW (see Slide 12). In other words, all clouds add some SW atmospheric warming but the smaller negative value in the LW is actually the overall result of a lot of compensation between warming and cooling. So, on a planetary scale, clouds do not really affect clear-sky radiative cooling, with the compensation needed to achieve energy equilibrium coming mainly from latent heating (precipitation).

Slide 8 Attempts were made about 2.5 decades ago to find out how different cloud classes affect the Earth's radiation budget. Hartmann *et al.* (1992) combined ERBE and ISCCP observations to break down the CRE at the TOA of five broad cloud types in terms of seasonal zonal averages. These results have been recently updated by Stephens *et al.* (2018) using CloudSat and CALIPSO data. The ISCCP has previously also tried to address the same problem using different cloud types and the same methodology central to Stephens and colleagues' analysis, namely, obtaining flux estimates from supplying cloud property retrievals into a radiative transfer code. From the perspective of the planet as a whole (TOA), these studies find that, with the exception of high-thin or cirrus clouds, all other cloud classes exhibit a negative total CRE overall, i.e., a cooling effect on the earth-atmosphere system. These types of CRE breakdowns are also possible with other cloud classifications, as shown in the next slide.

Slide 9 One such alternate cloud classification is the Cloud Regime (CR) classification. The most recent set of MODIS CRs was derived by Oreopoulos *et al.* (2016) (represented by the 12 mean joint histograms on the right), and their daily occurrences around the globe were matched with CERES CREs in order to assess the radiative effect of each regime. Here, we show again the global annual mean results. This plot summarizes the mean radiative effect of the MODIS CRs after being weighted by their frequency of occurrence. The SW and LW CRE were combined to estimate the total CRE at both the TOA and SFC, and these two total CREs were plotted against each other in order to visually infer the total CRE divergence, which is the total CRE of the atmosphere. All MODIS CRs cool the planet, with the effect being stronger for low-cloud-dominated CRs, which have much stronger negative SW CREs than positive LW CREs. All of them also cool the SFC (negative total CREs), with stronger values for high-cloud-dominated CRs. When the cooling of the SFC exceeds that at TOA, the CR exerts warming on the ATM (below diagonal), while a radiative cooling contribution to the ATM occurs when the negative TOA CRE exceeds the negative SFC CRE. In short, CRs dominated by high clouds warm the atmosphere, while those with predominantly low clouds cool the atmosphere.

Slide 10 Another cloud classification can be achieved using a joint CloudSat/CALIPSO product identifying cloud vertical locations. Oreopoulos *et al.* (2017) devised a scheme that uses the standard ISCCP layers separating clouds into low (*L*), mid-level (*M*), and high (*H*) using the 440 and 680 hPa atmospheric levels and numerous simplifications/assumptions to assign cloud layers to either one, two or three of the standard ISCCP layers. The end result was ten cloud vertical configurations which they called Cloud Vertical Structures (CVSs), and which mirror those of Tselioudis *et al.* (2013) but come from a different cloud profile assignment algorithm. Clouds can either occupy a single standard layer, two standard layers as a single entity, three standard layers as a single entity, and either two or three standard layers simultaneously as separate entities. Another CloudSat/CALIPSO product allowed them to calculate the CRE (TOA, SFC/BOA, ATM) of each configuration, both SW (blue bars) and LW (red bars). In the middle of the graph, the total atmospheric CRE is shown, calculated as the difference between the TOA total (= SW + LW) and SFC total CRE. The numbers shown are weighted by the frequency of occurrence of each CVS class. All CVS classes except two have negative TOA total CRE (blue bars bigger than red bars), so most CVS classes cool the planet. All CVS classes have negative BOA (SFC) total CRE, and so these cool the surface. Six CVS classes warm the atmosphere (positive difference between TOA and BOA total CRE), three cool it, and one (isolated mid-level clouds) is near neutral. To warm the atmosphere, a CVS has to include *H* clouds; to cool it, it has to have *L* clouds. The "*L*" CVS class has a huge CRE contribution to TOA net cooling (followed by "*H* × *M* × *L*") but a much smaller contribution to surface cooling because its positive LW SFC CRE is also large. Note that even the "*H*" CVS class cools the SFC instantaneously, so the planetary warming effect of these clouds is realized through the warming of the atmosphere.

Slide 11 Another possible cloud classification accounts only for their thermodynamic phase. This undertaking was documented by Matus and L'Ecuyer (2017), who used the same datasets as in Slide 10. SW, LW and total CREs were derived for the TOA, SFC and ATM, including the geographical distribution. In their representation, atmospheric columns containing multi-layer clouds cannot be assigned a phase and are presented separately from the "mixed" class, which contains contiguous cloud layers where both phases may co-exist in the same cloud volume. The phase classification indicates that only ice clouds warm the planet and clouds of all phases cool the surface, while the atmosphere is warmed by all clouds other than liquid clouds. The combined effect for the atmosphere exhibits large spatial variability with large swaths of negative and positive values, depending on the geographical preference of clouds of particular phases.

Slide 12 A more detailed look at the atmospheric (ATM) CRE requires a separate examination of the SW and LW components. One can see that the spatial variability of the total CRE (right most column) is driven by the LW (middle column), which has very strong negative values for liquid clouds. The overall LW ATM CRE is almost zero but comes as a result of widespread cancellation. The SW ATM CRE (left most column) is uniformly positive, with all clouds enhancing the SW atmospheric absorption but with substantial cloud phase dependence.

Slide 13 Summary

References

Chen T, Rossow WB, Zhang YC. (2000) Radiative effects of cloud-type variations, *Journal of Climate* **13**:264–86. doi:10.1175/1520-0442(2000)013<0264:REOCTV>2.0.CO;2

Hartmann DL, Ockert-Bell ME, Michelsen ML. (1992) The effect of cloud type on Earth's energy balance: Global analysis. *Journal of Climate* **5**:1281–304.

Matus AV, L'Ecuyer TS. (2017) The role of cloud phase in Earth's radiation budget, *Journal of Geophysical Research: Atmospheres* **122**:2559–78. doi:10.1002/2016JD025951

Oreopoulos L, Cho N, Lee D, Kato S. (2016) Radiative effects of global MODIS cloud regimes, *Journal of Geophysical Research: Atmospheres* **121**:2299–317. doi:10.1002/2015JD024502

Oreopoulos L, Cho N, Lee D. (2017) New insights about cloud vertical structure from CloudSat and CALIPSO observations, *Journal of Geophysical Research: Atmospheres* **122**:9280–300. doi:10.1002/2017JD026629

Stephens GL, O'Brien D, Webster PJ, Pilewski P, Kato S, Li J-l. (2015) The albedo of Earth, *Reviews of Geophysics* **53**:141–63. doi:10.1002/2014RG000449

Stephens GL, Winker D, Pelon J, Trepte C, Vane D, Yuhas C, L'Ecuyer T, Lebsock M. (2018), CloudSat and CALIPSO within the A-Train: Ten years of actively observing the Earth system. *Bulletin of the American Meteorological Society* https://doi.org/10.1175/BAMS-D-16-0324.1

Tselioudis G, Rossow WB, Zhang Y, Konsta D. (2013) Global weather states and their properties from passive and active satellite cloud retrievals. *Journal of Climate* **26**:7734–46. doi:10.1175/JCLI-D-13-00024.1

LECTURE 9

Calculation, Evaluation and Application of Long-term, Global Radiative Flux Datasets at ISCCP: Past and Present

Yuanchong Zhang[1,3], *William B. Rossow*[2], *Andrew A. Lacis*[3] *and Valdar Oinas*[1]

[1]*SciSpace, LLC, New York, New York, USA*

[2]*CREST at the City College of New York, NY, USA*

[3]*NASA Goddard Institute for space Studies, New York, NY, USA*

Yuanchong Zhang is a retired Research Scientist from Columbia University and NASA Goddard Institute for Space Studies (GISS). His research activities/interests include calculating global, 3-hourly and long-term radiative flux profiles that have resulted in the World Climate Research Program's International Satellite Cloud Climatology Project (ISCCP) FC, FD and FH products, validating flux results, studying global radiation/energy budgets, general circulation/energy transports for the whole atmosphere-earth system and its separate atmospheric and oceanic components, and cloud-radiation interactions as well as studying Earth Tides, Seismology, etc.

William B. Rossow is Emeritus Distinguished Professor of Remote Sensing at The City College of New York and former Senior Research Scientist at NASA GISS. His research covered cloud physics and dynamics, atmospheric general circulations, atmospheric radiative transfer and radiation budgets, satellite remote sensing of all components of the Earth's climate system, and advanced analysis of climate and climate feedbacks. He was the Head of the Global Processing Center for the World Climate Research Program's ISCCP from 1982 to 2017.

Andrew A. Lacis is a Senior Research Scientist at the NASA GISS in New York City. He has been a member of the GISS climate modeling group since the mid-1970s. His area of expertise is the development of fast and accurate radiative transfer techniques to model solar and thermal radiation with applications to the study of global climate change and the remote sensing of planetary atmospheres. He is the principal architect of the radiation modeling methodology that is used in the GISS climate GCM.

Valdar Oinas is a Senior Research Scientist at NASA GISS in New York. Now fully retired from teaching physics at the CUNY Queensborough College in New York, he has also been a member of the GISS climate modeling group since the mid-1970s. His chief area of expertise is the development of fast and accurate radiative transfer techniques to model solar and thermal radiation in the study of global climate change with applications also to remote sensing. He has been a key developer of the GISS GCM radiation modeling methodology.

Introduction

It should never be overemphasized that the tempo-spatial variations of all the forms of energy in the atmosphere and at the surface are vitally important to any branch of earth sciences as well as human lives. Among all the forms of energy exchanges, radiative fluxes play a fundamental role as this is virtually the only way for energy exchange between the earth-atmosphere system and outer space, and it is the primary driving force in the general circulation of the whole atmosphere-ocean system.

There are three main inference methods for estimating global radiative fluxes at the top of atmosphere (TOA), surface and in the atmosphere, namely, (1) regression methods based on satellite observations (see, e.g., [Cess and Vulis, 1989]), often with the assistance of some radiation model, (2) so-called "look-up table (LUT) approaches" or "matching" methods that match radiative transfer computations and satellite observations to infer the desired radiative fluxes at TOA and surface under various scenarios (see, e.g., [Ma and Pinker, 2012]) and (3) detailed flux calculation methods that use observation-based physical parameters as realistic as possible for all the associated atmospheric and surface properties as inputs to a detailed, relatively accurate and efficient radiative transfer model to calculate fluxes. This third method is probably the most comprehensive and more physically meaningful method that we have employed since the beginning of the ISCCP flux-calculation project. This project was pioneered by Rossow and Lacis (1990), but the production of an actual global flux product had to wait for a global, comprehensive cloud data product to appear, which is the first generation of the International Satellite Cloud Climatology Project (ISCCP) datasets, the C-series product, or ISCCP-C (Schiffer and Rossow, 1983; Rossow and Schiffer, 1991). This is because all of the global cloud climatologies prior to the ISCCP-C usually report on cloud amount only and do not contain sufficient information on cloud-top temperature (or altitude) and/or on the optical properties of clouds required by climate modelers to calculate the first-order effects of clouds on the Earth's radiation budget or the climate feedbacks produced by cloud variations (WMO, 1975). The ISCCP-C product for the first time made it possible for us to use the detailed radiation model of the NASA Goddard Institute for Space Studies (GISS) GCM Model II (Hansen *et al.*, 1983) to calculate global, satellite-observation-based radiative fluxes at TOA and surface, which is our first-generation global flux product, the ISCCP-FC (Zhang *et al.*, 1995), where the acronym FC means Flux calculated (mainly) using C data (ISCCP-C, or C-series, and similarly for FD and FH for the second and third-generation flux products as described below).

In the following years, ISCCP has developed two more generations of its products, D-series (Rossow and Schiffer, 1999) and the current H-series products (Young *et al.*, 2018). The ISCCP-D product reported separate cloud properties for liquid and ice forms of low and middle-level clouds along with other many important improvements (Rossow and Schiffer, 1999) and, as the third-generation product, the ISCCP-H has been more advanced. Instead of using every-30-km-sampled B3 for ISCCP-D, ISCCP-H uses 10-km-sampled B1. There are also other refinements in radiance quality control (QC), calibration, cloud detection (especially high, thin and polar clouds), cloud and surface property retrievals with more modern and more homogeneous ancillary datasets, e.g., more accurate surface type and topography, snow/ice datasets, reprocessed ozone data, etc. The cloud and surface retrievals are now based on more realistic atmospheric properties, with MACv1 aerosols (Kinne *et al.*, 2013) included so that the uncertainties are reduced as all the ancillary datasets have been improved. The temperature and humidity profiles with increased vertical resolution have been statistically adjusted to have diurnal variation wherever suitable. The new ISCCP H-series has also increased sub-data product categories (e.g., five for L3) with new globally gridded pixel-level (L2) data.

Accordingly, we have developed and produced the next two generations of radiative profile flux products, namely, the ISCCP-FD (Zhang *et al.*, 2004) and the current ISCCP-FH, which are based on improved NASA GISS radiation models of the GCM ModelD, or SI2000 (Hansen *et al.*, 2002) and ModelE (Schmidt *et al.*, 2006), respectively. All three generations of the flux products are global, with the same temporal and spatial resolution as their corresponding ISCCP products. Moreover, FD and FH have added fluxes at three levels in the atmosphere (between TOA and the surface), as a Cloud Vertical Structure (CVS) model is introduced to construct realistic vertical cloud layers instead of using single, fixed 100-hPa-thick cloud layers as in ISCCP-FC. The temporal coverage is the Earth Radiation Budget Experiment (ERBE) period for ISCCP-FC, while ISCCP-FD covered July 1983 to December 2009. The ISCCP-FH products now cover the period from July 1983 to June 2017 (34 years) and will be extended whenever the ISCCP-H is completed. Both the ISCCP-FH (version 0.0) and ISCCP-FD products (version 0.0 except 2001 of v. 0.01) are publicly available online at: https://isccp.giss.nasa.gov/projects/flux.html

The ISCCP-FH product may also become publicly available at NOAA National Centers for Environmental Information (NCEI).

The core of all the NASA GISS radiation models is the correlated *K*-distribution method as described in (Hansen *et al.*, 1983) for GCM Models I and II. Since then, the spectral resolution has been increased by increasing the number of the noncontiguous correlated spectral resolution intervals, *K*, to obtain higher accuracy, from 12 and 25 to the current 16 and 33 for shortwave (SW) and longwave (LW), respectively, paralleling the evolution of the GISS GCM models I and II to SI2000 or ModelD (Hansen *et al.*, 2002), and ModelE (Schmidt *et al.*, 2006). The radiation part of the GCM Model II, code-named RadII, was used for the ISCCP-FC production, while ModelD's radiation part, code-named RadD, for the ISCCP-FD flux profile product (Zhang *et al.*, 2004). Based on the radiation part of the GISS GCM ModelE (of 2011), code-named RadE, we have developed the current radiation calculation code, RadH. RadE (and therefore RadH) has an accuracy of 1 W/m^2 for cooling rates (in degree/day) throughout the troposphere and most of the stratosphere as compared with line-by-line calculations (Lacis and Oinas, 1991) for LW and is close to 1% for SW for RadE/RadH now. The major advanced features of RadH include: (1) reformulation of the SW line absorption for H_2O, O_2, CO_2, CH_4, N_2O, etc., using the latest HITRAN2012 atlas (Rothman *et al.*, 2013) with additional weak-SW-absorption values for H_2O, O_2 and CO_2, (2) corrections of some H_2O deficiencies, especially for large water vapor amounts (see, e.g., [DeAngelis *et al.*, 2015]), so it is improved in terms of accuracy, spectral coverage, additional absorption phenomena, and validity, (3) use of a new global aerosol data MAC-v2, based on AeroCom's advances (Kinne, 2013), (4) improved LW modeling of the H2O continua, CFC absorption cross-sections, SO_2 line absorption, CH_4 and N_2O overlap treatment and polar region (conditions) profile calculations,

(5) improved vertical cloud layer construction using the Vertical Cloud Layer Configuration (VCLC) algorithm that uses 5-year CloudSAT-CLIPSO climatology (Stephens *et al.*, 2002), and (6) new TSI (total solar irradiance) is introduced, which is a self-consistent daily time series based on SORCE V-15, Davos WRC composite and RMIB (from Dr. Shashi Gupta). The RadH is equivalent to the radiation code of the current NASA GISS ModelE2.1 (Kelley *et al.*, 2020).

All the three generations of the radiation models (RadII, RadD and RadH) that we adopted for ISCCP-FC, FD and FH production have the following positive features: (1) the atmosphere from the surface to the TOA can be physically divided into any number of layers at any pressure level so that, e.g., physical cloud layers can be positioned precisely at any altitude and interleaved with clear air layers just like in the real world, (2) all the input parameters are physical quantities that can be as realistic as possible, and naturally and realistically variable in each cloud or air layer as well as at the surface (e.g., each cloud layer has its own top and base temperatures and pressures, optical thickness, and microphysical model specified by the phase, particle shape, effective particle size, and size distribution variance. All air/cloud layers can also have their own aerosol mixtures with different optical properties, i.e., all the constituents and their physical properties in the model can be specified vertically independently), (3) the corresponding output fluxes were calculated for all the specified interfaces of the air/cloud layers, including TOA and the surface, for all the downwelling and upwelling broadband shortwave (SW of 0.2 to 5.0 μm) and longwave (LW of 5.0 to 200 μm) component fluxes (with additional downward SW direct and diffuse fluxes at surface but for ISCCP-FH only) for all-sky, clear-sky and overcast scenes, (4) the model is detailed and complete (i.e., not for a "bulk" atmosphere, etc.) and is physically self-consistent at all wavelengths, i.e., SW and LW are all treated using a correlated *K*-distribution (CKD) method, atmospheric and surface properties are from consistent data sources, so they can be used to diagnose/determine the radiative heating/cooling effects of any specific physical parameter(s) and/or structure(s) of the atmosphere, clouds and surface in a consistent way (the self-consistency may help reduce some flux errors), and (5) with improvements of the accuracy and knowledge of all input physical quantities, as well as the radiative transfer theory itself, the model can be relatively easily updated to incorporate any newly available information (e.g., when ice cloud detection became feasible and better information about their properties became available in ISCCP-D data that had new, separated ice and liquid clouds, RadD was able to use this information to produce ISCCP-FD flux profile products with better cloud information for both the ice and liquid clouds).

Various methods have been used to validate/evaluate our flux products (ISCCP-FC, FD and FH), mainly using observations of ERBE and the Clouds and the Earth's Radiant Energy System (CERES) (Wielicki *et al.*, 1996) at TOA and the Global Energy Balance Archive (GEBA) (Ohmura and Gilgen, 1991) and the Baseline Surface Radiation Network (BSRN) (Ohmura *et al.*, 1998) at surface and other flux products (see [Rossow and Zhang, 1995] and [Zhang *et al.*, 2004], respectively, for the validation of the ISCCP-FC and FD, and see accompanying slides for ISCCP-FH). Based on our previous validation and for regional, monthly mean fluxes, the ISCCP-FC has overall uncertainties of 10–15 W/m^2 at TOA and 20–25 W/m^2 at surface, while the ISCCP-FD has been improved to smaller 5–10 W/m^2 at TOA and 10–15 W/m^2 at surface. (The latter is probably underestimated; see Slides 2, 18 and 19). The preliminary uncertainties of the ISCCP-FH are overall slightly better than FD (especially for atmospheric

SW absorption and surface fluxes) with FH's increased spatial resolution of 110 km from FC's and FD's 280 km.

Since the release of ISCCP-FC and FD products, they have been used worldwide by researchers in numerous research studies and have inspired/promoted a number of new avenues in cloud-radiation research. The following are a few examples for the most important applications of ISCCP flux data in several research areas based on work by various authors.

(1) Validation/evaluation and comparison studies for radiative flux datasets

All datasets are subject to some kind of validation/evaluation before they can become useful. As stated above, the uncertainty estimates for ISCCP flux products are obtained through various validation/evaluation procedures. Similarly, other radiative flux products need to go through such processing by comparing them with various observed fluxes (such as GEBA, BSRN) and observation-based-calculated flux products such as ISCCP and GEWEX-SRB (Stackhouse *et al.*, 2011). Indeed, there have been extensive uses of the ISCCP flux datasets by various authors that have promoted the improvements of both climatological data production and radiative models, e.g., Oreopoulos *et al.* (2012) have found that our RadD-based atmospheric SW absorption has ~5 W/m^2 low bias via inter-comparison of about a dozen of radiation models with identical input datasets; the finding led to our improvement for RadH, Another example is that CERES has improved its Angular Distribution Models (ADMs) for TOA radiative flux estimation (Loeb *et al.*, 2007) over the ADM of ERBE (see, e.g., [Barkstrom and Smith, 1986]), after inaccuracies of ERBE's ADM were pointed out by many authors, including Rossow and Zhang (1995). The most comprehensive flux assessment activities were organized by WCRP GEWEX Radiative Flux Assessment (RFA) conducted over several years, with more than a dozen of institutes and their datasets involved (see [Raschke *et al.*, 2012a; 2012b]). The RFA activities have benefited all the participated institutes in improving their flux datasets.

(2) Sensitivity study of flux calculation to atmospheric and surface physical properties or models

In radiative flux calculation, there are many physical parameters to represent, such as the Sun's spectral incoming flux intensity at TOA, radiative properties of atmospheric constituents (gases, aerosols, etc.), atmospheric temperature and humidity, macrophysical and microphysical properties of clouds (morphology, particle size, phase, etc.), and surface properties (albedo, temperature, etc.) that are imported to a radiation model to produce total flux results (broadband SW and LW fluxes). This means that we can relatively easily quantify the radiative flux changes due to changes of individual (or a group of) properties after changing one or several physical input parameters or even the atmospheric structure and the model's parameterization in the flux calculations to diagnose their respective influences on flux results that would help understand their separate roles in the radiative energy distribution in the earth-atmosphere system. In the real world, because we can only measure the total flux results, it is difficult to separate the radiative flux components according to their individual contributions and properties. Sensitivity studies have provided a powerful tool to estimate the roles of individual contributions of physical properties and how they affect the radiative fluxes or cooling/heating rates in the atmosphere, ocean and land. Furthermore, the calculated fluxes are not without errors or uncertainties since there are always uncertainties/errors coming with input datasets and even the radiation model itself. Accordingly, the sensitivity test is also useful to break down how and how much these errors/uncertainties are caused by different components, either from input variables or from radiation modeling, so as to improve the input data and radiation model in order to have more accurate flux results. We have systematically conducted various sensitivity tests on ISCCP radiation fluxes, e.g., Table 2 in (Zhang *et al.*, 1995) shows a list of typical sensitivity tests for ISCCP-FC; also see (Zhang *et al.*, 2004; 2006; 2007a; Chen *et al.*, 2000a; 2000b).

The methodology of (1) and (2) has been substantially generalized and widely used over the past two decades in Model Inter-comparison Projects (MIPs) for different global circulation models (GCMs), as well as for other climate and radiation models such as RFMIP (for Radiation Forcing, see, e.g., [Pincus *et al.*, 2016]), CMIPS (for Climate, see, e.g., [Eyring *et al.*, 2015]), CFMIP (for Cloud Feedback, see, e.g., [Webb *et al.*, 2017]), etc. These inter-comparisons have been invaluable for model and climatological feedback evaluation and for a better understanding of climate modeling and improving its prediction capabilities.

(3) Earth radiation balance and budget at TOA and surface

Our earth-atmosphere system has been in a quasi-equilibrium state for a geologically long-time period with fluctuation within a moderate temperature range (e.g., 0.8 million years, as described in [Hansen and Sato, 2011], https://www.giss.nasa.gov/research/briefs/hansen_15/PaleoImplications.pdf). It is under such quasi-equilibrium conditions that human beings and other living things have been able to survive. Since the earth-atmosphere system's energy source is the Sun (while the system's energy sink is only through thermal radiation at TOA [OLR for Outgoing Longwave Radiation] emitted to outer space), the TOA incoming SW and OLR fluxes must be approximately balanced in order to maintain such a quasi-equilibrium state. In other words, the earth radiation budget's balance at TOA is fundamentally vital for all life on Earth and thus has long drawn scientific interest since Simpson (1929). As the radiation budget at TOA and the surface of the Earth are closely related, it is no wonder that the ISCCP-FD results have been widely used for radiation budget studies at both TOA and surface, see, e.g., (Ohmura, 2014; Trenberth *et al.*, 2009; Wong *et al.*, 2006; Loeb *et al.*, 2009). Furthermore, how can the earth-atmosphere system be maintained to have such status for such a long period that has not been possible on other planets? Could this status be destroyed by today's increasing human activities or by natural disasters? Such important scientific questions are what climatologists and other scientists are now working on in order to address and answer them.

(4) Global energy transports and the derivation of surface turbulence fluxes.

The differential distribution of the TOA total net flux with latitude can be used to derive the required global meridional total energy transport (see, e.g., [Carissimo *et al.*, 1985]). The total energy transport of the earth-atmosphere system can be partitioned into the atmosphere and ocean as shown in (Peixoto and Oort, 1992). Zhang and Rossow (1997) suggested using the boundary fluxes, i.e., surface radiative and turbulent fluxes, to do such partitioning instead of direct measurement in the atmosphere or/and ocean, which prompted

SeaFlux projects, in which turbulent fluxes are produced (see, e.g., [Curry *et al.*, 1999; 2004; Yu and Weller, 2007]). Our earth-atmosphere system is similar to a (low-efficiency) heat engine that takes input energy from the Sun (SW flux) to drive all kinds of dynamic processes in the system and expel the energy through OLR, which is exactly what the GCM has been developed for in order to simulate and study all the dynamic processes in the system, including global energy transport and how turbulence affects the planetary boundary layer.

(5) Long-term energy trends and variations of the atmosphere-earth system

As human beings, we are concerned about how long the current climate system can be maintained in its quasi-equilibrium status that enables human beings to survive. The temporal variations of the TOA radiation budget are what climatologists have been working on in order to study long-term energy trends and variations of the atmosphere-earth system and to refine the ability to predict the future climate change, see, e.g., (Hansen *et al.*, 1997; Romanou *et al.*, 2007; Zhang *et al.*, 2007b; Wong *et al.*, 2006).

In short, the above incomplete list of research has shown how important a good flux product can be. The ISCCP-based flux products like ISCCP-FH (and GEWEX-SRB) are especially valuable because there is no other product that has global coverage and minimum-required diurnal sampling (eight times a day) with a temporal coverage period longer than the ISCCP period of 34 years.

In the following slides, we focus on the fundamental features of the current ISCCP-FH Radiative Profile Flux Production and its validation/evaluation. We provide some necessary comparisons with other products, past and present, so that readers can understand how it has been developed and what its current status and uncertainty estimates are.

The slides begin with an overall background introduction (Slides 2–3) and outline (Slides 4–9) for the ISCCP-FH product, and then describe overall improvements of ISCCP-FH (Slides 7–11) and specific, important improvements (Slides 10–12). Slides 13–16 compare fluxes of ISCCP-FH and FD with CERES at TOA and in the atmosphere, respectively, and Slides 17–18 are for surface validation for ISCCP-FH (and FD and CERES) against BSRN. The last two slides (19–20) provide an overall uncertainty estimate of ISCCP-FH and conclusions, respectively.

Acknowledgments

We would like to give special thanks to Dr. Reto Reudy, who helped separate the radiation code (RadE) from NASA GISS GCM ModelE that made it possible to initiate the ISCCP-FH project. We thank Dr. Stefan Kinne, who has provided us with Aerocom, MACv1 and MACv2 aerosol datasets. We also thank Dr. Paul W. Stackhouse and Dr. Shashi K. Gupta, who provided us with TSI datasets. Computer facilities are supplied by NOAA's National Centers for Environmental Information and NASA GISS/Columbia University. The development of the ISCCP-FH production code was funded by the NOAA Climate Data Record Project (Grant NA11NES4400002 for 2011–2014). Finally, we thank the CERES, BSRN, CloudSat and CALIPSO teams for providing us with their datasets for the ISCCP-FH project's development and validation work.

References

Barkstrom BR, Smith GL. (1986) The Earth Radiation Budget Experiment: Science and implementation. *Reviews of Geophysics* **24**(2):379–90.

Carissimo BC, Oort AH, Vonder Haar TH. (1985) Estimating the meridional energy transports in the atmosphere and ocean. *Journal of Physical Oceanography* **15**:82–91.

Cess RD, Vulis IL. (1989) Inferring surface solar absorption from broadband satellite measurements. *Journal of Climate* **2**(9):1989.

Chen T, Rossow WB, Zhang Y-C. (2000a) Radiative effects of cloud type variations. *Journal of Climate* **13**:264–86.

Chen T, Zhang Y-C, Rossow WB. (2000b) Sensitivity of atmospheric radiative heating rate profiles to variations of cloud layer overlap. *Journal of Climate* **13**:2941–59.

Curry JA, Bentamy A, Bourassa MA, *et al.* (2004) SeaFlux. *Bulletin of the American Meteorological Society* **85**:409–24.

Curry JA, Clayson CA, Rossow WB, *et al.* (1999) High-resolution satellite-derived dataset of the ocean surface fluxes of heat, freshwater and momentum for the TOGA COARE IOP. *Bulletin of the American Meteorological Society* **80**:2059–80.

DeAngelis AM, Qu X, Zelinka MD, Hall A. (2015) An observational radiative constraint on hydrologic cycle intensification. *Nature* **528**:249–53. doi:10.1038/nature15770

Eyring V, Bony S, Meehl GA, *et al.* (2015) Overview of the Coupled Model Intercomparison Project Phase 6 (CMIP6) experimental design and organization. *Geoscientific Model Development* **8**:10539–83. doi:10.5194/gmdd-8-10539-2015

Hansen J, Russell G, Rind D, *et al.* (1983) Efficient three-dimensional global models for climate studies: Model I and II. *Monthly Weather Review* **111**:609–62.

Hansen J, Sato M. (2011) Paleoclimate implications for human-made climate change, Accepted for publication in "Climate Change at the Eve of the Second Decade of the Century: Inferences from Paleoclimate and Regional Aspects: Proceedings of Milutin Milankovitch 130th Anniversary Symposium" (A. Berger, F. Mesinger, and D. Šijači, Eds.)

Hansen, J, Sato M, Ruedy R, *et al.* (2002) Climate forcings in Goddard Institute for Space Studies SI2000 simulations. *Journal of Geophysical Research* **107**(D18):4347. doi:10.1029/2001JD001143

Hansen J, Sato M, Ruedy R. (1997) Radiative forcing and climate response. *Journal of Geophysical Research* **102**:6831–6864.

Kelley M, Schmidt GA, Nazarenko LS, *et al.* (2020) GISS — E2.1: Configurations and climatology. *Journal of Advances in Modeling Earth Systems* **12**:e2019MS002025. https://doi.org/10.1029/2019MS002025

Kinne S, O'Donnel D, Stier P, *et al.* (2013) MAC-v1: A new global aerosol climatology for climate studies. *Journal of Advances in Modeling Earth Systems* **5**:704–40. doi:10.1002/jame.20035

Lacis AA, Oinas V. (1991), A description of the correlated k distributed method for modeling nongray gaseous absorption, thermal emission, and multiple scattering in vertically inhomogeneous atmospheres. *Journal of Geophysical Research* **96**:9027–63. doi:10.1029/90JD01945

Loeb NG, Kato S, Loukachine K, *et al.* (2007) Angular distribution models for top-of-atmosphere radiative flux estimation from the Clouds and the Earth's Radiant Energy System instrument on the Terra satellite. Part II: Validation. *Journal of Atmospheric and Oceanic Technology* **2007**(24):564–84.

Loeb NG, Wielicki WBA, Doelling DR, *et al.* (2009) Toward optimal closure of the Earth's top-of-atmosphere radiation budget. *Journal of Climate* **22**:748–66.

Ma Y, Pinker RT. (2012) Modeling shortwave radiative fluxes from satellites. *Journal of Geophysical Research* **117**:D23202. doi:10.1029/2012JD018332

Ohmura A. (2014) The development and the present status of energy balance climatology. *Journal of the Meteorological Society of Japan* **92**(4):245–85. doi:10.2151/jmsj.2014-401

Ohmura A, Gilgen H. (1991) The GEBA data base, interactive applications, retrieving data, Report 2. Global Energy Balance Archive, World Climate Program Water Project A7, Zurich, 60 pp.

Ohmura A, Dutton EG, Forgan B, *et al.* (1998) Baseline Surface Radiation Network (BSRN/WCRP): New precision radiometry for climate research. *Bulletin of the American Meteorological Society* **79**:2115–36.

Oreopoulos L, Mlawer E, Delamere J. (2012) The continual intercomparison of radiation codes: Results from phase I. *Journal of Geophysical Research* **117**:D06118. doi:10.1029/2011JD016821

Peixoto JP, Oort AH. (1992) *Physics of Climate.* American Institute of Physics, 520 pp.

Pincus R, Forster PM, Stevens B. (2016) The Radiative Forcing Model Intercomparison Project (RFMIP): Experimental protocol for CMIP6. *Geoscientific Model Development* **9**:3447–60. doi:10.5194/gmd-9-3447-2016

Raschke E, Kinne S, Stackhouse PW, *et al.* (2012a) GEWEX Radiative Flux Assessment (RFA) volume 1: Assessment, a project of the World Climate Research Programme Global Energy and Water Cycle Experiment (GEWEX) radiation panel, WCRP Report No. 19/2012.

Raschke E, Kinne S, Stackhouse PW, *et al.* (2012b) GEWEX Radiative Flux Assessment (RFA) volume 2: Supplementary information, a project of the World Climate Research Programme Global Energy and Water Cycle Experiment (GEWEX) radiation panel, WCRP Report No. 19/2012.

Romanou A, Liepert B, Schmidt GA, *et al.* (2007) 20th century changes in surface solar irradiance in simulations and observations. *Geophysical Research Letters* **34**:L05713. doi:10.1029/2006GL028356

Rossow WB, Lacis AA. (1990) Global, seasonal cloud variation from satellite radiance measurements. 2, Cloud properties and radiative effects. *Journal of Climate* **3**:1204–53.

Rossow WB, Schiffer RA. (1991) ISCCP cloud data products. *Bulletin of the American Meteorological Society* **72**:2–20.

Rossow WB, Schiffer RA. (1999) Advances in understanding clouds from ISCCP. *Bulletin of the American Meteorological Society* **80**:2261–88. https://doi.org/10.1175/1520-0477(1999)080<2261:AIUCFI>2.0.CO;2.

Rossow WB, Zhang Y-C. (1995) Calculation of surface and top of atmosphere radiative fluxes from physical quantities based on ISCCP data sets: 2. Validation and first results. *Journal of Geophysical Research* **100**:1167–97.

Rothman LS, Gordon IE, Babikov A. (2013) The HITRAN2012 molecular spectroscopic database. *Journal of Quantitative Spectroscopy and Radiative Transfer* **130**(2013):4–50.

Schiffer RA, Rossow WB. (1983) The International Satellite Cloud Climatology Project (ISCCP): The first project of the World Climate Research Program. *Bulletin of the American Meteorological Society* **64**:779–84.

Schmidt GA, Ruedy R, Hansen JE, *et al.* (2006) Present day atmospheric simulations using GISS ModelE: Comparison to in-situ, satellite and reanalysis data. *Journal of Climate* **19**:153–92. doi:10.1175/JCLI3612.1

Simpson GC. (1929) The distribution of terrestrial radiation. *Memoirs of the Royal Meteorological Society* **3**:53–78.

Stackhouse Jr PW, Gupta SK, Cox SJ, *et al.* (2011) The NASA/GEWEX surface radiation budget release 3.0: 24.5-year dataset. *GEWEX News* **21**(1):10–12.

Stephens GL, Vane DG, Boain RJ, *et al.*, and the CloudSat Science Team. (2002) The CloudSAT mission and the A-train, a new dimension of space-based observations of clouds and precipitation. *Bulletin of the American Meteorological Society* **83**(12):1771–90.

Trenberth KE, Fasullo JT, Kiehl J. (2014) Earth's global energy budget. *Bulletin of the American Meteorological Society* **90**(3):311–23.

Webb MJ, Andrews T, Bodas-Salcedo A, *et al.* (2017) The Cloud Feedback Model Intercomparison Project (CFMIP) contribution to CMIP6. *Geoscientific Model Development* **10**:359–84. doi:10.5194/gmd-10-359-2017

Wielicki BA, Barkstrom BR, Harrison EF, *et al.* (1996) Clouds and the Earth's Radiant Energy System (CERES): An Earth Observing System experiment. *Bulletin of the American Meteorological Society* **77**:853–68.

Wong T, Wielicki BA, Lee III RB. (2006) Reexamination of the observed decadal variability of the earth radiation budget using altitude-corrected ERBE/ERBS nonscanner WFOV data. *Journal of Climate* **19**:4028–40.

World Meteorological Organization (1975) The physical basis of climate and climate modeling. Report of the International Study Conference in Stockholm, 29 July to 10 August 1974, GARP Publications Series No. 16, Global Atmospheric Research Programme, World Meteorological Organization, Geneva.

Young AH, Knapp KR, Inamdar A, *et al.* (2018) The International Satellite Cloud Climatology Project H-Series climate data record product. *Earth System Science Data* **10**:583–93. https://doi.org/10.5194/essd-10-583-2018

Yu L, Weller RA. (2007) Objectively analyzed air–sea heat fluxes for the global ice-free oceans (1981–2005). *Bulletin of the American Meteorological Society* **88**:527–39.

Zhang Y-C, Rossow WB. (1997) Estimating meridional energy transports by the atmospheric and oceanic general circulations using boundary fluxes. *Journal of Climate* **10**:2358–2373. doi:10.1175/1520-0442(1997)010

Zhang Y-C, Rossow WB, Lacis AA. (1995) Calculation of surface and top of atmosphere radiative fluxes from physical quantities based on ISCCP data sets, 1. Method and sensitivity to input data uncertainties. *Journal of Geophysical Research* **100**:1149–65.

Zhang Y-C, Rossow WB, Lacis AA, *et al.* (2004) Calculation of radiative fluxes from the surface to top-of-atmosphere based on ISCCP and other global datasets: Refinements of the radiative transfer model and the input data. *Journal of Geophysical Research* **109**:D19105. doi:10.1029/2003JD004457 (1-27 + 1-25)

Zhang Y, Rossow WB, Stackhouse Jr PW. (2006) Comparison of different global information sources used in surface radiative flux calculation: Radiative properties of the near-surface atmosphere. *Journal of Geophysical Research* **111**:D13106. doi:10.1029/2005JD006873

Zhang Y, Rossow WB, Stackhouse Jr PW. (2007a) Comparison of different global information sources used in surface radiative flux calculation: Radiative properties of the surface. *Journal of Geophysical Research* **112**:D01102. doi:10.1029/2005JD007008

Zhang Y, Rossow WB, Stackhouse Jr. PW, *et al.* (2007b) Decadal variations of global energy and ocean heat budget and meridional energy transports inferred from recent global data sets. *Journal of Geophysical Research* **112**:D22101. doi:10.1029/2007JD008435

Slide 1

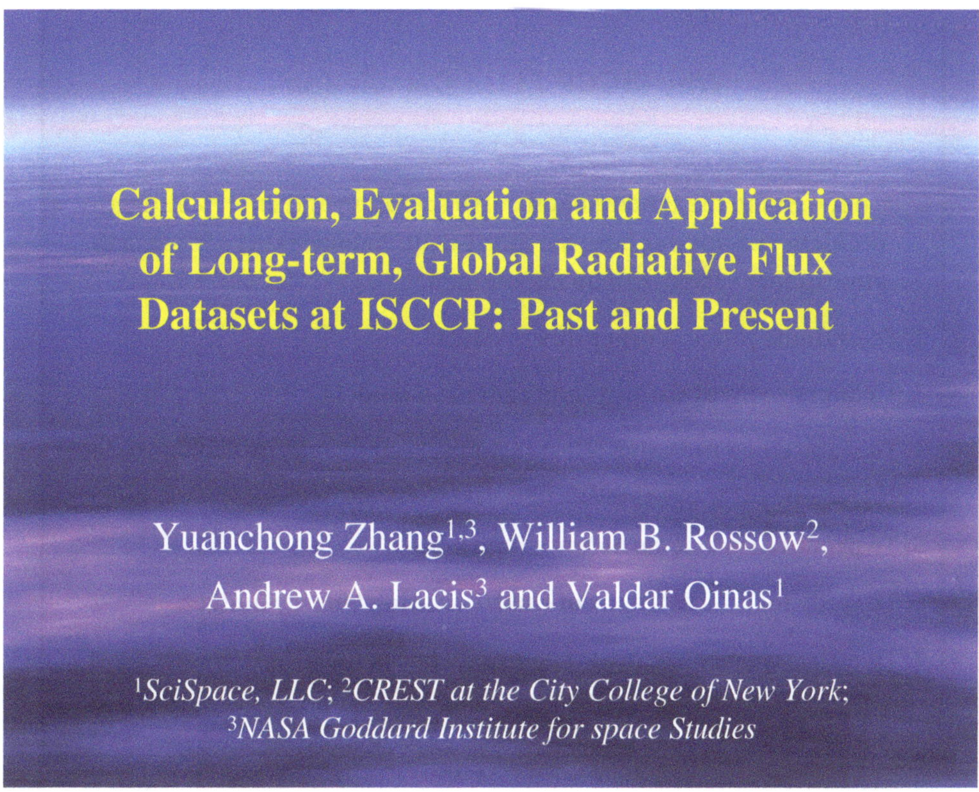

Slide 2

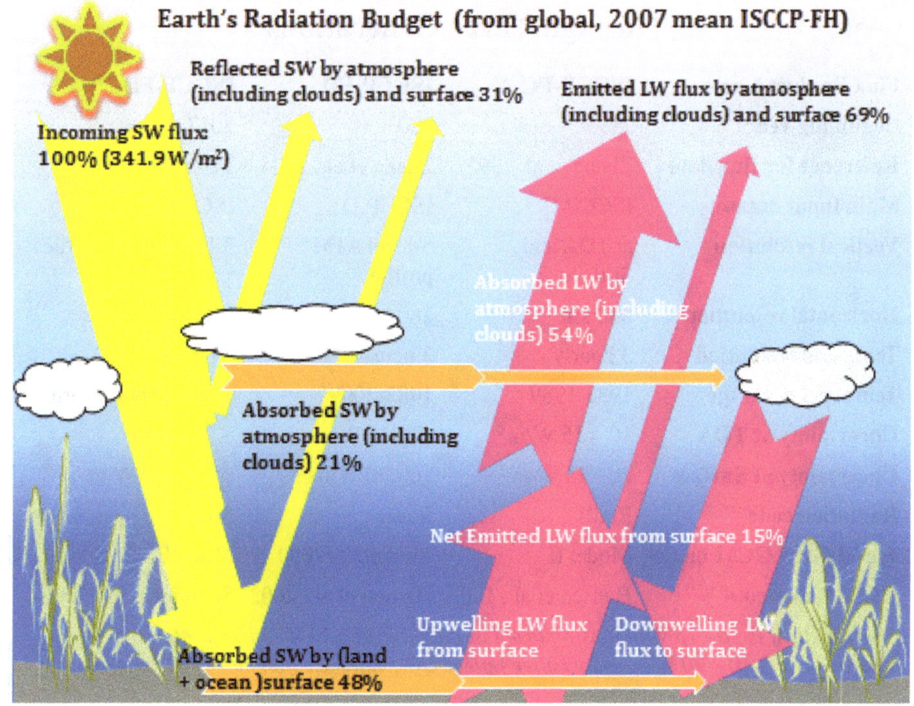

Slide 3

Why ISCCP FLUX Calculation?

1. To understand the fundamental earth radiation and energy budget problem and answer how the quasi-equilibrium status of the earth-atmosphere system can be maintained.
2. To evaluate this and other flux products.
3. To improve input datasets and radiation models.
4. To study cloud radiative effects (cloud forcing), energy transport, local or regional radiation/energy balance, etc.
5. To produce surface turbulence fluxes.
6. To study long-term trends of earth radiation/energy budget.
7. To estimate solar energy for solar-electric panels.

Slide 4

Evolution of the ISCCP Flux Products Through Three Generations

Flux Product Name	ISCCP-FC	ISCCP-FD	ISCCP-FH
Beginning Year	1995	2003	2017
Reference for flux data	Zhang et al. 1995	Zhang et al., 2004	TBD
Main input datasets	ISCCP-C	ISCCP-D	ISCCP-H
Vertical resolution	at TOA and Surface	5-level ATM profile	5-level ATM profile
Horizontal resolution	280 km	280 km	110 km
Temporal resolution	3 hourly	3 hourly	3 hourly
Temporal coverage	1985-1989	1983-2009	1983-2009-onward
Uncertainty at TOA	10 – 15 W/m^{-2}	5 – 10 W/m^{-2}	≲ 10 W/m^{-2}
Uncertainty at Surface	20 - 25 W/m^{-2}	10 - 20 W/m^{-2}*	≲ 10 - 20 W/m^{-2}
Radiation code	RadII	RadD	RadH
Based GISS GCM model	Model II	ModelD/SI2000	ModelE
GCM References	Hansen et al., 1983	Hansen et al., 2002	Schmidt et al., 2006

**Originally it is 10 - 15 W/m^{-2}, as reported in Zhang et al. (2004) based on limited validation; but 10 – 20 W/m^{-2} may better reflect the truth based on more validation.*

Slide 5

Main Radiation-model Improvements of ISCCP-FH

- **Based on** RadH, improved/revised from RadE, the radiation code of the GISS ModelE (of 2011), vs. RadD, revised from GISS Model SI2000 or ModelD (of 2002)
- **Spectral resolution** in k's (for Correlated K-distribution method): Improved
  reformulated/updated 16 k's for SW (0.2 - 5.0 μm) [though same 16 k's in FD]
  reformulated/updated 33 k's LW (5.0 -200.0 μm) [though same 33 k's in FD]
- **Spatial resolution**: 110 km [vs. 280 km in FD]
- **Accuracy**: same 1 W/m^2 and 1% cooling rates at TOA and SRF for LW and SW, respectively, but with significant reformulation and updates, especially atmospheric gas absorption and elaboration of LW calculation w.r.t. ISCCP-FD.
- **Reformulation of Atmospheric Gases for SW calculation**:
  Added weak line absorption for H2O, O2 and CO2, and updated line absorption for CH4, N2O, etc., which has virtually removed low bias of atmospheric absorption.
- **Reformulation/Refining for LW calculation**:
  RadH has several improvements for LW flux calculation over RadD, including additional Ma2008 option and MT-CKD H2O continua options (vs. RadD's sole Ma2000 scheme), CFC absorption cross-section, SO2 line absorption and better treatment of CH4 and N2O overlap with major absorbers with HITRAN2012 atlas, if possible.

  In addition, RadH increases the base atmospheric vertical resolution using a 43-layer standard atmosphere (old 24 layers), and now takes into account of amount of water vapor above and below a given layer as well as the water vapor gradient.

Slide 6

Contents of the ISCCP-FH Product

► It is a SuRFace (SRF)-to-TOA, 5-level, flux profile product:

- FH stands for: Flux profile data calculated (mainly) using ISCCP H series data
  to replace its precursor = ISCCP-FD (2003, final coverage: 8307-0912)
- Spectral coverage: 0.2 – 200 μm (SW: 0.2 – 5.0 and LW: 5.0 – 200)
- Spatial resolution: horizontal: 110km equal-area (1.0° on equator)
  vertical: 5 levels (SRF-680mb-440mb-100mb-TOA)
- Temporal resolution: 3-houly (UTC = 0, 3, … 21)
- Spatial coverage: virtually fully global
- Temproal coverage: July 1983 – June 2017 (and onwards)

► Outputs are compiled into five sub-products:

(1) FH-**TOA** Top-Of-Atmosphere radiative fluxes (23 variables)
(2) FH-**SRF** SuRFace Radiative Fluxes (34 var's)
(3) FH-**PRF** 5-level PRoFile Radiative Fluxes (including TOA and SRF, 91 var's))
(4) FH-**MPF** Monthly mean of FH-PRF (same 91 var's)
(5) FH-**INP** Complete INPut dataset (up to a maximum of 335 var's)

-- All are available at: https://isccp.giss.nasa.gov/projects/flux.html
in NetCDF except FH-INP only in Binary for July 1983 to June 2017.

Slide 7

Summary of Input Data for ISCCP-FH Production

(1) Atmospheric Gases: Climatology from NASA GISS radiation code of ModelE

(2) Atmospheric temperature/humidity Profile: nnHIRS (in HGG) vs. TOVS in D1

(3) Atmospheric aerosol climatology: MACv2 (Stefan Kinne, MPI-Meteorology)

(5) Clouds: ISCCP-HGG (18 types vs. 15 types in ISCCP-D1)

(6) Particle size of liquid/ice clouds based on (Han *et al.*, 1994) climatology

(7) Surface air temperature: from ISCCP-HGG (of nnHIRS); in addition, RadH makes cloud-caused, diurnal adjustment on it for land areas (> 1/3 fraction) using climatology from NCEP and NMC Surface Weather station reports

(8) Surface skin temperature: from ISCCP-HGG; RadH also makes additional cloud-caused, diurnal-adjustment (for land)

(9) Surface albedo: MACv2-aerosol-corrected reflectance for 0.55 μm from non-aerosol-corrected (processed based on ISCCP-HXG), modulated using VIS/NIR of revised RadE to have broadband albedo (for six wavebands)

(10) O3, snow/ice, vegetation and other surface characteristic (type, topography, land ice, etc.) data: from ISCCP-H Ancillary data

(11) TSI (total solar irradiance): Self-consistent daily time series based on SORCE V-15, Davos WRC composite and RMIB (from Dr. Shashi Gupta).

Slide 8

Summary of Output Variables in ISCCP-FH Product

(1) Radiative Flux Profile:

Full-sky	SW↑, SW↓, LW↑, LW↓ (and direct/diffuse downward at SRF)
Clear-sky	SW↑, SW↓, LW↑, LW↓ (and direct/diffuse downward at SRF)
100% overcast	SW↑, SW↓, LW↑, LW↓ (and direct/diffuse downward at SRF)

at 5 levels:

TOA ——— (~ 100 km high)
100 mb ———
440 mb ———
680 mb ———
Surface ——— (Ground ≤ 1100 mb)

} **Flux Profile**

(2) Input data Variables:

- Summary input variables for **TOA, SRF, PRF** and **MPF** sub-products
- Complete inputs for **INP** sub-product that may be used to reproduce FH

Slide 9

Comparison of main Global, Long-term flux products

Feature	CERES (Level 3) (SYN1deg Edition3A)	GEWEX-SRB (v3.1LW/3.0SW)	ISCCP-FH (v 0.00)
Period	2000 – current	1983 – ISCCP-D/H's current	1983 – ISCCP-H's current
Spatial Reso	1° x 1°	1° x 1°	1° x 1° (110 km EQ)
Temporal Reso	3-hourly	3-hourly	3-hourly
TOA flux	yes (observed + calculated)	yes (calculated)	yes (calculated)
SRF flux	yes (calculated)	yes (calculated)	yes (calculated)
In-Atmosphere Flux (Profile)	Yes, 3 levels: 70, 200 and 500 mb	No	Yes, 3 levels: 100. 440 and 680 mb
SW: algorithm based on	Various (http://ceres.larc.nasa.gov/atbd.php)	Pinker and Laszlo (1992)	Correlated K-distribution (Schmidt et al., 2006)
LW: algorithm based on		Fu et al. (1997)	
PAR/UV index	Yes	PAR	No

Slide 10

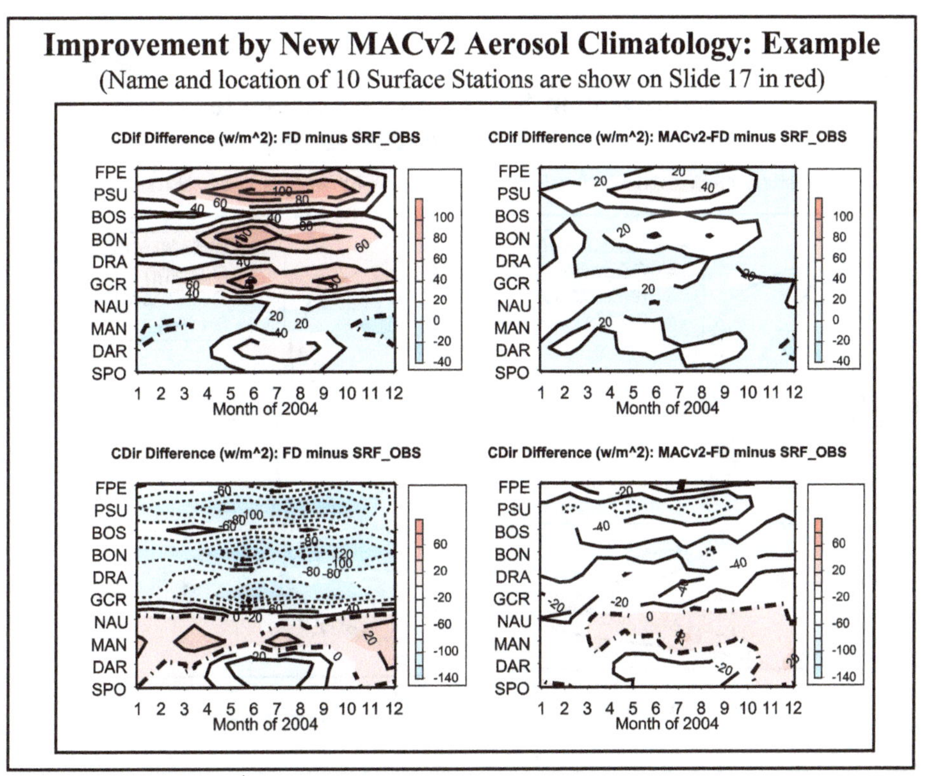

Slide 11

Improvement by New Vertical Cloud Layer Configuration (VCLC)

- **VCLC consists of (1) CVS and (2) CLTC**

(1). CVS (Cloud Vertical Structure): Model B for FH (18 cloud types)

Layer	ISCCP-H Cloud Class	Sub-class	Vertical structure	How to construct
HC	Ci		1H	= single layer cloud
	Cs	Thin	HM*	Radiatively reconstructed
		Thick	HML	ISCCP Clim reconstructed
	Cb		1 H-M-L	ISCCP Clim reconstructed
MC	Ac	Thin	1M	= single layer cloud
		Thick	HL*	Radiatively reconstructed
	As	Thin		
		thick	ML	ISCCP Clim reconstructed
	Ns			
LC	Cu		1L	= single layer cloud
	Sc			
	St			

(2). CLTC (Cloud Layer Thickness Configuration):

-- Function of Cloud optical thickness (τ), Longitude, Latitude & Ocean/Land

Based on 20-yr Rawinsonde and 5-yr CloudSat-CALIPSO climatology

Slide 12

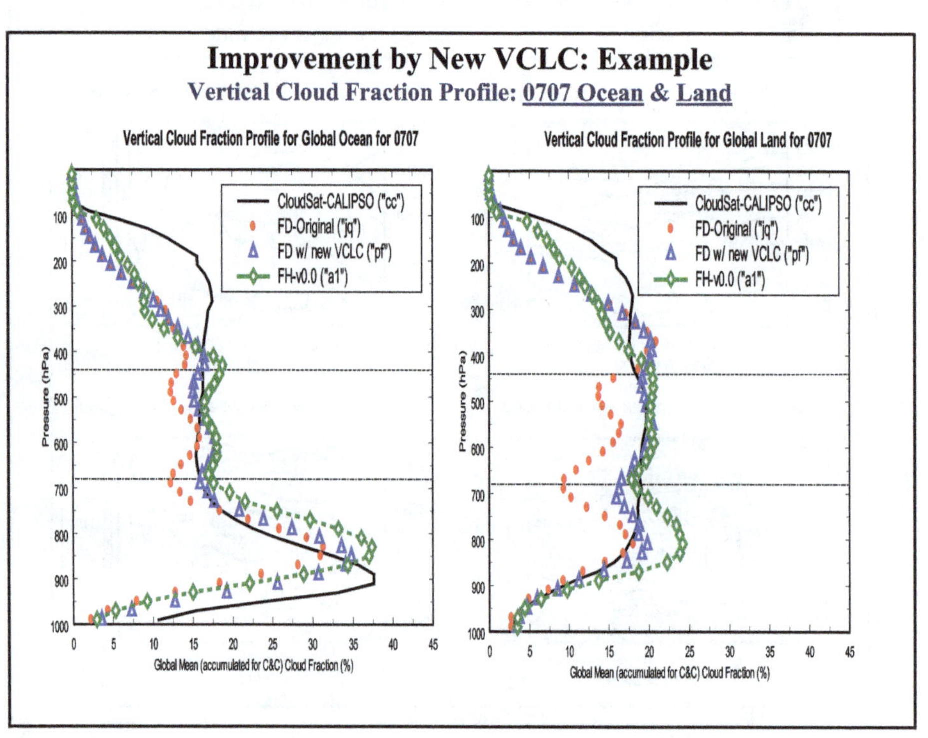

Slide 13

Preliminary validation against CERES for 2017 Monthly means: All-sky at TOA

All-sky ISCCP-FH vs CERES at TOA

Variable	FH mean	CERES mean	Mean diff	Stdv	cor coef	Slope	intrcept	Nrm dev	Eq cell #
ALBEDO (%)	33.64	31.48	2.153	3.413	0.9691	0.95	-0.60	2.43	485291
SW_net (W/m^2)	235.56	242.61	-7.054	8.455	0.9970	1.00	6.66	5.97	494824
LW_net (W/m^2)	-232.11	-239.18	7.071	4.560	0.9912	1.00	-6.62	3.22	494824
SW_ce (W/m^2)	-49.52	-47.16	-2.359	10.261	0.9572	0.92	-1.45	7.27	494599
LW_ce (W/m^2)	26.95	26.76	0.189	6.124	0.9305	0.98	0.37	4.37	485960

All-sky ISCCP-FD vs CERES at TOA

Variable	FD mean	CERES mean	Mean diff	Stdv	cor coef	Slope	intrcept	Nrm dev	Eq cell #
ALBEDO (%)	34.13	31.64	2.489	3.718	0.9627	1.00	-2.47	2.63	77842
SW_net (W/m^2)	234.57	242.57	-8.000	8.577	0.9969	1.01	5.54	5.98	79152
LW_net (W/m^2)	-236.10	-239.16	3.061	4.876	0.9899	1.01	-0.50	3.42	79152
SW_ce (W/m^2)	-52.78	-47.15	-5.631	8.182	0.9763	0.90	0.13	5.40	79152
LW_ce (W/m^2)	26.88	26.94	-0.059	4.884	0.9574	0.93	2.00	3.47	79063

Slide 14

Preliminary validation against CERES for 2017 Monthly means: Clear-sky at TOA

Clear-sky ISCCP-FH vs CERES at TOA

Variable	FH mean	CERES mean	Mean diff	Stdv	cor coef	Slope	intrcept	Nrm dev	Eq cell #
ALBEDO (%)	18.95	17.16	1.789	4.325	0.9519	0.90	0.12	3.04	484911
SW_net (W/m^2)	285.15	289.85	-4.698	10.977	0.9961	0.99	7.75	7.75	494671
LW_net (W/m^2)	-259.45	-266.32	6.877	6.183	0.9828	1.03	1.74	4.23	486037

Clear-sky ISCCP-FD vs CERES at TOA

Variable	FD mean	CERES mean	Mean diff	Stdv	cor coef	Slope	intrcept	Nrm dev	Eq cell #
ALBEDO (%)	18.56	17.30	1.261	3.620	0.9630	0.98	-0.81	2.58	77752
SW_net (W/m^2)	287.35	289.72	-2.369	8.191	0.9979	0.99	5.15	5.76	79152
LW_net (W/m^2)	-262.97	-266.10	3.126	6.474	0.9802	0.99	-5.58	4.60	79063

Slide 15

Preliminary validation against CERES for 2017 Monthly means: In atmosphere all-sky

All-sky ISCCP-FH vs CERES in Atmosphere

Variable	FH mean	CERES mean	Mean diff	Stdv	cor coef	Slope	intrcept	Nrm dev	Eq cell #
SW_net (W/m^2)	76.60	78.02	-1.412	8.008	0.9687	1.02	0.26	5.61	494824
LW_net (W/m^2)	-178.25	-183.40	5.157	17.801	0.8823	0.69	-59.78	11.27	494824
SW_ce (W/m^2)	2.61	2.15	0.460	7.456	0.3515	0.78	0.12	5.85	494599
LW_ce (W/m^2)	5.91	1.10	4.816	11.248	0.8835	1.01	-4.89	7.90	485960

All-sky FISCCP-FD vs CERES in Atmosphere

Variable	FD mean	CERES mean	Mean diff	Stdv	cor coef	Slope	intrcept	Nrm dev	Eq cell #
SW_net (W/m^2)	71.76	78.00	-6.240	10.167	0.9512	1.08	0.34	6.72	79152
LW_net (W/m^2)	-185.08	-183.39	*-1.690*	18.118	0.7889	0.84	-28.06	13.48	79152
SW_ce (W/m^2)	3.32	2.15	1.173	6.406	0.3724	0.61	0.13	5.30	79152
LW_ce (W/m^2)	-4.48	0.95	-5.429	8.455	0.9367	0.94	5.18	6.07	79063

Slide 16

Preliminary validation against CERES for 2017 Monthly: In-atmosphere clear-sky

Clear-sky ISCCP-FH vs CERES in Atmosphere

Variable	FH mean	CERES mean	Mean diff	Stdv	cor coef	Slope	intrcept	Nrm dev	Eq cell #
SW_net (W/m^2)	74.02	75.89	-1.873	8.002	0.9671	1.03	-0.03	5.56	494671
LW_net (W/m^2)	-184.18	-184.59	0.407	17.024	0.9254	0.78	-40.86	11.05	486037

Clears-sky ISCCP-FD vs CERES in Atmosphere

Variable	FD mean	CERES mean	Mean diff	Stdv	cor coef	Slope	intrcept	Nrm dev	Eq cell #
SW_net (W/m^2)	68.44	75.85	**-7.412**	9.173	0.9611	1.12	-0.97	5.70	79152
LW_net (W/m^2)	-180.55	-184.30	3.752	18.951	0.8571	0.95	-13.56	13.71	79063

Slide 17

BSRN Stations used for Surface-flux Validation (39 are Data-available for 2007)

Station	Lat	Lon	Station	Lat	Lon
1 ALE	82.49	297.58	21 DRA	36.626	243.982
2 EUR	79.989	274.060	22 BIL	36.605	262.484
3 NYA	78.925	11.930	23 E13	36.605	262.515
4 LER	60.139	358.815	24 TAT	36.058	140.126
5 TOR	58.254	26.462	25 GCR	34.255	270.127
6 LIN	52.210	14.122	26 BER	32.267	295.333
7 CAB	51.971	4.927	27 SBO	30.860	34.779
8 REG	50.205	255.287	28 TAM	22.790	5.529
9 CAM	50.217	354.683	29 KWA	8.720	167.731
10 PAL	48.713	2.208	30 NAU	-0.521	166.917
11 FPE	48.317	254.900	31 MAN	-2.058	147.425
12 PAY	46.815	6.944	32 COC	-12.193	96.835
13 CAR	44.083	5.059	33 DAR	-12.425	130.891
14 SXF	43.730	263.380	34 ASP	-23.798	133.888
15 PSU	40.720	282.067	35 LAU	-45.045	169.689
16 BOS	40.125	254.763	36 SYO	-69.005	39.589
17 BON	40.067	271.633	37 GVN	-70.650	351.750
18 BOU	40.050	254.993	38 DOM	-75.100	123.383
19 XIA	39.754	116.962	39 SPO	-89.983	335.201
20 CLH	36.905	284.287			

(Red-colored also for Slide 10 MACv2 test)

Slide 18

Preliminary validation against BSRN: for 2017 Monthly Mean

FLux	FH	BSRN	M. diff	Stdv	Cr coef	Slope	intrcept	Nrm dev	Stn #
SWdn (W/m²)	170.89	172.26	-1.369	20.911	0.9746	0.98	4.14	14.87	441
LWdn (W/m²)	302.52	310.00	-7.481	17.317	0.9770	0.96	20.34	12.26	458
SWup (W/m²)	56.92	67.87	-10.949	26.354	0.9543	1.09	6.01	17.29	89
LWup (W/m²)	287.18	289.15	-1.973	20.389	0.9755	1.09	-23.09	13.00	96

FLux	FD	BSRN	M. diff	Stdv	Cr coef	Slope	intrcept	Nrm dev	Stn #
SWdn (W/m²)	168.49	172.26	-3.772	23.820	0.9667	1.00	4.23	16.87	441
LWdn (W/m²)	320.25	310.00	10.249	21.508	0.9657	1.08	-37.45	14.00	458
SWup (W/m²)	44.53	67.87	-23.344	32.712	0.9281	1.11	18.56	21.31	89
LWup (W/m²)	289.68	289.15	0.534	19.080	0.9762	1.02	-7.71	13.25	181818 1818

FLux	CERES	BSRN	M. diff	Stdv	Cr coef	Slope	intrcept	Nrm dev	Stn #
SWdn (W/m²)	176.65	172.26	4.396	16.183	0.9848	1.00	-4.89	11.43	441
LWdn (W/m²)	304.98	310.00	-5.018	11.280	0.9899	0.99	6.95	7.99	458
SWup (W/m²)	57.27	67.87	-10.598	24.017	0.9598	1.02	9.69	16.82	89
LWup (W/m²)	282.13	289.15	-7.017	11.472	0.9923	1.04	-5.66	7.5	96

Slide 19

Preliminary Uncertainty Estimate for monthly/regional ISCCP-FH

(1) At TOA: for *Single flux component*: if taking CERES as 'truth'.

Bias ≲ 7 W/m² **STDV ≲ 11 W/m²**

Corr coefficient ≥ 0.95 **Normal Deviation ≲ 7 W/m²**

▶ Uncertainty ≲ 10 W/m²

FH is comparable with but slightly worse than FD (but FH with higher spatial resolution) at TOA.

(2) At Surface: for *single flux component* based on Comparisons with BSRN:

Bias ≲ 11 W/m² **STDV ≲ 26 W/m²**

Corr coefficient ≥ 0.95 **Normal Deviation ≲ 17 W/m²**

▶ Uncertainty ≲ 15 - 20 W/m²

-- FH is overall better than FD but slightly worse than CERES.

(3) In Atmosphere: for *Net and CE*,

Difference ≲ 5 W/m² **STDV ≲ 18 W/m²**

Corr coefficient: 0.35 – 0.88 **Normal Deviation ≲ 11 W/m²**

▶ Uncertainty ≲ 15 W/m²

ISCCP-FH seems slightly better than FD, especially substantially improved in atmospheric SW-Net (absorption) based on comparison with CERES.

Slide 20

Conclusions

1. **RadH represents the most recent improvements of the radiation code of the NASA GISS ModelE, especially in atmospheric gas absorption by SW and polar-region LW calculation, i.e., ISCCP-FH flux profile product is an improvement over its precursor, RadD-based ISCCP-FD.**
2. **Besides increasing spatial resolution (from 280 km to 110 km), ISCCP-H has many substantial improvements. For temperature and humidity profiles, new nnHIRS may be better than previous ISCCP-D's TOVS in temporal homogeneity as well as others.**
3. **New aerosol data MACv2 seems an improvement as validated using 2004 high-quality surface observations.**
4. **Our cloud-type-dependent statistical VCLC model may be slightly better than previous one for ISCCP-FD; however, because there is no unique solution even with CloudSat-CALIPSO data, it remains to be further improved.**
5. **The new ISCCP-FH product seems overall slightly better than FD but slightly worse than CERES at surface. The preliminary validation shows that for monthly, regional mean, ISCCP-FH has**
   uncertainties ≲ 10 W/m² for TOA and ≲ 20 W/m² for Surface fluxes
6. **It may imply a LIMIT we encounter now under the current status of input parameters and, secondarily, radiation modeling. The limit is largely caused by the restriction of the accuracy of the atmospheric, cloud and surface properties, i.e., UNLESS we make substantial improvements on some major input datasets that cause leading errors, substantial reduction on flux calculation uncertainties may not be achievable.**

Slide Captions

Slide 1 Cover

Slide 2 Earth's radiation budget (from global, 2007 mean ISCCP-FH)

This earth-atmosphere budget cartoon is created from the actual global, annual mean ISCCP-FH product for 2007 (after downgrading to a 280-km resolution for a fully covered global map of 110-km resolution). It shows that the energy at TOA is perfectly balanced (sum = zero): the incoming SW of 100% for 341.9 W/m^2 is balanced by reflected SW (31%) and OLR (69%), i.e., the earth-atmosphere system neither loses to nor gains energy from outer space in the global, 2007 annual mean. The non-reflected SW flux is absorbed partly by the atmosphere and the (land + ocean) surface, while all the emitted/reflected LW flux from the surface and atmosphere are combined into the 69% of OLR flux with a net (= upward minus downward) LW flux of 15% emitted to the atmosphere from the surface. The atmosphere has various constituents such as clouds, aerosols, gases, and water vapors that all contribute to the flux redistribution, including also the surface albedo and temperature for different terrains. Even in this cartoon picture, it shows how complicated the radiation calculation can be.

Slide 3 Why ISCCP FLUX calculation?

Listed are the seven most important reasons why we need flux datasets such as this ISCCP-FH product as we have explained for the ISCCP-FC and FD products in the introduction.

Slide 4 Evolution of the ISCCP flux products through three generations

Columns 2, 3 and 4 (left to right) outline the basic characteristics of the three generations of the ISCCP flux products to illustrate their evolution history, from the earliest ISCCP-FC to ISCCP-FD and then to the current ISCCP-FH with its corresponding three generations of the NASA GISS GCM radiation models (RadII, RadD and RadH), which are what the ISCCP flux products are based on. The ISCCP-FH data are now available for online access for July 1983 to June 2017 at:

https://isccp.giss.nasa.gov/projects/flux.html

and may become available to the public at NOAA NCEI. The widely used ISCCP-FD product for July 1983 to December 2009 is still available online at the above same site.

Slide 5 Main radiation-model improvements of ISCCP-FH

The radiation code for ISCCP-FH production, RadH, is based on the GISS ModelE's radiation code (of 2011), RadE. It has been revised to add several new improvements, of which the most important items are listed here. Among these listed features, adding new values of SW weak line absorption for H_2O, O_2 and CO_2, and the update of line absorption for CH_4, N_2O, etc., has virtually removed the low bias (about 4–5 W/m^2 in global average) of atmospheric SW absorption of ModelE, as found by Oreopoulos *et al.* (2012). Thus, this is a significant improvement (see Slide 16 for more specific information). In the meanwhile, the increases of the base atmospheric vertical resolution to a 43-layer standard atmosphere (from an original 24 layers) and the treatment for water vapor amount above and below a given layer as well as the water vapor gradient have improved the LW accuracy, especially in the Polar Regions. There are also a few other improvements that will be mentioned later. The RadH code is equivalent to the radiation code of the current NASA GISS ModelE2.1 (Kelley *et al.*, 2020).

Slide 6 Contents of the ISCCP-FH product

Beginning with ISCCP-FD, the content of the ISCCP flux product was extended from TOA and SRF only (as in ISCCP-FC) to the whole atmospheric flux profile (PRF), TOA and SRF inclusive, as we were able to climatologically construct the Cloud Vertical Structure (CVS) based on a 20-year rawinsonde dataset (Wang *et al.*, 2000; also see Slide 11 for ISCCP-FH). Moreover, we have added two more subproducts, MPF and INP, to meet current research needs: there are a total of five subproducts, simplified as TOA, SRF, PRF, MPF and INP, as shown here. Among them, MPF is more convenient for those users who are only interested in monthly mean values, while INP is a virtually complete dataset that can be used to reproduce profile flux products; this is especially useful for in-depth study on cloud-radiation interaction as well as other radiative effects. In addition, beginning with ISCCP-FH, four of the subproducts (excluding INP that remains in binary format) are now available in NetCDF-4 format for even wider access in response to the requests from ISCCP-FD users.

Slide 7 Summary of input data for ISCCP-FH production

Here we give a list of all the important input datasets that are used for ISCCP-FH production; all of them reflect the research progress recently achieved since ISCCP-FD was produced. Among them, a new atmospheric temperature and humidity profile (nnHIRS) is produced using a neural network by NOAA to replace the old TOVS profile for ISCCP-D and ISCCP-FD production, which shows inhomogeneities in the temporal LW flux variations as explained in (Zhang *et al.*, 2006). This change reduces inhomogeneity. ISCCP-H produces aerosol-corrected surface reflectance based on MACv1 aerosol climatology at 0.65 microns, but RadH is based on 0.55 microns and MACv2, so ISCCP-FH has to first remove the aerosol effects on surface reflectance (from ISCCP-H) and then apply the 0.55-micron aerosol effects based on MACv2 to the processed reflectance. In addition, the ISCCP-FH now uses more accurate TSI data that is based on the best SORCE daily solar constant as well as the instantaneous earth-sun distance that removes the leap-year errors in the ISCCP-FD solar constant that is based on the climatology of 365 days a year.

Slide 8 Summary of output variables in ISCCP-FH Product

Besides all downwelling and upwelling, SW and LW fluxes at five atmospheric levels for all-, clear- and overcast-sky scenes. The ISCCP-FH product now includes additional surface direct and diffuse downward SW fluxes that are useful for surface validation and other applications (see, e.g., [Zhang *et al.*, 2010]).

Slide 9 Comparison of main global, long-term flux products

As CERES, GEWEX-SRB and ISCCP-FH (and FD) are the principal and most widely used long-term, global flux products, it is necessary to understand their basic similarities and differences as shown here for people to select the right one(s) they need to explore.

Slide 10 Improvement by New MACv2 aerosol climatology: Example

As mentioned in (3) and (9) of Slide 7, introducing new MACv2 aerosol data has overall improved our SW flux results. Here, we compare surface clear-sky downwelling diffuse and direct flux difference with surface observation ("SRF_OBS") for the original ISCCP-FD ("FD") that uses NASA GISS ModelD's monthly aerosol climatology and revised FD ("MACv2-FD"), with aerosol-only change to MACv2 to see how much the improvement could be. The surface observational data are from ten high-quality controlled surface stations that are climate-representative and arranged from the south-most (lowest) to north-most (uppermost): from the South Pole (SPO) to the North Polar Region (FPE) (Zhang *et al.*, 2010). The comparison (differences with SRF_OBS) is for all 2004 monthly means. The left column is for original FD and the right column revised FD; the upper row is for clear-sky diffuse flux ("Cdif") and the lower row clear-sky direct flux ("Cdir"). Because the NASA GISS ModelE's aerosol input data is not substantially different from ModelD's, this slide may be taken as the demonstration of the advantage of new MACv2 aerosol data used in ISCCP-FH production. From this slide, we can see substantial improvements for essentially the northern hemisphere (all stations above NAU station on the Y-axis) but not much change for the southern hemisphere. Note that the total SW downward flux is also improved somewhat (up to 10 W/m^2) but not as much as the separated diffuse and direct fluxes because these two fluxes somewhat compensate each other when they are added up to be the total SW flux.

Slide 11 Improvement by new Vertical Cloud Layer Configuration

In developing the ISCCP-FD production code, we have introduced a CVS model to construct climatologically realistic vertical profile clouds so as to calculate a realistic flux profile through the entire vertical atmospheric column for the first time on a global scale (Zhang *et al.*, 2004). For ISCCP-FH production, we improved the vertical cloud profile construction by building up the VCLC algorithm. As shown here, it consists of two parts: (1) CVS, essentially the same as the one used for FD but now for 18 original cloud types from ISCCP-H data, and (2) Cloud Layer Thickness Configuration (CLTC) using a combination of a previous 20-year rawinsonde climatology and new 5-year CloudSat-CALIPSO climatology. In (1) of this slide, all the 18 original ISCCP-H cloud types are categorized into one of the nine cloud classes, from Ci, Cs, ... to St, as shown in Column 2, based on cloud top pressure of the three ranges of surface — 680, 680–440 and 440–0 hPa for Low, Middle and High Cloud, denoted as LC, MC and HC, respectively, and cloud optical thickness, τ, with three ranges of 0–3.6, 3.6–23 and 23–450 (see [Rossow *et al.*, 1996], for ISCCP-D, but whose largest τ is 379 instead of the current 450 for ISCCP-H). There may be some Sub-cloud classes, Thin and Thick (Column 3), based on τ values subdivided at $\tau = 1.3$ for low τ of 0–3.6, and $\tau = 9.4$ for middle τ of 3.6–23, for MC and HC classes, respectively, if necessary. Column 4 shows such cloud-class-based, eight CVS types, e.g., 1H for one high cloud layer, HML for three separated High, Middle and Low cloud layers, and 1 H-M-L for one cloud layer from the high through middle to low pressure layer. The original total column τ will be redistributed, proportional to the sum of the new cloud layers' thickness if there are two or more cloud layers for a determined CVS type. The total column τ is conserved except for the two special CVS types, HM* and HL*, for which we have to redistribute the original 1-layer-based τ to a new two-cloud-layer-based, consistent with assumed satellite detection. The cloud top position of all the cloud layers for a determined CVS type is based on either the current cloud top temperature (if available) or climatological cloud information from ISCCP-H. In (2), the cloud thickness (therefore, cloud base) is now a function of cloud optical thickness τ, longitude, latitude, ocean/land, and month, where τ and longitude are new variables that ISCCP-FD did not use because of a lack of cloud profile data (like CloudSat-CALIPSO) at that time. By these steps, we are able to construct a vertical profile of clouds with all cloud layers interleaved with possible clear-air layers for a determined CVS, for which the cloud phase and particle size are also determined based on all available information from ISCCP-H and climatology before calculating the flux profile.

Slide 12 Improvement by the new VCLC: Example — Vertical cloud fraction profile: 0707 Ocean and land

Here is an example of a monthly-mean vertical profile of cloud amounts for the ocean (left) and land (right), respectively, for July 2007, to see how the new VCLC algorithm performs versus the old one. The slide compares the monthly vertical profiles of cloud fraction (in %) for the CloudSAT-CALIPSO, original FD, revised FD using the new VCLC, and new FH (also using the new VCLC), respectively, with their symbols shown in the legend box. The pressure level is arranged such that the Y-axis is from the ground surface (1,000 hPa) to TOA (~0 hPa). Compared with CloudSat-CALIPSO (black solid line), both the revised FD (blue upward empty triangle) and FH (brown empty diamond with dashed line) overall outperform the original FD's cloud profile (red solid circle), especially for middle clouds and, in addition, FH has a better high cloud profile than the revised FD, showing the improvement in high-cloud detection in the ISCCP-H product. The CALIPSO lidar has much higher sensitivity to detect high clouds; thus, CloudSat-CALIIPSO's high cloud amount is higher than FH's, but those high clouds (mainly thin cirrus, see [Rossow and Zhang, 2010]) may have very low τ value that may not affect the flux much. For low clouds, CloudSAT-CALIPSO may have some difficulty in determining an accurate low cloud profile because of attenuation, so its uncertainty is larger. Given all of these issues for both ISCCP and CLoudSat-CALIPSO datasets, the current VCLC is in need of further improvement.

Slide 13 Preliminary validation against CERES for 2017 Monthly means: All-sky at TOA

As preliminary validation, we use all 2017 monthly means to compare the ISCCP-FH, ISCCP-FD with CERES, and the Baseline Surface Radiation Network (BSRN) for TOA and surface validation, respectively. We also compare with CERES and in-atmosphere for ensemble-based validation (Zhang *et al.*, 2006) since CERES is also a flux product for atmospheric fluxes. This slide shows the comparison for all-sky net ("net") fluxes, cloud effects ("ce") and albedo at TOA, where "Stdv", "cor coef", "intercept", "Nrm dev" and "Eq cell #" stand for standard deviation, correlation coefficient, intercept, normal standard deviation calculated based on the distance of a reference dataset (here FH or FD) to their regression line (with CERES) that is a measure of rms scattering for the reference dataset, and total equal-area cell number of all available 2017 monthly means. (If it is different in the original spatial resolution, a higher-resolution equal-area map is downgraded to the same as a low-resolution dataset.) From the table, we can see that FH is better than FD by a fraction of a percent for all bias ("mean difference"), standard deviation (Stdv), and Nrm dev for albedo, while only slightly better for SW_net. For SW_ce, although FH has ~3 W/m^2 lower bias than FD, its variation is worse than FD by ~2 W/m^2. For LW fluxes, FH is actually worse than FD by up to ~4 W/m^2 in LW_net bias (but FH's LW_net has a slightly smaller Stdv and Nrm dev than FD). Note that FH now has the same higher spatial resolution of 110 km as CERES while FD has a 280-km resolution, so its comparison with CERES is for a downgraded 280-km resolution that usually would result in a better agreement through the average for CERES.

Slide 14 Preliminary validation against CERES for 2017 Monthly means: Clear-sky at TOA

For clear-sky fluxes at TOA, FH seems slightly worse than FD by about half a percent and ~2 W/m^2 for albedo and SW_net, respectively. For LW_net, FH is nearly ~4 W/m^2 worse than FD in bias, but again its Stdv and Nrm dev are slightly better than FD. Note, however, that the meaning of "clear sky scene" is different between CERES and the calculated global clear-sky scene, which is the same as an all-sky scene except that all cloud τ is set to zero in clear-sky calculation, i.e., there are environmental differences between the two clear-sky concepts.

Slide 15 Preliminary validation against CERES for 2017 Monthly means: In-atmosphere all-sky

For all-sky SW fluxes in the atmosphere, FH is closer (than FD) to CERES by up to ~5 W/m^2 (in SW_net bias) except Stdv and Nrm dev of SW_ce, for which FD has a better agreement with CERES. For LW_net, FD is closer to CERES by 3.5 W/m^2 in bias, but FH has a smaller Stdv and Nrm dev than FD. For LW_ce, FH has a smaller bias with CERES than FD, but FD has smaller Stdv and Nrm dev. The fact that the FH's SW_net is ~5 W/m^2 closer to CERES than FD demonstrates much improved SW's atmospheric absorption in the RadH code, as mentioned above.

Slide 16 Preliminary validation against CERES for 2017 Monthly: In-atmosphere clear-sky

For atmospheric clear-sky fluxes, FH has a better agreement with CERES than FD, with substantial improvement for SW_net by 5.6 W/m^2 as a demonstration of improved atmospheric absorption of the RadH code.

Slide 17 BSRN stations used for surface-flux validation (39 are data-available for 2007)

We now compare all the monthly mean surface fluxes from CERES, FD and FH for 2007 with ground-based measurements, the observed data from BSRN. Here is the list of the 39 BSRN stations whose 2007 data is used for surface validation here; the ten red-colored stations are what were used in Slide 10 for demonstrating the effects of aerosol data change from NASA GISS' climatology to MACv2.

Slide 18 Preliminary validation against BSRN: For 2017 Monthly Mean

With BSRN as the ground "truth", CERES seem to have the best overall performance, though ISCCP-FH outperforms CERES by 3 W/m^2 for SWdn in bias and 5 W/m^2 for LWup in bias. ISCCP-FD has the worst performance. ISCCP-FH is overall improved over ISCCP-FD by up to ~13 W/m^2 for SWup bias and ~2–3 W/m^2 for SWdn and LWdn bias, and Stdv and Nrm dev.

Slide 19 Preliminary uncertainty estimate for monthly/regional ISCCP-FH

We summarize the above validation results. Overall, ISCCP-FH is slightly better than FD, but their uncertainties are virtually about the same.

Slide 20 Conclusions

Based on the above presentation, we make several concluding remarks. We mention a "limit" based on the current error estimates for all the input variables of the atmospheric and surface physical properties, which we think are unlikely to have substantial improvement in the near future. Accordingly, the current flux uncertainty may not change much in a short time.

References

Fu Q, Liou KN, Cribb MC, *et al.* (1997) Multiple scattering parameterization in thermal infrared radiative transfer. *Journal of the Atmospheric Sciences* **54**:2799–812.

Han, Q, Rossow, WB, and Lacis, AA. (1994) Near-global survey of effective droplet radii in liquid water clouds using ISCCP data. *Journal of Climate* **7**(4): 465–97. https://doi.org/10.1175/1520-0442(1994)007<0465:NGSOED>2.0.CO;2.

Hansen J, Russell G, Rind D, et al. (1983) Efficient three-dimensional global models for climate studies: Model I and II. *Monthly Weather Review* **111**:609–62.

Hansen, J, Sato M, Ruedy R, et al. (2002) Climate forcings in Goddard Institute for Space Studies SI2000 simulations. *Journal of Geophysical Research* **107**(D18):4347. doi:10.1029/2001JD001143

Kelley M, Schmidt GA, Nazarenko LS, et al. (2020) GISS — E2.1: Configurations and climatology. *Journal of Advances in Modeling Earth Systems* **12**:e2019MS002025. https://doi.org/10.1029/2019MS002025

Oreopoulos L, Mlawer E, Delamere J. (2012) The continual intercomparison of radiation codes: Results from phase I. *Journal of Geophysical Research* **117**:D06118. doi:10.1029/2011JD016821

Pinker RT, Laszlo I. (1992) Modeling surface solar irradiance for satellite application on a global scale. *Journal of Applied Meteorology* **31**:194–211. doi:10.1175/1520-0450(1992)031<0194:MSSIFS>2.0.CO;2

Rossow WB, Walker AW, Bueschel D, Roiter M. (1996) International Satellite Cloud Climatology Project (ISCCP) documentation of new cloud datasets, WMO/TD-737, World Climate Research Programme, Geneva, 115 pp.

Schmidt GA, Ruedy R, Hansen JE, *et al.* (2006) Present day atmospheric simulations using GISS ModelE: Comparison to in-situ, satellite and reanalysis data. *Journal of Climate* **19**:153–92. doi:10.1175/JCLI3612.1

Wang J, Rossow WB, Zhang Y-C. (2000) Cloud vertical structure and its variations from 20-yr global rawinsonde dataset. *Journal of Climate* **12**:3041–56.

Zhang Y-C, Rossow WB, Lacis AA. (1995) Calculation of surface and top of atmosphere radiative fluxes from physical quantities based on ISCCP data sets, 1. Method and sensitivity to input data uncertainties. *Journal of Geophysical Research* **100**:1149–65.

Zhang Y-C, Rossow WB, Lacis AA, *et al.* (2004) Calculation of radiative fluxes from the surface to top-of-atmosphere based on ISCCP and other global datasets: Refinements of the radiative transfer model and the input data. *Journal of Geophysical Research* **109**:D19105. doi:10.1029/2003JD004457 (1-27 + 1-25)

Zhang Y, Rossow WB, Stackhouse Jr PW. (2006) Comparison of different global information sources used in surface radiative flux calculation: Radiative properties of the near-surface atmosphere. *Journal of Geophysical Research* **111**:D13106. doi:10.1029/2005JD006873

Zhang Y-C, Rossow WB, Long CN, Dutton EG. (2010) Exploiting diurnal variations to evaluate the ISCCP-FD flux calculations and Radiative-Flux-Analysis-Processed Surface Observations from BSRN, ARM and SURFRAD. *Journal of Geophysical Research* **115**:D15105, (1–21). doi:10.1029/2009jd012743

Lecture 10

Cirrus Clouds Observed From Himawari-8

Toshiro Inoue[1], *Hiroshi Ishimoto*[2], *Masahiro Hayashi*[2] *and Johannes Schmetz*[3]

[1]*Atmosphere and Ocean Research Institute, The University of Tokyo, Kashiwa, Japan*

[2]*Meteorological Research Institute, Japan Meteorological Agency, Tsukuba, Japan*

[3]*European Organization for the Exploitation of Meteorological Satellites (EUMETSAT), Darmstadt, Germany*

Toshiro Inoue is a researcher at the University of Tokyo after retiring from the Meteorological Research Institute/Japan Meteorological Agency (MRI/JMA). He first showed the effectiveness of two infrared data (within 10 μm atmospheric window region) for analysis of cirrus cloud properties. He is interested in the comparison of the life cycle of deep convective clouds observed in satellite observations and simulated in global cloud-resolving models (Nonhydrostatic Icosahedral Atmospheric Model (NICAM)).

Hiroshi Ishimoto has been with the Meteorological Research Institute, Japan Meteorological Agency, since 1998. His research interests include satellite remote sensing and light scattering by nonspherical atmospheric particles.

Masahiro Hayashi is a researcher at the Meteorological Research Institute, Japan Meteorological Agency. His research focuses on satellite remote sensing, especially for retrievals of atmospheric parameters.

Johannes Schmetz retired as the Chief Scientist from the EUMETSAT (European operational satellite agency for monitoring weather, climate and the environment from space). His career started at the University of Cologne as a scientist working on cloud-radiation interaction. He continued this work as a research scientist at the Max-Planck Institute for Meteorology in Hamburg. At the ESA (European Space Agency), he created the first viable operational calibration for the Meteosat satellites and developed satellite products used in Numerical Weather Prediction. As Division Head at the EUMETSAT, he started and developed the Meteorological Division which, inter alia, supported and guided the development of future satellite programs.

Introduction

Climate change has been a big issue in recent decades. Clouds play an essential role in climate change through radiative processes. Clouds reflect shortwave radiation to space (solar albedo effect) and warm the Earth by reducing outgoing longwave radiation (greenhouse effect). The balance of the greenhouse effect and solar albedo effect varies depending on cloud type. Manabe and Strickler (1964) showed that high-level clouds could warm the

Earth's surface, while middle and low-level clouds cool the surface and lower atmosphere. Liou (1986) recognized that cirrus clouds, which are high, cold, and optically thin, play an important role in modifying the Earth's radiation budget because of the global coverage with cirrus clouds. However, Stephens *et al.* (1990) pointed out that mankind lacks an understanding of how the gross radiative properties of cirrus are controlled by cloud microphysical properties, i.e., the size and shape of ice crystals.

In situ observations of cirrus clouds are not an easy task to perform because of their high altitudes. Heymsfield (1975) reported characteristics of ice crystals for cirrus clouds from aircraft observations. The International Cirrus Experiment (ICE) was conducted (Raschke *et al.*, 1990) to study the optical properties of cirrus clouds. However, we still do not have enough information on the properties of cirrus clouds.

We generally know that cirrus clouds are ubiquitous; however, we do not know where and how much cirrus clouds exist globally. Satellite observations are the only possible tool to consistently observe cirrus clouds globally. Motivated by these problems, the International Satellite Cloud Climatology (ISCCP) was initiated (Schieffer and Rossow, 1983) using visible and infrared data to classify cloud types and optical properties, and the retrieval methods of cloud properties from space were reviewed by Rossow (1989).

Besides conventional visible and infrared data used in the ISCCP, some channels are noted to be just as effective in retrieving cloud properties. Inoue (1985) showed the feasibility of detecting and retrieving the temperature of semi-transparent cirrus clouds using two channels in the 10 μm atmospheric window region. The technique can be applied to both day and night observations since visible data is not required in the method. Inoue (1989) showed the feature of cloud distributions, including a cirrus cloud type over the tropical Pacific, as well as day-to-night cloudiness differences (Inoue, 1997) based on the cloud type classification using the split window data (Inoue, 1987). Infrared sounding data can also be used (Wylie *et al.*, 1994, Stubenrauch 2006) to describe the frequency, geographical distribution, and seasonal changes of upper tropospheric clouds, though some ambiguity in determining cirrus clouds arises from the field of view of the sounder, which is larger than that of imagers.

Data from the Earth radiation budget satellites also brought helpful information to quantify the cirrus cloud impact on the radiation budget. A simple estimation was performed to study cirrus cloud radiative properties, using a collocated split window and Earth Radiation Budget Experiment (ERBE) observations (Inoue and Ackerman, 2002). They showed that some cirrus clouds observed by a split window indicate a warming effect. However, a severe issue in such a radiation budget study is that satellite passive methods in retrieving ice cloud properties have difficulties obtaining vertical profiles and detecting optically thin cirrus (so-called sub-visible cirrus). Therefore, further investigations are needed to better characterize the radiative effects of cirrus clouds using global and vertically resolved observations of cirrus clouds.

It is decisive, therefore, to utilize the data from active satellite sensors, as the advent of the Cloud Profiling Radar (CPR) on-board CloudSat and Cloud-Aerosol Lidar with Orthogonal Polarization (CALIOP) on-board Cloud-Aerosol Lidar and Infrared Pathfinder Satellite Observations (CALIPSO) as part of the A-Train of formation flying satellites (Stephens *et al.*,

2002; Winker *et al.*, 2003), and the future EarthCARE satellite to simultaneously carry CPR and LIDAR (Illingworth *et al.*, 2015).

The vertical (horizontal) resolutions in the upper troposphere of these active remote sensing satellite pioneers are 480 m (1.4 km) for the CPR and 60 m (1.0 km) for the CALIOP green (0.532 μm) channel. However, observations by CALIPSO and CloudSat are limited in time and space. Because the repetition cycle of CALIPSO and CloudSat is 16 days, the CALIOP and CPR observations have difficulty retrieving information on life cycles and diurnal cycles of atmospheric features such as cirrus clouds.

Recent studies identified several cooling/warming effects using multi-satellite data. Sun *et al.* (2011), for example, studied ice clouds (optical thickness smaller than 0.3) using a synergy of measurements from the Clouds and the Earth's Radiant Energy System (CERES), the Moderate Resolution Imaging Spectroradiometer (MODIS), and CALIOP. They demonstrated that these thin ice clouds have a diurnal mean SW radiative effect of -2.5 Wm^{-2} and that the LW radiative warming effect can reach 15 Wm^{-2} for ice clouds with an optical thickness of about 0.1. Berry and Mace (2014) found that ice clouds of the Asian summer monsoon produce a net warming radiative effect of 21 Wm^{-2} at the top of the atmosphere (TOA), and those ice clouds with ice water paths (IWP) around 20 gm^{-2} contribute the most significant heating to the Earth based on the CloudSat and CALIPSO data.

Hong *et al.* (2016) investigated the radiative effects of all ice clouds (including low-level ice clouds) by performing radiative transfer modeling using ice cloud properties retrieved from CloudSat and CALIPSO measurements as inputs. Their results for the 2008 period show that the warming effect (~21.8 ± 5.4 Wm^{-2}) was induced by ice clouds trapping longwave radiation that exceeded their cooling effect (~ −16.7 ± 1.7 Wm^{-2}) caused by shortwave reflection, resulting in a net warming effect (~5.1 ± 3.8 Wm^{-2}) globally on the earth-atmosphere system. The net warming is over 15 Wm^{-2} in the tropical deep convective regions, whereas cooling occurs in the mid-latitudes, less than 10 Wm^{-2} in magnitude. However, further studies are still required to retrieve cirrus cloud properties and their spatial and temporal variation to improve radiative effects evaluation.

The observations of cloud properties of cirrus clouds and their diurnal variations are essential for climate studies and for constraining climate models. Kox *et al.* (2014) developed a method to retrieve optical properties of thin cirrus clouds based on a neural network model applying to infrared channels of the Spinning Enhanced Visible and Infrared Imager (SEVIRI) aboard the geostationary METEOSAT Second Generation (MSG) satellite (Schmetz *et al.*, 2002). The network was trained by coincident cirrus optical properties derived from CALIOP and infrared data. They showed a diurnal variation of the optical thickness over the METEOSAT coverage.

Himawari-8 (Japanese geostationary meteorological satellites) carry an optical sensor, Advanced Himawari Imager (AHI), with significantly higher radiometric, spectral, and spatial resolution than those previously available in the geostationary orbit. They have 16 observation bands, and their spatial resolution is 0.5 and 1 km for visible and near-infrared bands, respectively, and 2 km for infrared bands (Bessho *et al.*, 2016).

Optimal Cloud Analysis (OCA) has been developed by the European Organization for the Exploitation of Meteorological Satellites (EUMETSAT) (Watts *et al.*, 1998). With the cooperation of the EUMETSAT, the OCA was modified by Hayashi (2018) for Himawari-8. The OCA retrieves cloud optical properties from multi-band radiance data by the optimal estimation (OE) technique based on the Radiative Transfer Model (RTM) simulation of radiances. In the OCA, the ice crystal shape is assumed as Voronoi non-spherical aggregates. Its single-scattering properties are computed by Ishimoto *et al.* (2012). The OCA has two versions for day and night. The daytime version uses all the channels, and the nighttime version uses only infrared channels of Himawari-8. The nighttime version uses only infrared data, including the "split window" data. We could retrieve the optical properties of cirrus clouds throughout the day with consistent accuracy using only the nighttime version of OCA, although the daytime version is superior in retrieving optical thickness and an effective radius.

In this lecture, using hourly Himawari-8 infrared data, we introduce some results toward showing the diurnal variation of cirrus clouds over the western Pacific (140°E–190°E, 40°N–20°S). We first introduce the characteristic features of brightness temperature differences in the infrared data of 10 μm, 12 μm and 8.6 μm for the detection of cirrus clouds, then some results of cirrus clouds analyzed by the infrared version of OCA over the western Pacific. Then, we show the diurnal variation of cloud amount, cloud optical thickness (COT), cloud effective particle radius (Re), and cloud top pressure (CTP) for thin cirrus clouds over the western Pacific. Next, we show additional features of temporal variation of anvil clouds associated with deep convective clouds. We then present a case study of rainfall rate at various life stages of deep convection.

References

Berry E, Mace GG. (2014) Cloud properties and radiative effects of the Asian summer monsoon derived from A-Train data. *Journal of Geophysical Research* **119**:9492–508.

Bessho, K, Date K, Hayashi M, *et al.* (2016) An introduction to Himawari-8/9 — Japan's new-generation geostationary meteorological satellites. *Journal of the Meteorological Society of Japan* **94**:151–83.

Hayashi M. (2018) Introduction to the computation method for cloud radiative processes and its application for the advanced Himawari imager onboard Himawari-8. *Meteorological Satellite Center Technical* Note **63**:1–38.

Heymsfield AJ. (1975) Cirrus uncinus generating cells and the evolution of cirriform clouds. Part I: Aircraft observations of the growth of the ice phase. *Journal of Atmospheric Sciences* **32**:799–808.

Hong Y, Liu G, Li J-LF. (2016) Assessing the radiative effects of global ice clouds based on CloudSat and CALIPSO measurements, *Journal of Climate* **29**:7651–74. doi:10.1175/JCLI-D-15-0799.1

Illingworth AJ, Barker HW, Beljaars A, *et al.* (2015) The EarthCARE Satellite: The next step forward in global measurements of clouds, aerosols, precipitation, and radiation. *Bulletin of the American Meteorological Society* **96**:1311–32. doi:10.1175/BAMS-D-12-00227.1

Inoue T. (1985) On the temperature and effective emissivity determination of semi-transparent cirrus clouds by bi-spectral measurements in the 10 μm window region. *Journal of the Meteorological Society of Japan* **63**:88–99.

Inoue T. (1987) A cloud type classification with NOAA 7 Split-Window measurements. *Journal of Geophysical Research* **92**:3991–4000.

Inoue T. (1989) Features of clouds over the tropical Pacific during northern hemispheric winter derived from split window measurements. *Journal of the Meteorological Society of Japan* **67**:621–37.

Inoue T. (1997) Day-to-night cloudiness change of cloud types inferred from split window aboard NOAA polar-orbiting satellite. *Journal of the Meteorological Society of Japan* **75**:59–66.

Inoue T, Ackerman S. (2002) Radiative effect of various cloud types as classified by the split window technique over the eastern sub-tropical Pacific derived from collocated ERBE and AVHRR data. *Journal of the Meteorological Society of Japan* **80**:1383–94.

Ishimoto H, Masuda K, Mano Y, *et al.* (2012) Irregularly shaped ice aggregates in optical modeling of convectively generated ice clouds. *Journal of Quantitative Spectroscopy and Radiative Transfer* **113**:632–43. https://doi.org/10.1016/j.jqsrt.2012.01.017

Kox S, Buglianro L, Ostler A. (2014) Retrieval of cirrus cloud optical thickness and top altitude from geostationary remote sensing. *Atmospheric Measurement Techniques* **7**:3233–46. doi:10.5194/amt-7-3233-2014

Liou KN. (1986) Influence of cirrus clouds on weather and climate processes: A global perspective. *Monthly Weather Review* **114**:1167–99. doi:10.1175/1520-0493(1986)114,1167:IOCCOW.2.0.CO;2

Manabe S, Strickler RF. (1964) Thermal equilibrium of atmosphere with a convective adjustment. *Journal of the Atmospheric Sciences* **21**:361–85.

Raschke E, Schmetz J, Heintzenberg J, *et al.* (1990) The International Cirrus Experiment (ICE) — A joint European effort. *European Space Agency Journal* **14**:193–99.

Rossow WB. (1989) Measuring cloud properties from space: A review. *J. Climate* **2**:201–13.

Schiffer RA, Rossow WB. (1983) The International Satellite Cloud Climatology Project (ISCCP): The first project of the World Climate Research Programme. *Bulletin of the American Meteorological Society* **64**:779–84.

Schmetz J, Pill P, Tiemkes S, *et al.* (2002) An introduction to Meteosat Second Generation (MSG). *Bulletin of the American Meteorological Society* **83**:977–92.

Stephens GL, Tsay SC, Stackhouse JPW, Flatau P. (1990) The relevance of the microphysical and radiative properties of cirrus clouds to climate and climatic feedback. *Journal of the Atmospheric Sciences* **47**:1742–53. doi:10.1175/ 1520-0469(1990)047,1742:TROTMA.2.0.CO;2

Stephens GL, Vane DG, Boain RJ, *et al.* (2002) The CloudSat mission and the A-train:A new dimension of space-based observations of clouds and precipitation. *Bulletin of the American Meteorological Society* **83**:1771–90. doi:10.1175/BAMS-83-12-1771

Stubenrauch CJ, Chédin A, Rädel G, N. A. Scott, and S. Serrar (2006) Cloud properties and their seasonal and diurnal variability from TOVS Path-B, *Journal of Climate* **19**:5531–53. https://doi.org/10.1175/JCLI3929.1

Sun W, Videen G, Kato S, *et al.* (2011) A study of subvisual clouds and their radiation effect with a synergy of CERES, MODIS, CALIPSO, and AIRS data. *Journal of Geophysical Research* **116**:D22207. doi:10.1029/2010JG001573

Watts PD, Mutlow CT, Baran A.J, Zavody AM. (1998) Study on cloud properties derived from Meteosat Second Generation observations, Final Report, EUMETSAT ITT no. 97/181.

Winker DM, Pelon J, McCormick MP. (2003) The CALIPSO mission: Spaceborne lidar for observation of aerosols and clouds. In *Lidar Remote Sensing for Industry and Environment Monitoring III*. Eds. Singh UN, Itabe T, Liu Z. International Society for Optical Engineering. doi:10.1117/12.466539

Wylie DP, Menzel WP, Strabala KI. (1994) Four years of global cirrus cloud statistics using HIRS. *Journal of Climate* **7**:1972–86.

Slide 1

Slide 2

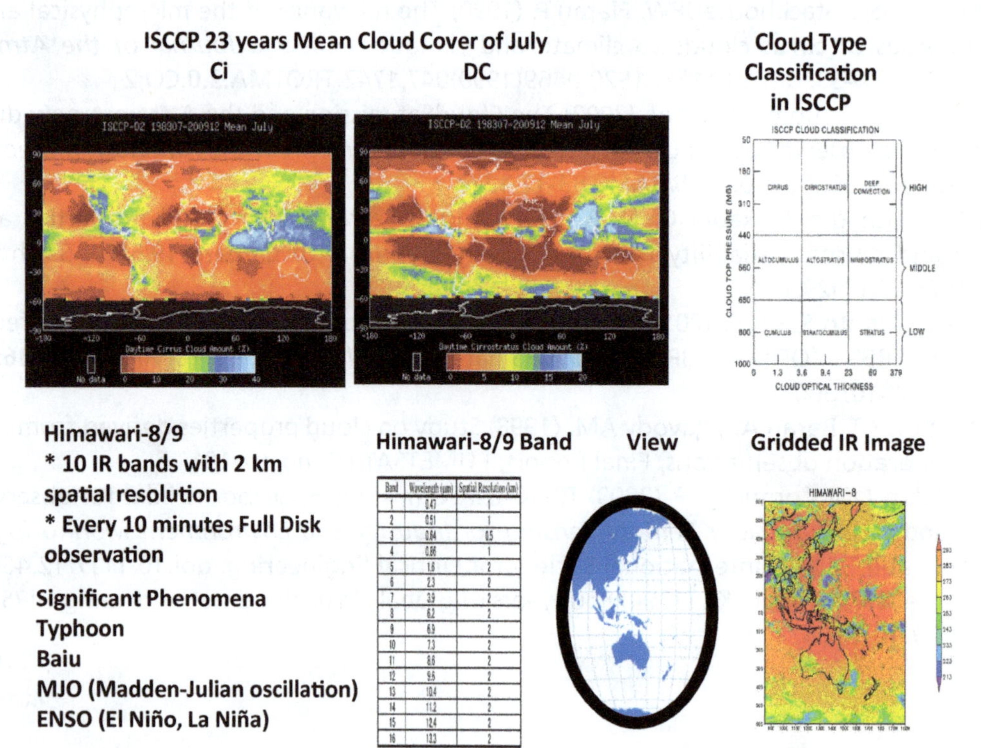

Band	Wavelength (μm)	Spatial Resolution (km)
1	0.47	1
2	0.51	1
3	0.64	0.5
4	0.86	1
5	1.6	2
6	2.3	2
7	3.9	2
8	6.2	2
9	6.9	2
10	7.3	2
11	8.6	2
12	9.6	2
13	10.4	2
14	11.2	2
15	12.4	2
16	13.3	2

Slide 3

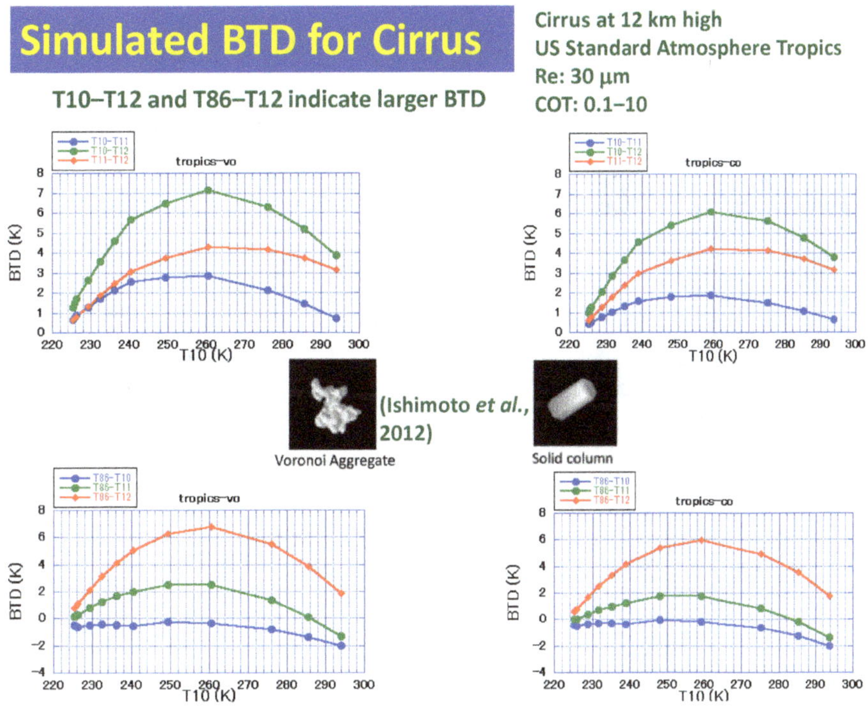
Simulated BTD for Cirrus
T10–T12 and T86–T12 indicate larger BTD
Cirrus at 12 km high
US Standard Atmosphere Tropics
Re: 30 µm
COT: 0.1–10
T10-T11
T10-T12
T11-T12
tropics-vo
tropics-co
BTD (K)
T10 (K)
(Ishimoto et al., 2012)
Voronoi Aggregate
Solid column
T86-T10
T86-T11
T86-T12

Slide 4

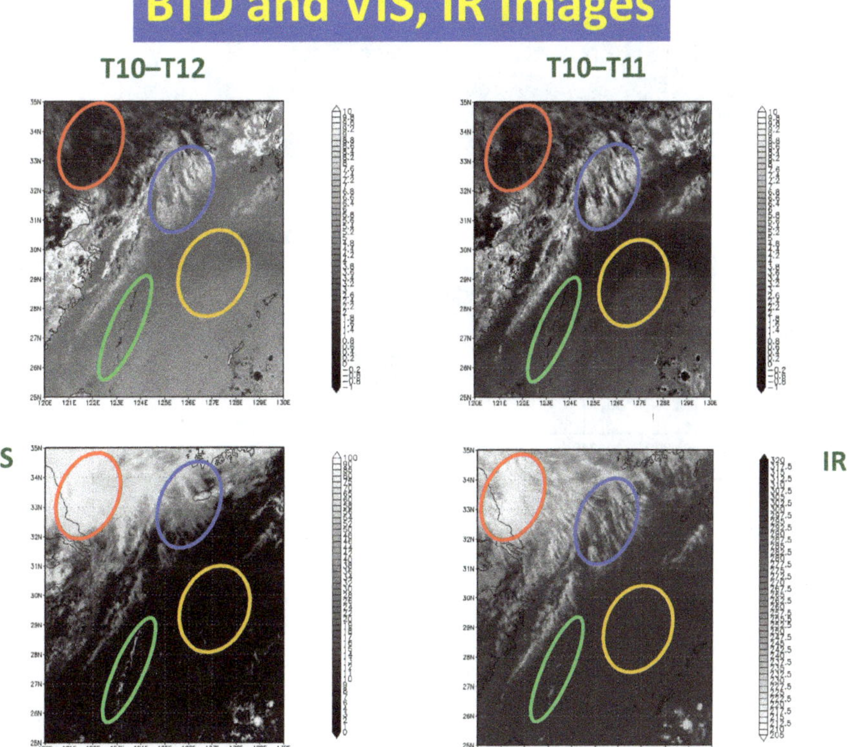
BTD and VIS, IR Images
T10–T12
T10–T11
VIS
IR

Slide 5

RGB Composite and CALIOP

R: IR G: T10–T12 B: T86–T12

03 UTC 01 July 2016

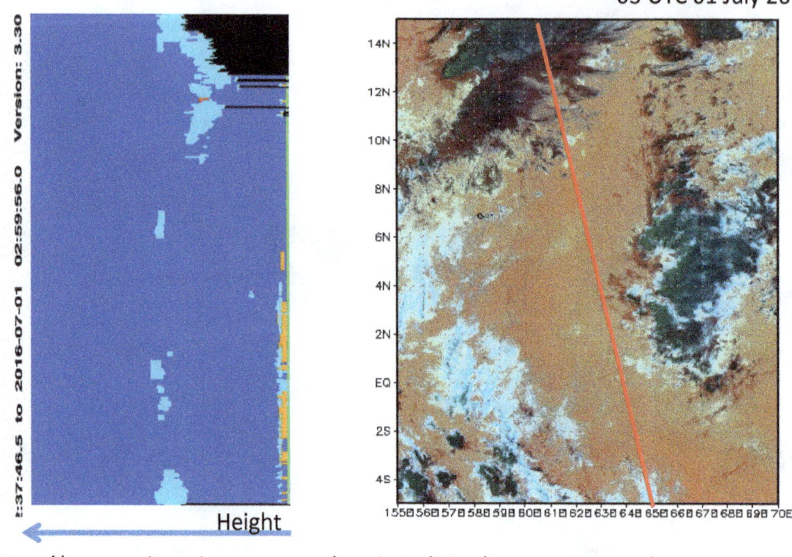

(https://www-calipso.larc.nasa.gov/products/lidar/browse_images/)

Slide 6

BTD for CALIOP Signal Seen at Higher than 8 km

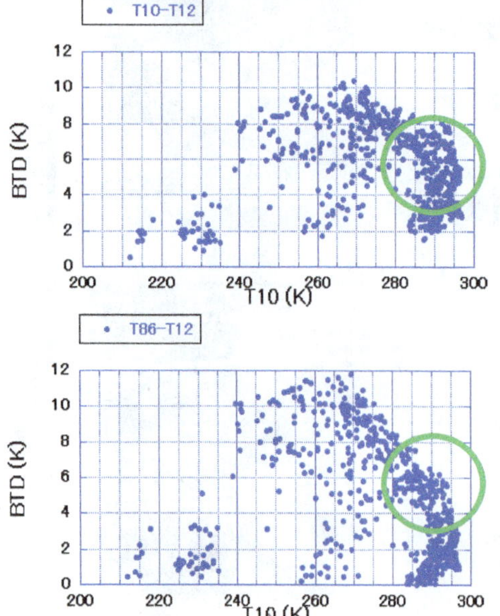

T10–T12, T86–T12 indicates larger BTD For T10 warmer than 273 K cloud

Slide 7

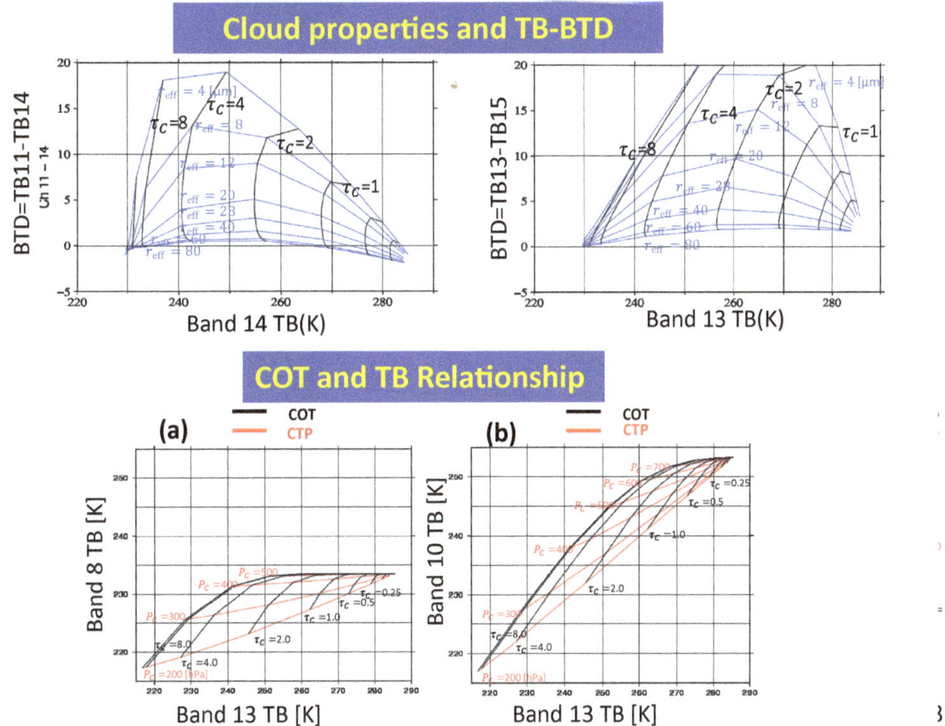

Slide 8

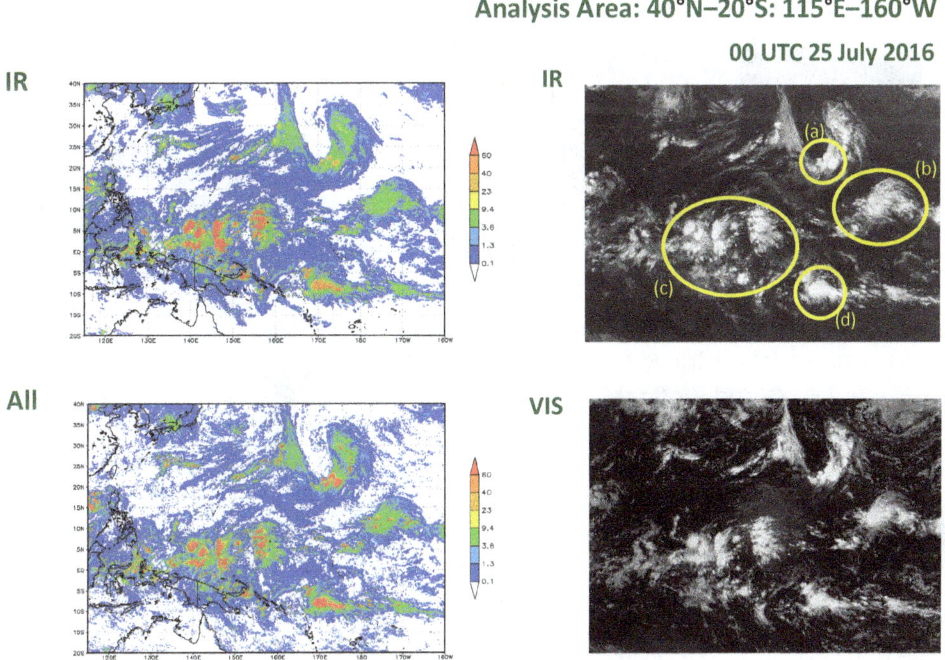

Slide 9

Difference of OCA-IR and OCA-ALL (Re, CTP)

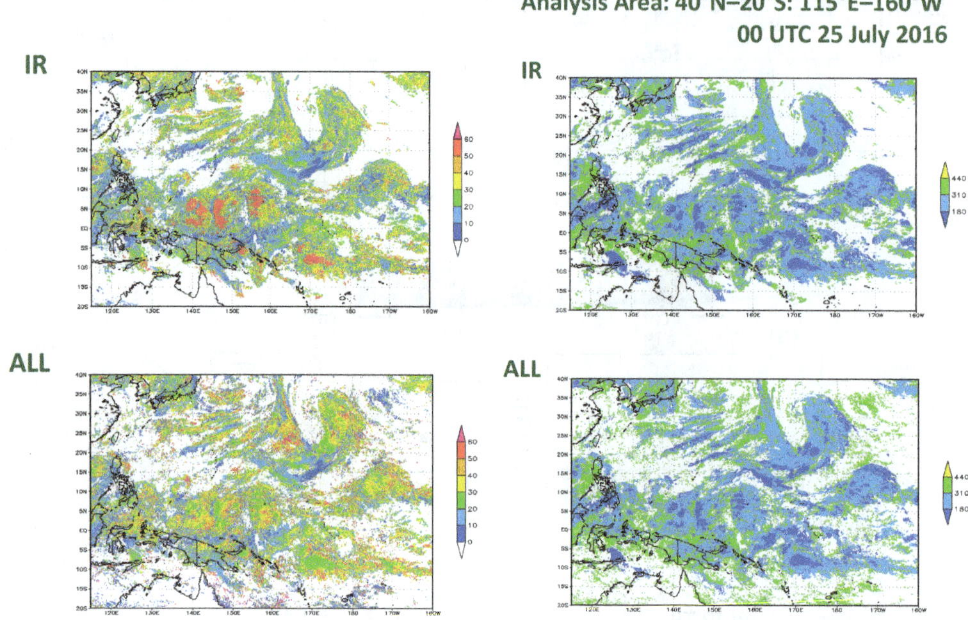

Slide 10

Comparison with CALIOP

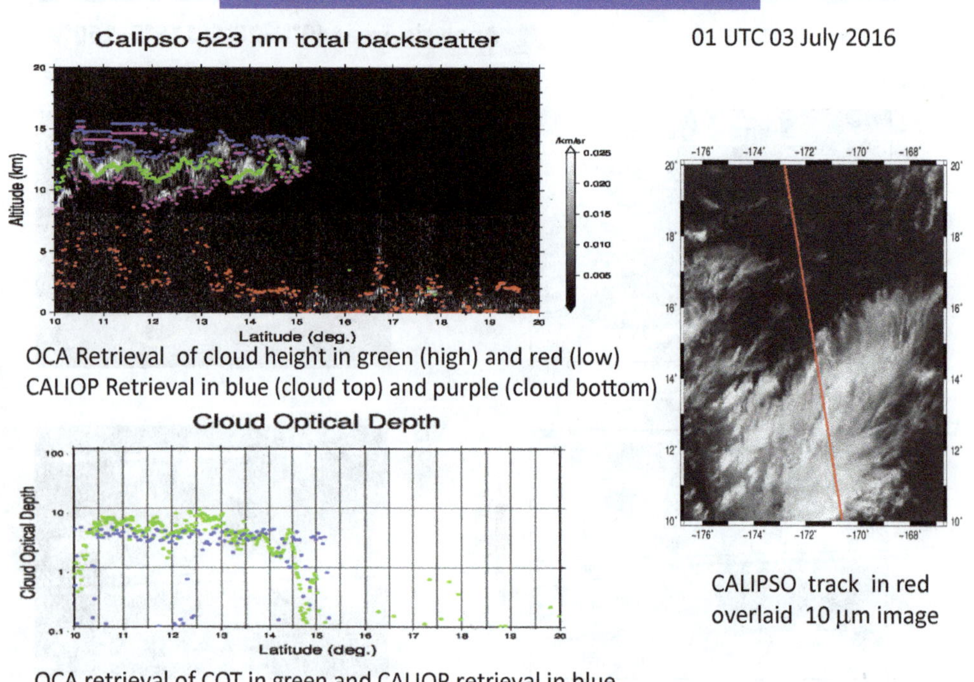

OCA Retrieval of cloud height in green (high) and red (low)
CALIOP Retrieval in blue (cloud top) and purple (cloud bottom)

CALIPSO track in red overlaid 10 μm image

OCA retrieval of COT in green and CALIOP retrieval in blue

Slide 11

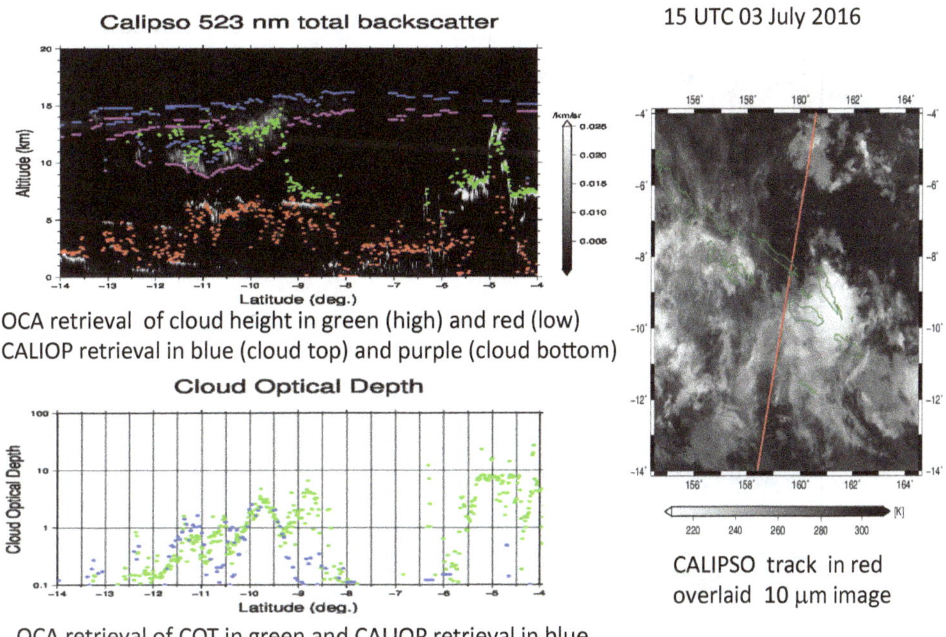
Comparison with CALIOP
Calipso 523 nm total backscatter
15 UTC 03 July 2016
Altitude (km)
Latitude (deg.)
OCA retrieval of cloud height in green (high) and red (low)
CALIOP retrieval in blue (cloud top) and purple (cloud bottom)
Cloud Optical Depth
Cloud Optical Depth
Latitude (deg.)
OCA retrieval of COT in green and CALIOP retrieval in blue
CALIPSO track in red
overlaid 10 μm image

Slide 12

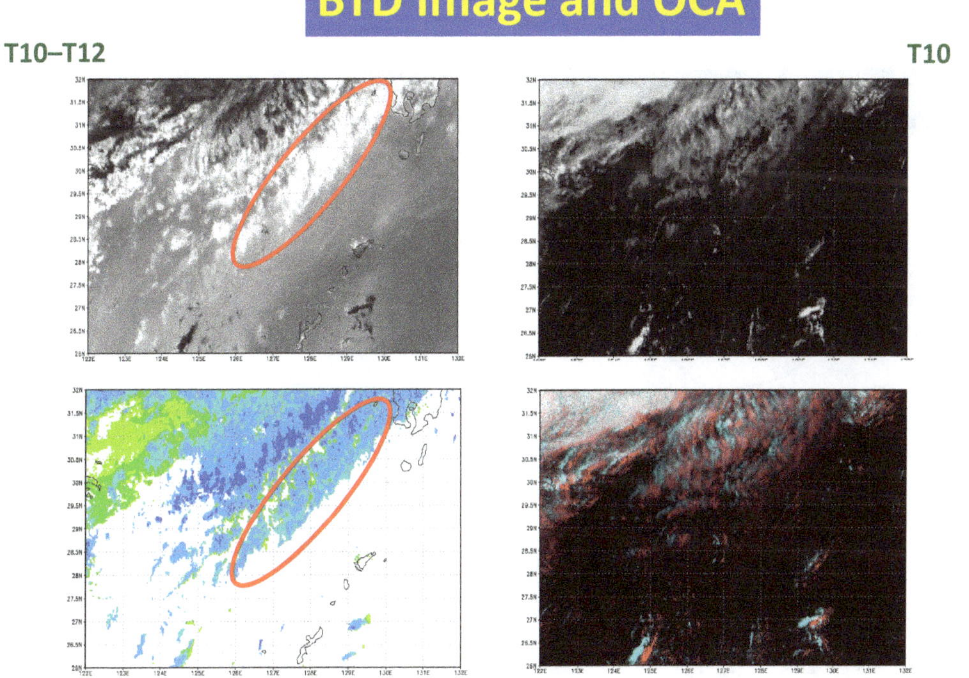
BTD Image and OCA
T10–T12
T10
Cloud height (hPa) analyzed by the OCA
Projecting image (red and green glasses)

Slide 13

Monthly Mean Cloud Cover of High Cloud

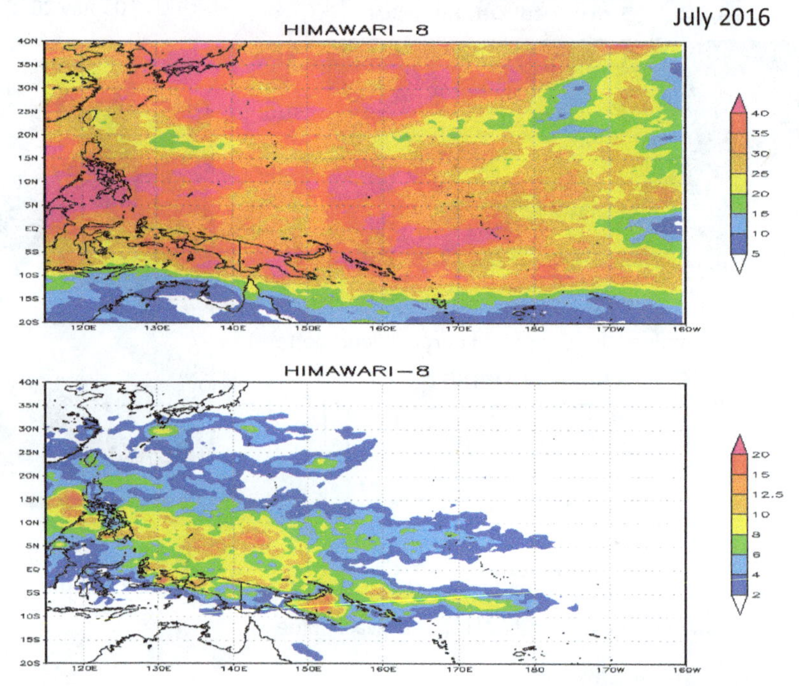

Slide 14

Diurnal Variation of Cirrus Cloud Cover (COT < 1.3)

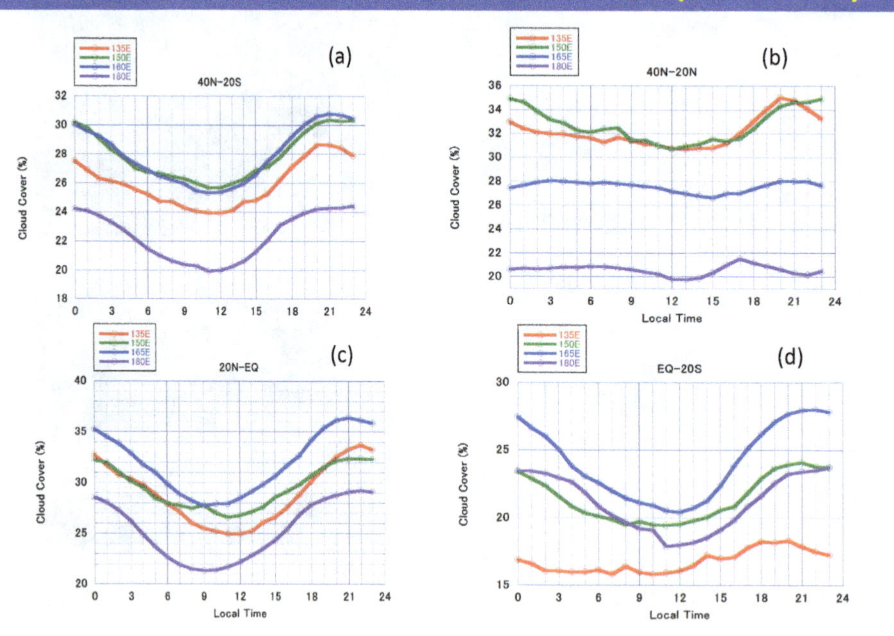

Slide 15

Diurnal Variation of Cloud Properties of Cirrus (COT < 1.3: Red and 1.3 < COT < 3.6: Green)

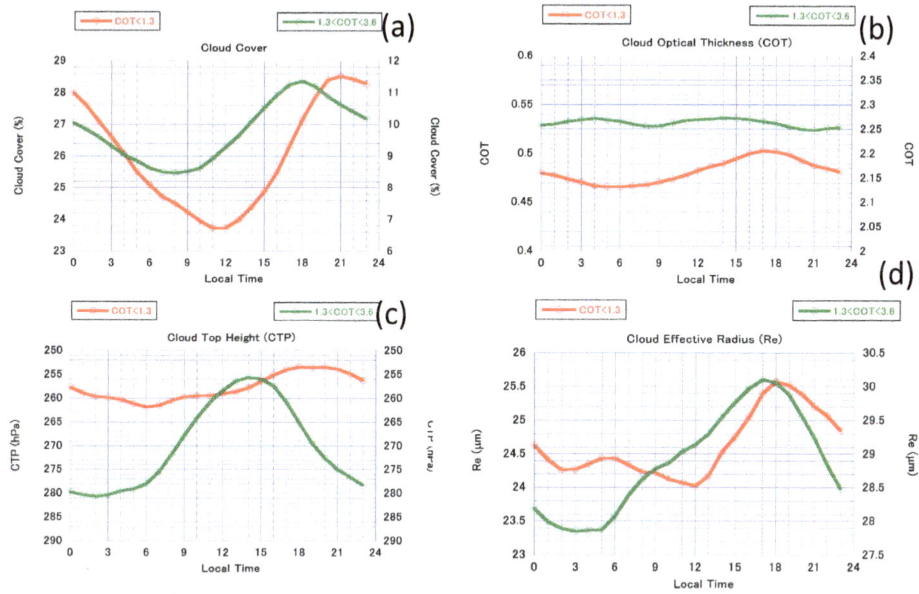

Slide 16

Diurnal Variation of Optical Properties for Thin Cirrus (COT < 1.3) over (20°N–EQ:Red) and (40°N–20°N:Green)

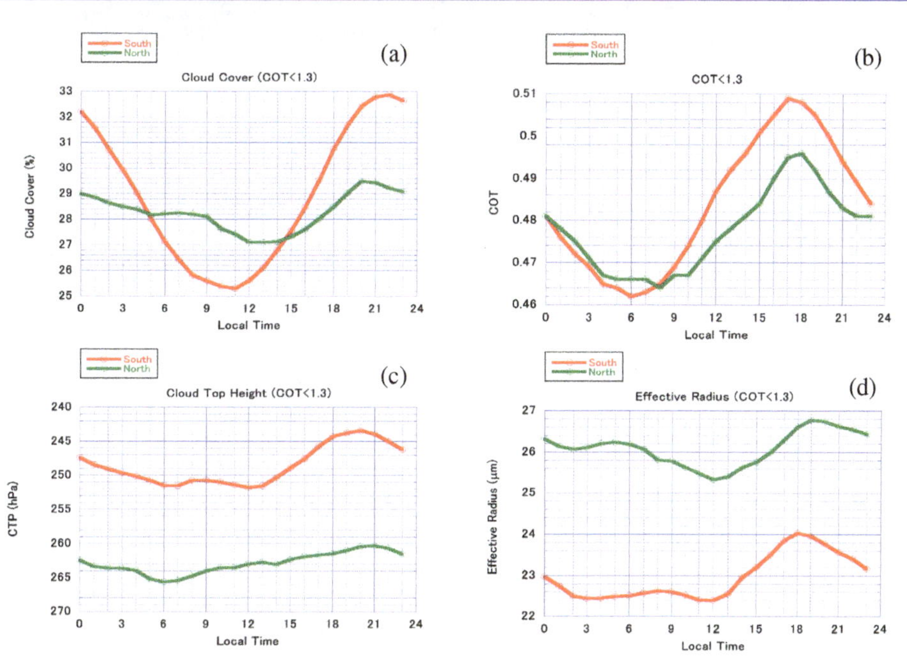

Slide 17

Developing stage : T86–T12 is small. Decaying stage : T86–T12 becomes larger

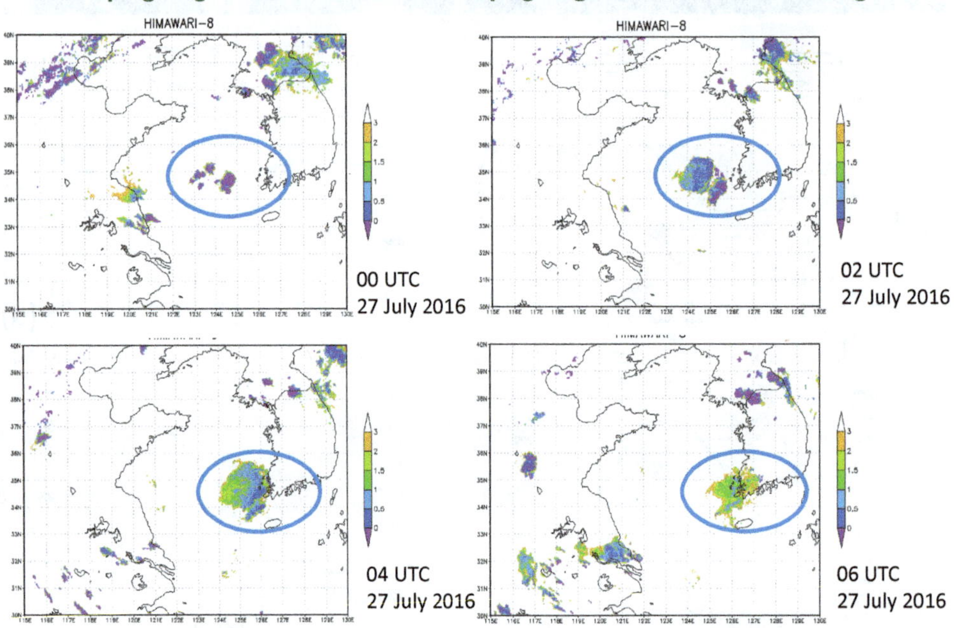

Slide 18

Evolution of Optical Properties within Deep Convection (<253 K)

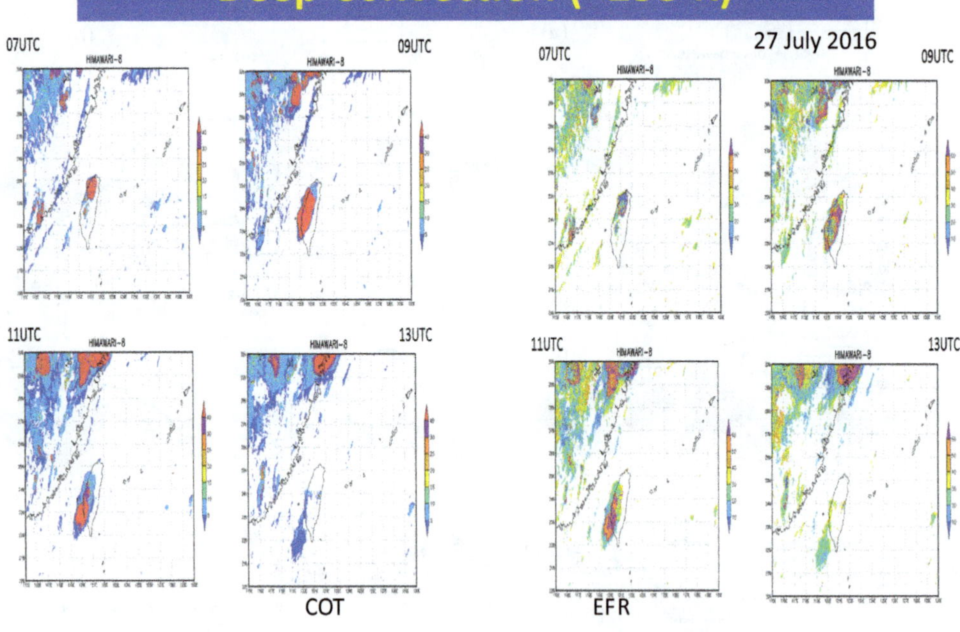

Slide 19

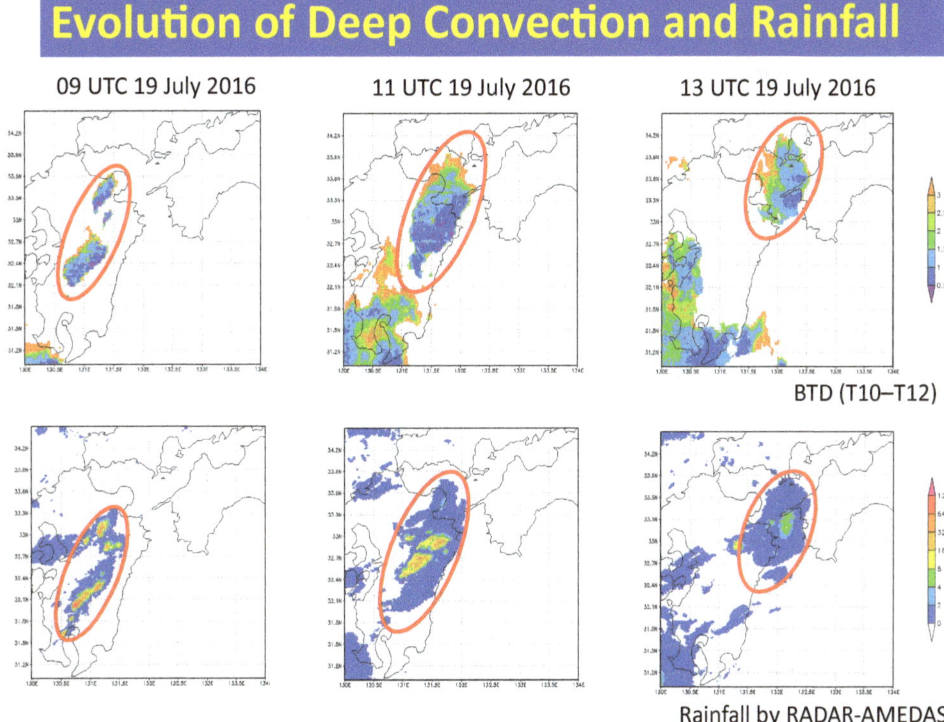

Slide 20

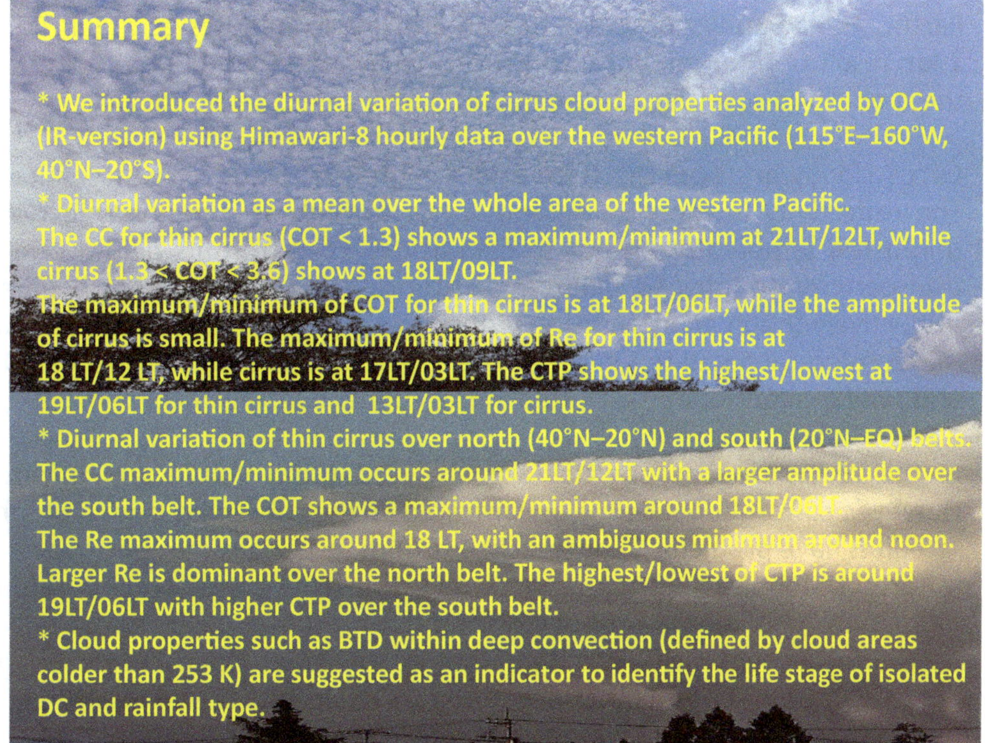

Summary
* We introduced the diurnal variation of cirrus cloud properties analyzed by OCA (IR-version) using Himawari-8 hourly data over the western Pacific (115°E–160°W, 40°N–20°S).
* Diurnal variation as a mean over the whole area of the western Pacific.
The CC for thin cirrus (COT < 1.3) shows a maximum/minimum at 21LT/12LT, while cirrus (1.3 < COT < 3.6) shows at 18LT/09LT.
The maximum/minimum of COT for thin cirrus is at 18LT/06LT, while the amplitude of cirrus is small. The maximum/minimum of Re for thin cirrus is at 18 LT/12 LT, while cirrus is at 17LT/03LT. The CTP shows the highest/lowest at 19LT/06LT for thin cirrus and 13LT/03LT for cirrus.
* Diurnal variation of thin cirrus over north (40°N–20°N) and south (20°N–EQ) belts.
The CC maximum/minimum occurs around 21LT/12LT with a larger amplitude over the south belt. The COT shows a maximum/minimum around 18LT/06LT.
The Re maximum occurs around 18 LT, with an ambiguous minimum around noon.
Larger Re is dominant over the north belt. The highest/lowest of CTP is around 19LT/06LT with higher CTP over the south belt.
* Cloud properties such as BTD within deep convection (defined by cloud areas colder than 253 K) are suggested as an indicator to identify the life stage of isolated DC and rainfall type.

Slide Captions

Slide 1 Cover

Slide 2 The International Satellite Cloud Climatology Project (ISCCP) (Rossow and Schiffer, 1999) uses visible and infrared data from a combination of polar-orbiting and geostationary weather satellites for a globally complete long-term cloud data record with a temporal sampling of 3 hours and spatial scale of 25 km. The top right diagram shows the cloud type classification in the ISCCP. Cloud types are defined by cloud top pressure (CTP) and cloud optical thickness (COT). A cirrus cloud is defined as a cloud higher than 440 hPa with a COT thinner than 3.6. A deep convective cloud is defined as a cloud higher than 440 hPa with a cloud optical thickness larger than 23. The top left (middle) figure shows the 23-year-mean cirrus (deep convection) cloud amount of July analyzed by the ISCCP. The high cloud amount area is widely spread over the western tropical Pacific, although the high cloud amount area of deep convection is confined to the north of Australia.

Himawari-8/9 (Japanese geostationary meteorological satellites) carries an optical sensor, AHI (Advanced Himawari Imager), with higher radiometric, spectral, and spatial resolution. The bottom left table shows 16 bands of Himawari-8/9 with central wavelength and spatial resolution. The bottom middle shows the Earth's view from Himawari-8/9, and the effective observation area is shown in the bottom right figure as an infrared image.

Significant meteorological phenomena such as typhoons, Baiu, Madden-Julian Oscillation, and ENSO (El Niño and La Niña) occur over this western tropical Pacific. Himawari-8/9 observes this area every 10 minutes as a Full Disk with a 2-km resolution for ten infrared bands. Himawari-8/9 observation depict well for the high cloud cover area of cirrus and deep convective clouds.

Slide 3 Cirrus clouds are optically thin, and a high-level cloud consists of ice crystals. The reflectivity in the visible band is not so high as cumulus-type clouds, and brightness temperature is not so cold as expected from its height due to semi-transparency on infrared radiation. Historically, a cloud type has been analyzed subjectively (neph-analysis) using visible and infrared images by specialists at the Meteorological Satellite Center at the Japan Meteorological Agency (MSC/JMA) and similar agencies in the United States (US) and EUMETSAT.

In the ISCCP, cloud properties are analyzed using visible and infrared data. However, conventional visible and infrared images observed from meteorological satellites have problems to detect cirrus clouds, especially at night. We need other infrared channels to retrieve cirrus cloud properties consistently during day and night since visible data cannot be used during the night. Inoue (1985) showed how effective emissivity and temperature of cirrus clouds could be retrieved objectively using two channels (centered at 10 μm and 12 μm) at a 10-μm window region (later called a split window) of Advanced Very High Resolution Radiometer

(AVHRR) onboard NOAA-7. He noted cirrus clouds indicate a larger brightness temperature difference (BTD = TB10 μm − TB12 μm). The large BTD is caused by the effective emissivity difference between the split windows.

Cirrus cloud is composed of ice crystals with different shapes like solid columns, rosettes, plates, and aggregation of various types. Here, the BTD is theoretically computed for cirrus clouds consisting of ice crystals of Voronoi aggregates (Ishimoto *et al.*, 2012) and a solid column for a four-band (8.6, 10, 11, 12 μm) combination of AHI. The cirrus cloud is assumed to have a 30-μm effective radius (Re) and is located at the 12-km level of the US standard atmosphere. Single scattering properties are computed based on (Ishimoto *et al.*, 2012).

Figures show the computed BTD by changing the COT from 0.1 to 10 for ice crystals assumed as the Voronoi aggregate (left) and solid column (right). Figures on top show the combination of T10, T11 and T12, while figures at bottom show the combination of T86, T10, T11 and T12. The BTD for thin cirrus with a COT of 0.1 is shown at right most, and the thicker cirrus of COT as 10 is shown at left most in each figure. The BTD indicates larger values with an increase of optical thickness and shows a maximum value of around COT 3. It then decreases with the increase of COT. The cirrus cloud shows a larger BTD for 10, 11 and 12-μm bands of AHI, as Inoue (1985) noted.

Very thick clouds indicate smaller BTD since those clouds can be considered as a blackbody. Using an 8.6-μm band of AHI also indicates larger BTD values with an increase of around COT 3. Then the BTD decreases with increasing COT. BTDs between 10 and 12 μm (T10–T12) and 8.6–12 μm (T86–T12) are adequate to identify cirrus clouds since the BTD indicates the largest value among the combinations of four channels.

Slide 4 Examples of BTD images of T10–T12 (top left), T10–T11 (top right), and corresponding visible (bottom left) and infrared (10 μm: bottom right) over the western Pacific at 03 UTC 20 July 2016 are shown in the figure.

Optically thicker clouds are shown whiter in the visible image (bottom left). The colder cloud is more white in the infrared image (bottom right). The larger BTD is shown in bright white in the BTD images. We can see a higher, colder (bright white in the infrared image) and optically thicker (bright white in the visible image) cloud area at the north-west corner (red oval).

Over the southeast edges of this optically thick and high cloud area, we can see gray clouds in both visible and white infrared images in the BTD image (blue oval). These bright white areas in the BTD images can be subjectively (so-called neph-analysis) considered a cirrus cloud consisting of ice crystals based on the former figures. White dotted line shape clouds in visible are shown as black dotted line in the BTD images (green oval area). These clouds are hardly seen in the infrared image. Therefore, we can classify these clouds as shallow, optically thick clouds (cumulus).

The cloud-free area over the southeast part in the images is relatively smooth in visible and infrared images. Nevertheless, we note a gradually changing pattern over this cloud-free ocean in the BTD images (yellow oval). This pattern reflects the change of total column water vapor abundance over this area (Inoue and Smith, 1994), since the absorption characteristics by water vapor are different over the 10 μm window region.

Slide 5 This slide shows an example of a BTD image and Cloud-Aerosol Lidar with Orthogonal Polarization (CALIOP) observation at 03 UTC on 01 July 2016. Here we construct the RGB composite image using infrared data of T10, T10–T12, and T86–T12. The figure on the right shows an RGB composite image as infrared (in red), T10–T12 (in green), and T86–T12 (in blue) with the CALIPSO track as the red line.

The reddish color corresponds to a warmer brightness temperature with smaller BTDs, while green or blue corresponds to the larger BTD. Thus, the light blue cloud area (left middle and edge of the darker gray cloud) corresponds to cirrus clouds that have larger BTDs (darker blue and green) with slightly colder brightness temperature (lighter red).

The scattered reddish clouds (top and right side) are warmer (darker red) and optically thicker clouds (lighter blue and green). On the other hand, the dark gray cloud area is optically thicker (smaller BTDs: lighter blue and green) and higher cloud (lighter red). Thus, we can analyze these clouds as higher optically thick clouds from this RGB composite image.

The figure on the left shows a corresponding CALIOP analysis (https://www-calypso.larc.nasa.gov) from the north to the south. Ice cloud is shown as light blue in CALIOP analysis. Ice clouds are analyzed at the northernmost of the CALIOP observation with the black shade underneath that indicates that the LIDAR cannot observe below the thick ice cloud. Those cloud areas can be understood as very thick clouds with ice on the top from the CALIOP analysis. These clouds are black (low value of red, green and blue) clouds in the RGB composite image.

The CALIOP analysis shows that thicker high clouds change their height from high to lower toward the south. The RGB composite image shows the change from dark gray to a dark reddish (slightly lower) cloud. We can see the consistency of the analysis between RGB composite (BTDs and infrared data) and CALIOP analysis for this cloud area, in that the cloud top becomes lower from the north to the south.

The ice cloud at 4°S in the CALIOP observation shows a slightly thicker high-level cloud corresponding to the whitish (high value of red, green, blue) cloud in the RGB composite image. The whitish area at 6°N and near the Equator in an RGB composite image corresponds to the ice cloud at 6°N and the Equator by the CALIOP analysis. The slightly whitish area against the orange background in the RGB image indicates the limitation of infrared observation.

Slide 6 The top figure in this slide shows scatter plots of the *x*-axis for T10 and *y*-axis for BTD (T10–T12), and the bottom figure shows the *x*-axis for T10 and *y*-axis for BTD (T86–T12) for a cirrus cloud (higher than 8 km) observed by CALIOP using the coincident (less than 10 minutes) and collocated data of CALIOP and Himawari-8. Here, a cirrus cloud is defined as the cloud bottom being higher than 8 km by CALIOP analysis (https:/eosweb.larc.nasa.gov/project/calipso_table).

We often used a single threshold of 273 K in T10 to identify water clouds. However, figures in this slide show that many clouds indicate larger BTD with T10 being warmer than 273 K. These clouds might be ice clouds because the cloud height is higher than 8 km by CALIOP observation. Thus, a single threshold to identify water clouds is not practical. These clouds might be ice clouds rather than water. The clouds showing larger BTD with T10 higher than 273 K in the figure might be very thin cirri (invisible cirrus) that cannot be detected by an infrared sensor or a time lag of observation time between CALIOP and AHI.

Slide 7 Infrared data is essential in retrieving diurnal variation of cloud properties with consistency between day and night. Here, we introduce some simulation results showing the relationship between cloud properties and infrared bands. In this slide, TB8, 10, 11, 13, 14, 15 mean brightness temperature at band 8, 10, 11, 13, 14 in slide 2. TB8:TB by 6.2 μm, TB10:TB by 7.3 μm, TB11:TB by 8.6 μm, TB13:TB by 10.4 μm, TB14:TB by 11.2 μm, TB15:TB by 12.4 μm.

The top figures show the dependency on and the BTD for various combinations of optical thickness and effective radius. The curve simulates for cirrus clouds composed of Voronoi-shaped ice crystals at 300 hPa in the US standard atmosphere. We set the satellite zenith angle as 40 degrees. The figure on the left shows the TB14 and BTD (= TB11–TB14), while the figure on the right shows TB13 and BTD (= TB13–TB15). We can see from the arch-shaped curve (blue) that the BTD increases with increasing the optical thickness (black), with the maximum value around the optical thickness of 4, and then decreasing for even larger values of optical thickness. The maximum value of BTD becomes smaller with the increase of the effective radius. These diagrams show the feasibility of retrieving cirrus cloud properties using infrared data.

The bottom figures show the dependency of cloud top height and optical thickness between the TBs for two infrared bands. The relationship between TB13 and TB8 (see [(a)]) and TB10 (see [(b)]) for various optical thicknesses is shown in this figure. The computation is performed assuming the effective radius of 20 μm for the US standard atmosphere with a satellite zenith angle of 30 degrees. The fat solid line shows the TB relationship between the two bands for a blackbody cloud. The red line shows the brightness temperature for the semi-transparent cloud with changing optical thickness. The top right point at each figure shows the cloud-free condition. With the increase of optical thickness, the TB decreases toward the left and reaches the fat solid line (blackbody temperature).

The radiative transfer computation shows the feasibility of retrieving cloud properties using the Himawari-8 infrared observation. In this study, we perform the optimal cloud analysis (OCA) developed by Hayashi (2018) with the cooperation of EUMETSAT (Watts *et al.*, 1998) to analyze the clouds observed by the AHI. In OCA, cloud optical properties are determined using the inversion method. For reducing the computational cost of radiative transfer calculation, the product's formulation involves using a pre-computed look-up table (LUT) representing the cloud radiative properties of reflection, transmission and emission. For ice cloud retrieval, a Voronoi aggregate is assumed. Outputs are cloud optical thickness, cloud effective radius, cloud phase, cloud top pressure, and surface temperature. In addition, single-scattering properties are computed based on (Ishimoto *et al.*, 2012).

Slide 8 The OCA has two versions of IR-only and all bands of the AHI. The retrieval with visible data is more reliable than IR-only, especially for retrieving cloud optical thickness (COT) and effective radius (Re). In contrast, the advantage of the IR-only version is the consistent retrieval for day and night. Since we are interested in the diurnal variation of cloud properties of cirrus clouds, we use the IR-only version in this study. We analyzed the cloud over the area of 40°N–20°S and 115°E–160°W using the hourly observation of the AHI with a 2-km spatial resolution during 1 July and 31 July in 2016.

This slide shows infrared (top right) and visible (bottom right) images. In both visible and infrared images, we can identify four deep convective areas (yellow ovals at the top) where both visible and infrared images indicate white clouds. The top left slide shows a map of retrieved COT for clouds higher than 440 hPa by IR version for 00 UTC 25 July 2016. The COT for the core of the deep convective cloud is retrieved as being more extensive than 23 COT surrounding with a COT smaller than 9.4 (greenish). A larger area is retrieved as a COT smaller than 3.6 (bluish).

The bottom left slide shows the COT retrieved by the band version. The distribution of the retrieved area is similar for both versions, although the band version retrieves more area than the IR-only version's significantly thinner cirrus. However, as expected, a considerable retrieval difference is seen over the optically thicker cloud area (orange), in that the COT value is more than 23 within (a), (b), (c) and (d). The IR retrieval is overestimated for area (b) and underestimated in areas (a), (c) and (d). Area (b) seems to be developing a stage of deep convection since a more significant percentage of the orange area is within the surrounding greenish area. This difference may be because infrared retrieval is less effective for optically thicker clouds.

The RMS difference of COT between IR-only and all-channel methods for clouds higher than 440 hPa and smaller than the 23 COT cloud area is 4.4. For clouds thinner than 3.6, the difference is 0.8.

Slide 9 Comparison of the effective radius (Re) retrieval for clouds higher than 440 hPa by the IR-only version at the top left and the all-bands version at the bottom left for 00 UTC 25 July 2016. The more considerable difference occurs over the larger COT cloud areas (a), (b), (c) and (d) in Slide 8. Area (b) indicates the underestimation by the IR-only version. The IR-only version overestimates the Re over the other three convective areas. The RMS difference of Re between an IR-only and all-channel versions for clouds higher than 440 hPa and smaller than a 23 COT cloud area is 11.0 μm, and 10.4 μm for a COT of less than 3.6 for cirrus clouds.

The panels at the top right and bottom right show, respectively, the map of cloud top height in hPa for the IR-only version and the all-band version. Generally, cloud top pressure retrieval shows good agreement over the core of the deep convective cloud area. However, the slightly larger difference appears in the surrounding area of the deep convective cloud, where thinner cirrus clouds cover. The RMS difference of CTP between IR-only and all-channel methods for clouds higher than 440 hPa and smaller than a 23 COT cloud area is 21.8 hPa, and 23.3 hPa for a COT of less than 3.6.

Slide 10 We compare the cloud top height and COT between OCA retrieval and CALIOP observations. This figure shows a comparison between the OCA retrieval of cloud top height and optical thickness for 01 UTC on 3 July 2016. The time difference of CALIOP and OCA observations is within five minutes. The figure on the right shows the infrared image with the CALIPSO track in red.

The top left figure shows the CALIOP total backscatter (grayscale from black to white) with the cloud top in blue and cloud bottom in purple for cirrus clouds analyzed by OCA and the CALIPSO team (https:/eosweb.larc.nasa.gov/project/). At the same time, cirrus clouds analyzed by the OCA are shown in green and the lower clouds in red.

The bottom left shows the optical depth (thickness) retrieval by OCR (in green) and CALIOP retrieval (in blue). The OCA retrieval (green) in the left half (10°N–15°N) reasonably coincides with the middle of the cloud top and cloud bottom analyzed by the CALIPSO team.

Slide 11 This slide shows another example of a comparison between OCA and CALIOP. This figure shows the case for 15 UTC 03 July 2016. The time difference is about 10 minutes, slightly more than the former slide figure. In this case, the OCA retrieval is similar to the former figure for the left half (south of 10°S) for both cloud top height and COT. We can see very thin cirrus from the backscatter intensity and cloud top and bottom analyzed by CALIOP over the northern half. OCA retrieval is not good except for the cloud at 5°S. As expected, the OCA could not retrieve very thin cirrus at 6°S–9°S, observed by CALIOP. CALIOP analysis shows only thin (about 1 km) cirrus cloud around 15 km high along 4°S–6°S. OCA retrieves cloud around 8-13 km high rather thin cirrus cloud around 15 km high retrieved by CALIOP. The OCA retrieval (in green) matches higher backscatter intensity (bright white) at 4°S–4.5°S and 5°S–5.5°S. The larger values of the total backscatter signal (bright white) by CALIOP also coincide with the OCA retrieval.

Slide 12 In this slide, we show the comparison between BTD and OCA analysis. The top left figure shows the T10–T12 image, infrared image (top right), cloud top height in hPa analyzed in OCA (bottom left), and artificial 3-D image (bottom right) for 03 UTC 02 July 2016. The former slides showed that the larger BTD area corresponds to the high-level cirrus cloud. The OCA analysis shows that the cloud height is higher than 350 hPa over the red oval area (bottom left), indicating a larger BTD (top left). We create an RGB composite image (bottom right) using the original visible image, and the visible image is displaced artificially in proportion to the cloud top altitude analyzed by OCA. We can see the higher clouds over the red oval area using 3-D glasses (red [left eye] and blue [right eye]).

Slide 13 Himawari-8 covers the western tropical Pacific where significant meteorological phenomena occur, such as typhoons, Baiu, Madden Julian Oscillation, and El Niño/La Niña. Monitoring clouds with high temporal resolution (10 minutes for a full disk and 2.5 minutes for sector regions) is essential for severe weather and climate study. Thus, we show examples of high cloud analysis using hourly data during July 2016.

Here, we offer mean features of cloud cover (%) over the Himawari-8 observation area (115°E–160°W: 40°N–20°S). The top figure shows the mean monthly cloud cover (%) defined over 0.5 degrees latitude/longitude for cirrus-type clouds defined as higher than 440 hPa and a COT thinner than 23. The bottom figure shows the cloud cover for deep convective clouds higher than 440 hPa and a COT larger than 23.

Convective activity is dominant over the area south of 20°N, as seen in the bottom figure. Cirrus-type clouds indicate a more superior value surrounding the active convective area and north of 20°N. Cirrus cloud is a significant cloud type in coverage over the western tropical Pacific. This feature is consistent with the ISCCP analysis in Slide 2.

Slide 14 Cirrus clouds are an essential cloud type in studying the radiation budget. Himawari-8/9 covers the western tropical Pacific where the cirrus-type cloud is dominant. Here, we examine the diurnal variation of cirrus clouds using the high temporal (hourly, although a full disk is possible every 10 minutes) and spatial (2 km) resolution data observed by Himawari-8.

First, we show the diurnal variation of cloud cover, then for cloud optical properties. This slide shows the cloud cover (%) of thin cirrus (COT < 1.3) over the latitudinal belt of (a) 40°N–20°S, (b) 40°N–20°N, (c) 20°N–EQ, and (d) EQ–20°S. Each figure consists of four lines over 15 longitudinal belts centered at 135°E, 150°E, 165°E and 180°E.

Figure (a) shows the monthly mean value over the area of (127.5°E–142.5°E:40°N–20°S) in red, (142.5°E–157.5°E:40°N–20°S) in green, (157.5°E–172.5°E:40°N–20°S) in blue, and (172.5°E–187.5°E:40°N–20°S) in purple. The diurnal variation of cloud cover indicates similar temporal variation with the minimum at 12 LT and maximum at 21 LT. The cloud cover values are similar over the area centered at 150°E (green) and 165°E (blue), although those are smaller over the area centered at 135°E and 180°E. When we averaged the area of 40°N–20°S, the diurnal variation is similar at each longitudinal belt, even over the far-east belt, where convective activity is weak.

Similarly, (b) shows the monthly mean value of slightly narrower longitudinal belts over the northernmost area of (127.5°E–142.5°E:40°N–20°N) in red, (142.5°E–157.5°E:40°N–20°N) in green, (157.5°E–172.5°E:40°N–20°N) in blue, and (172.5°E–187.5°E:40°N–20°N) in purple. The diurnal variation of the cloud cover is slightly different compared with (a). The areas centered at 150°E (green) and 135°E (red) show similar cloud cover values with smaller amplitude compared with (a). While in the areas centered at 165°E (blue) and 180°E (purple), the cloud cover value decreased with smaller amplitude.

The diurnal variation (c) over the 20°N–EQ belts shows similar variation with (a), although the area centered at 165°E (blue) and 180°E (purple) shows an earlier minimum. The diurnal variation over the area centered at 135°E (red) in (d) is different from other areas. This area might be affected by the Australian continent. Diurnal variation of thin cirrus cloud cover indicates a minimum around noon and a maximum around 21 LT. The amplitude is smaller over the higher latitude and eastern part of the analysis area.

Slide 15 Here we study the diurnal variation of cloud optical properties of thin cirrus (COT < 1.3) and cirrus ($1.3 < \text{COT} < 3.6$) over the western tropical Pacific. Diurnal variation of cirrus cloud optical properties is one of the essential parameters in affecting the effect of cirrus cloud on radiation budget.

Kox *et al.* (2014) studied the cirrus cloud optical thickness and cloud top altitude from infrared data of the Spinning Enhanced Visible and Infrared Imager (SEVIRI) aboard the geostationary Meteosat Second Generation (MSG) satellite. They developed the neural network method for retrieving cloud optical thickness and the cloud top pressure of cirrus cloud trained by co-incident measurements of CALIOP aboard the Cloud-Aerosol Lidar and Infrared Pathfinder Satellite Observations (CALIPSO). They showed the diurnal variation of cloud optical thickness with a maximum at 15 LT and minimum at 07 LT over southern Africa, with a maximum at 16 LT and minimum at 10 LT over the Mediterranean region.

Feofilov and Stubenrauch (2019) studied the diurnal variation of high opaque, cirrus, and thin cirrus clouds using two space-borne infrared sounders, the Atmospheric InfraRed Sounder (AIRS), and the Infrared Atmospheric Sounding Inter ferrometer, which observed the Earth four times per day from 2008 to 2015. The diurnal cycle of tropical thin cirrus seems similar over land and the ocean, with a minimum in the morning (09 LT) and a maximum during the night (01 LT).

Here, we show the diurnal variation of cloud properties of thin cirrus (COT < 1.3: red) and cirrus (1.3 < COT < 3.6: green) averaged over the area of (127.5°E–187.5°E: 40°N–20°S) as a monthly mean of July 2016. (a) (top left) shows the cloud cover for thin cirrus in red and cirrus in green. Cloud cover of thin cirrus (cirrus) indicates the maximum value at 21 LT (18 LT) and minimum value at 12 LT (08 LT). Optically, thicker cirrus shows an extreme value slightly earlier.

Similarly, (b) shows cloud optical thickness that indicates the maximum value at 17 LT and minimum value at 06 LT for thin cirrus (red); however, the amplitude for cirrus (green) is too small to assign a maximum or minimum. A cloud top height of cirrus (green) shows an LT maximum of 14 LT and a minimum of 02 LT, thin cirrus (red) indicates a late evening maximum and early morning minimum, although the amplitude of thin cirrus is small. Next, (d) (bottom right) shows the diurnal variation of the cloud effective radius. Cirrus (green) shows a maximum value of around 17 LT with a minimum of 03 LT, while the thin cirrus (red) shows the maximum value at 18 LT with an ambiguous minimum.

Slide 16 In Slide 13, the cirrus-type cloud cover shows a dip along 20°N latitude. The south of the 20°N area seems to have more active convection. Thus, we study the diurnal variation of cirrus cloud properties over the place where convective activity is a substantial (127.5°E–187.5°E: 20°N–EQ) and relatively weak area (127.5°E–187.5°E: 40°N–20°N). Diurnal variation of cloud cover (see [a]), cloud optical thickness (see [b]), cloud top height in pressure (see [c]), and cloud effective radius (see [d]) are shown over two latitudinal belts for thin cirrus (COT < 1.3).

Cloud cover (percentage within 20 degrees latitudinal belt) shows larger amplitude over the southern belt area (red) than the northern belt area (green). This larger amplitude of cloud cover is affected by the convective activity over the 20°N–EQ belt. The cloud cover indicates a maximum of around 21 LT and a minimum of around 12 LT for both longitudinal belts with a slight difference. The diurnal variation of COT for thin cirrus indicates a maximum of around 18 LT and a minimum of around 6 LT over both belts. The peak of the effective radius appears at 18 LT for the southern belt and 19 LT over the northern belt. The minimum of the effective radius appears at 12 LT over both belts, although a small secondary peak with different times is seen. Although further study is required, it is interesting that the effective radius is generally larger over the northern belt than the southern belt, where convective activity is significant. The cloud top is generally higher over the southern convectively active area than the northern belt. The maximum cloud top height appears around 21 LT for both belts, with a minimum early in the morning.

Comparison of cloud properties of thin cirrus between active convective area (southern belt) and less convective area (northern belt) shows that the timing of the maximum value is close for both belts in the evening (18–21 LT). In contrast, the timing of the minimum is slightly different for each property.

Slide 17 One of the sources of cirrus cloud is deep convection. Anvil cloud consists of ice crystal spreads surrounding the core of deep convection. Deep convection occurs regardless of whether it is day or night. Thus, infrared observation is inevitable to monitor the anvil cloud. Here, we show the evolution of deep convection in terms of BTD. We define deep convection as the area covered by clouds colder than 253 K brightness temperature (~400 hPa temperature). The BTD is an excellent indicator to classify the optical thickness for colder clouds (see Slides 3 and 7).

The figure shows the BTD (T86–T12) within clouds colder than 253 K every two hours from 00 UTC to 06 UTC on 27 July 2016. The thicker cloud areas are colored purple and blue, while the thinner clouds are colored green and yellow. With BTD, we can see the relatively thinner cirrus-type clouds within the deep convection. The figure shows that the smaller BTD area (purple) is initially dominant at the earlier stage of deep convection. The larger BTD (greenish color) then becomes dominant and shrinks the size. Inoue *et al.* (2009) showed that the DC area is initially covered mainly with smaller BTD before the percentage of the larger BTD area increases with time at the mature and decaying stage.

Slide 18 Here, we show the cloud properties in terms of the evolution of deep convection for another case over Taiwan on 27 July 2016. The OCA provides us with optical properties of clouds, including cirrus clouds. The set of four graphs on the left shows the evolution of optical thickness (COT) every two hours from 07 UTC to 13 UTC (top) and the effective radius (Re) (bottom).

The northern part of the convective area is covered mainly by a larger COT (reddish) with smaller Re (blue and green) at 07 UTC, while the tiny area to the south is covered with a smaller COT (bluish) and smaller Re (blue and green). The DC size expands by merging with a smaller area at 09 UTC. The DC area is covered chiefly with larger COT (reddish), with larger Re (reddish) over the northern part. The middle DC area is covered with smaller Re (greenish). The spatial distribution of the effective radius is complicated.

At 11 UTC, the size of the DC area is almost the same as 09 UTC. However, the larger Re (reddish) area is seen only over the middle and southern parts, while smaller COT (bluish) with smaller Re (greenish) area increases over the northern part.

Finally, the DC size shrinks, covered mainly by smaller COT (bluish) and smaller Re (greenish) clouds. We may summarize that the smaller COT (blue, greenish) and smaller Re (blue, greenish) area is dominant at the initial stage of DC, while larger COT (reddish) and larger Re (reddish) is dominant at the mature stage.

Slide 19 Deep convection is generally associated with heavy rainfall. Further, some deep convection naturally merges to become long-lived larger convective areas. Eventually, the convective area disappears. A pioneering study (Byers and Braham

Jr., 1949) shows from aircraft and radar observations that convective rain occurs at the mature stage and stratiform rain occurs at the decaying stage. Thus, identifying the stage of deep convection is a good indicator of convective/stratiform rain.

Historically, a specific cold brightness temperature (253 K, 233 K, 213 K, etc.) has been used to estimate rainfall rate based on the idea that high and tall deep convective clouds are associated with heavy rainfall. However, coincident observation of cloud areas and Radar observations show that rainfall occurs not only over cold cloud areas. The advent of the Tropical Rainfall Measuring Mission (TRMM) has shown more clearly that the colder cloud area does not necessarily correspond to heavy rainfall. Thus, considering the life stage of deep convection is essential. Inoue *et al.* (2009) show some results of the rainfall rate and life stage of deep convection using the TRMM and GOES-W.

The figures on top show a 2-hourly BTD distribution within the DC (cloud colder than 253 K) area over the western part of Japan from 09 UTC to 13 UTC 19 July 2016. The figures at the bottom show the rainfall rates analyzed by Radar-AMeDAS corresponding to the top figures. The three DC areas are covered mainly by the smaller BTD (T10–T12: bluish color) at 09 UTC, especially the smallest convective area among the three DCs. The three deep convections are merged and they expand the area size at 11 UTC with a relatively wide area of greenish color. The deep convective area slightly shrinks at 13 UTC with a rather major (minor) area of greenish (bluish) color. In the context of Slide 17, we may classify the DC as: developing stage at 09 UTC, mature stage at 11 UTC, and decaying stage at 13 UTC.

The Radar-AMeDAS (Automated Meteorological Data Acquisition System), data indicates that the three rainfall areas (orange) are corresponding to the bluish cloud part of the three DC areas in the BTD image. The higher rainfall area (yellow and red) within the rainfall area (colored area) at the bottom is significant at 09 UTC compared with other times. Still, a yellow area is seen at 11 UTC with a broader area of weak rain (blue). At 13 UTC, the reddish and yellow areas are gone, and stratiform rain (relatively weak) is dominant.

To study the areal coverage of optical properties such as an optically thick cloud area within the deep convective area may be a clue to find the relationship with the rainfall rate or convective/stratiform rain.

Radar-AMeDAS is a combined rainfall map around Japan obtained from the JMA's automatic observation facility, the AMeDAS, and ground-based weather radar.

Slide 20 Summary

References

Byers, H. R. and R. R. Braham, Jr., 1948: Thunderstorm structure and circulation, J. Meteor., Vol. 5, No. 3, 71–86.

Feofilov AG, Stubenrauch CJ (2019) Diurnal variation of high-level clouds from the synergy of AIRS and IASI space-borne infrared sounders. *Atmospheric Chemistry and Physics* **19**:13957–72. https://doi.org/10.5194/acp-19-13957-2019

Hayashi M. (2018) Introduction to the computation method for cloud radiative processes and its application for the advanced Himawari imager onboard Himawari-8. *Meteorological Satellite Center Technical Note* **63**:1–38.

Inoue T. (1985) On the temperature and effective emissivity determination of semi-transparent cirrus clouds by bi-spectral measurements in the 10 μm window region. *Journal of the Meteorological Society of Japan* **63**:88–99.

Inoue T, Smith WL. (1994) The feasibility of extracting low level wind by tracing low level moisture observed with GOES-7. *Journal of Applied Meteorology* **33**:594–604.

Inoue T, Vila D, Rajendran K, Hamada A, Wu X, Machado LAT. (2009) Life cycle of deep convective systems over the eastern tropical Pacific observed by TRMM and GOES-W. *J. Meteor. Soc. Japan* **87**:381–91. doi:10.2151/jmsj.87A.381

Ishimoto H, Masuda K, Mano Y, N. A. Scott, and S. Serrar (2012) Irregularly shaped ice aggregates in optical modeling of convectively generated ice clouds. *Journal of Quantitative Spectroscopy and Radiative Transfer* **113**:632–43. https://doi.org/10.1016/j.jqsrt.2012.01.017

Kox S, Buglianro L, Ostler A. (2014) Retrieval of cirrus cloud optical thickness and top altitude from geostationary remote sensing. *Atmospheric Measurement Techniques* **7**:3233–46. doi:10.5194/amt-7-3233-2014

Rossow WB, Schiffer RA. (1999) Advances in understanding clouds from ISCCP. *Bulletin of the American Meteorological Society* **80**:2261–87.

Watts PD, Mutlow CT, Baran A.J, Zavody AM. (1998) Study on cloud properties derived from Meteosat Second Generation observations, Final Report, EUMETSAT ITT no. 97/181.

Section 3

Cloud Modeling and Model Evaluations

Lecture 11

Organized Convection Parameterization for GCMs

Mitchell W. Moncrieff

National Center for Atmospheric Research, Boulder, CO, USA

Mitchell Moncrieff is a scientist at the National Center for Atmospheric Research (NCAR), Boulder, Colorado, US. Prior to 1986, he was a tenured faculty member of the Physics Department, Imperial College, London, UK. His research interests range from cloud-to-global scale and the process-to-parameterization methodology with a focus on convective organization and its large-scale role. His approach involves nonlinear dynamics, high-resolution numerical simulation, global weather prediction analyses, and verification of the results by satellite and field-campaign measurements.

Introduction

William Rossow played a leading role in the analysis of precipitating convective systems measured by instruments flown on geostationary and low-earth-orbit satellites. He is a leading light in international research programs, notably the Global Energy and Water-cycle Exchanges (GEWEX) and the International Satellite Cloud Climatology Project (ISCCP) (Rossow and Shiffer, 1999). His analyses of the vertical structure of precipitating tropical clouds revealed key categories of precipitating convection and weather states, which are salient to a convective organization and its parameterization in global climate models (GCMs), the principal themes of this chapter.

The unification of observations and models is a grand challenge. While global-scale field campaigns are logistically and fiscally out of the question, virtual global field campaigns in the form of analyses, short-term forecasts, and subgrid-scale information from high-resolution global weather models provide unique information on physical and dynamical interactions ranging from mesoscale to global. The inaugural virtual field campaign, the Year of Tropical Convection (YOTC) coordinated by the World Weather Research Programme (WWRP) and the World Climate Research Programme (WCRP), utilized the European Centre for Medium-Range Weather Forecast (ECMWF) Integrated Forecast System (IFS) with a 25-km computational grid. Organized moist convection was a primary research theme of the YOTC project. This chapter includes results from NASA-funded collaborative research on convective organizations utilizing the YOTC database with Rossow and this author as principal investigators.

The role of water in vapor, liquid and ice phases and the dominance of convection as a transport process for water, energy, mass and momentum is unique to the Earth's atmosphere. Of the three heat transport processes (convection, radiation, conduction), only convection is sufficiently efficient to equilibrate the gradient of heating between the tropical and polar regions and between the surface and the overlying atmosphere. The heat and momentum transport processes are modulated and controlled by atmospheric dynamics. Firstly, the meridional transport of heat between the Equator and the poles by quasi-horizontal convective conveyor belts in mid-latitude cyclone/anticyclone systems is resolved by GCMs, but small-scale vertical convection has to be parameterized. Secondly, precipitating convection in the tropics is structurally diverse, and in vertically sheared environments, extensively evolves upscale into larger-scale systems. There are notable consequences for convective parameterization. Upscaling in mid-latitudes occurs when ensembles of cumulus convection organized by vertical shear evolve into mesoscale convective systems (MCSs) that can travel hundreds of kilometers and beyond. In contrast, instead of being primarily controlled by the large-scale environment, tropical convection organizes upscale into coherent structures that interact strongly with atmospheric waves.

Organized convection processes have been explored in extensive international programs such as GEWEX, satellite missions and field campaigns, and a host of individual field studies and numerical and dynamical models. In particular, the WCRP GEWEX Cloud System Study (GCSS) focused on cloud systems spanning the mesoscale, such as boundary-layer clouds, cirrus, extratropical layer cloud, and precipitating convective systems. In scientific respects, these categories are not mutually exclusive. In the GCSS, field-campaign observations were utilized to develop and validate cloud-resolving numerical models that, in turn, were utilized for parameterization development for global models (GCSS Science Team, 1993; Moncrieff *et al*., 1997).

Despite laudable advances in understanding organized processes, longstanding shortcomings in parameterization seriously compromise the fidelity of GCMs of significantly lower resolution than global weather models. Furthermore, the mean-state of a GCM is not controlled by tens of millions of measurements, mostly satellite data, assimilated every few hours into weather models. It is unlikely that the variability of weather extremes in a warmer world can be understood and projected without serious attention to organized convection.

Global weather models will resolve mesoscale organization once their grid-spacing is about 1 km, say in a decade or beyond, as dictated by computational advancements. In GCMs, convective organization will need to be parameterized for the foreseeable future, especially in probabilistic frameworks employing large ensembles that require serious computational resources. The following questions set the scene for this chapter:

- What key processes are missing from contemporary convective parameterizations?
- Can new paradigms be devised to treat such deficiencies?
- Is a minimalist proof-of-concept available?

These questions will be shown to have affirmative answers. Firstly, organized moist convection is missing in GCMs because mesoscale systems occupy the assumed gap between cumulus and the resolved scales that underpins contemporary convection parameterizations. Secondly, nonlinear dynamical models that approximate these mesoscale systems have been verified against observations and convection-resolving numerical simulations. Thirdly, the effects of a minimalist parameterization implemented in a leading GCM are reasonably consistent with field-campaign and satellite observations. The fact that slow-manifold circulations (coherent structures) of scale much larger than embedded cumulus elements are widely observed and numerically simulated gives credence to the parameterization of organized convection as *multiscale coherent structures* and the utilization of analytic *slantwise layer overturning models* as transport modules. Multiscale examples include MCSs (cumulonimbus ensembles), tropical superclusters (MCS ensembles) and the Madden–Julian Oscillation (MJO), a remarkable multiscale ensemble of cumulonimbus, MCSs, superclusters, and convectively coupled waves.

Slantwise layer overturning has distinguished transport properties. While transient small cumulus clouds mix momentum down the mean-flow velocity gradient (downgradient transport), slantwise layer overturning can transport momentum counter to the velocity gradient (upgradient transport or negative viscosity) and bolster the mean-flow kinetic energy. This behavior occurs in large-scale mid-latitude cyclone/anticyclone systems and multiscale tropical convection, albeit in different ways. In mid-latitudes, the predominately dry large-scale meridional slope convection transports momentum upgradient to sustain the jet streams. In the tropics, convective transports are primarily moist. For example, MCSs that are aligned perpendicular to the vertical shear transport the momentum upgradient and sustain the large-scale flow against dissipation, especially in the tropics. Other categories of convective organization transport momentum downgradient, e.g., 3-D shear-parallel rainbands embedded in the Intertropical Convergence Zone (ITCZ) throughout the year and the Mei-Yu front during the Asian summer monsoon. It follows that global weather and climate are significantly affected by upscale transport mechanisms not taken into account by contemporary cumulus parameterizations in GCMs.

The ubiquity of organized convection has been amply demonstrated by observational, numerical and theoretical approaches. Numerical simulation of organized convection began in the 1970s utilizing cloud-resolving models (CRMs) with non-hydrostatic dynamical cores, O (1 km) grids, and O(10–100 km) computational domains. Dynamical analogs were developed in concert. Exponential advances in computer technology over the past half-century enabled CRMs with global computational domains that explicitly treat organized convection around the globe (see [Satoh *et al.*, 2019] for recent progress). Superparameterization, whereby 2-D CRMs replace convection and planetary boundary layer parameterizations, is another explicit approach, albeit with enormous computational overhead. Besides being an intellectual challenge in its own right, organized convection parameterization is identified as a scientific imperative, especially for long GCM integrations, large ensembles, and Earth System Models.

The new paradigm for organized convection parameterization, summarized in this chapter and dubbed "MCSP", is based on the dynamical principles of multiscale coherent structures and slantwise layer overturning. A coherent structure is a fundamental fluid property where ensembles of transient small-scale cumulus evolve into long-lived larger-scale systems, e.g., upscale evolution of a cumulus ensemble into MCSs. The MCSP acronym has a dual definition with cross-scale implications. Firstly, MCSP identifies *Mesoscale Convective System Parameterization* in the form of slantwise layer overturning driven by horizontal pressure gradients generated by an ensemble of embedded cumulonimbus in different stages of their life cycle. Secondly, an approach that utilizes the remarkable self-similar properties of slantwise layer overturning across scales (cumulonimbus, squall-lines, MCSs, tropical super-clusters, and convectively coupled equatorial waves) identifies the generalized form of MCSP, dubbed *Multiscale Coherent System Parameterization.*

The dynamical foundation and explicit numerical simulation of organized precipitating convection are described by Moncrieff and Green (1972), Moncrieff and Miller (1976), Thorpe *et al.* (1982), Moncrieff (1981; 1992; 2004; 2010), and Grabowski and Moncrieff (2001). International programmatic outreach on convective organization is addressed by Moncrieff *et al.* (2007; 2012), Brunet *et al.* (2010), Waliser *et al.* (2012), and Moncrieff and Waliser (2015). Finally, Moncrieff *et al.* (2017) designed a minimalist form of MCSP implemented in the NCAR CAM. The following slides summarize key aspects outlined in this introduction.

Acknowledgments

NASA-funded collaborative research with the City College of New York "Diagnostic Analysis & Cloud-System Modeling of Organized Tropical Convection in the YOTC-ECMWF Global Database". The National Center for Atmospheric Research is sponsored by the National Research Foundation.

References

Brunet G, Shapiro M, Hoskins B, *et al.* (2010) Collaboration of the weather and climate communities to advance subseasonal-to-seasonal prediction. *Bulletin of the American Meteorological Society* **91**:1397–412.

GEWEX Cloud System Science Team (A Betts, Browning KA, Jonas PR, Kershaw R, Manton M, Mason PJ, Miller M, Moncrieff MW, Sundqvist H, Tao WK, Tiedtke M). (1993) The GEWEX Cloud System Study (GCSS). *Bulletin of the American Meteorological Society* **73**:387–99.

Grabowski WW, Moncrieff MW. (2001) Largescale organization of tropical convection in two-dimensional explicit numerical simulations. *Quarterly Journal of the Royal Meteorological Society* **127**:445–68.

Majda, AJ, Stechmann SN. (2009) The skeleton of tropical intraseasonal oscillations. *Proceedings of the National Academy of Sciences* **106**:8417–22.

Moncrieff MW. (1981) A theory of organized steady convection and its transport properties. *Quarterly Journal of the Royal Meteorological Society* **107**:29–50.

Moncrieff MW. (1992) Organized convective systems: Archetypal models, mass and momentum flux theory, and parameterization. *Quarterly Journal of the Royal Meteorological Society* **118**:819–50.

Moncrieff MW. (2004) Analytic representation of the large-scale organization of tropical convection. *Journal of the Atmospheric Sciences* **6**:1521–38.

Moncrieff MW. (2010) The multiscale organization of moist convection and the intersection of weather and climate. In *Climate Dynamics: Why Does Climate Vary*? Eds. Sun D-Z, Bryan F. American Geophysical Union, Washington, DC, pp. 3–26. doi:10.1029/2008GM000838

Moncrieff MW, Green JSA. (1972) The propagation and transfer properties of steady convective overturning in shear. *Quarterly Journal of the Royal Meteorological Society* **98**:336–52.

Moncrieff MW, Krueger SK, Gregory D, *et al.* (1997) GEWEX Cloud System Study (GCSS) Working Group 4: Precipitating Convective Cloud Systems. *Bulletin of the American Meteorological Society* **78**:831–45.

Moncrieff MW, Miller MJ. (1976) The dynamics and simulation of tropical cumulonimbus and squall-lines. *Quarterly Journal of the Royal Meteorological Society* **102**:373–94. doi:10.1002/qj.49710243208373-394

Moncrieff MW, Liu C, Bogenschutz P. (2017) Simulation, modeling and dynamically based parameterization of organized tropical convection for global climate models. *Journal of the Atmospheric Sciences* **74**:1363–80. doi:10.1175/JAS-D-16-0166.1

Moncrieff MW, Waliser DE. (2015) Organized convection and the YOTC project. In *Seamless Prediction of the Earth-System: From Minutes to Months*. Eds. Brunet G, Jones S, Ruti, PM. WMO-No. 1156, World Meteorological Organization, pp. 283–309. http://library.wmo.int/pmb_ged/wmo_1156_en.pdf

Moncrieff MW, Waliser DE, Miller MJ, *et al.* (2012) Multiscale convective organization and the YOTC virtual global field campaign. *Bulletin of the American Meteorological Society* **93**:1171–87. doi:10.1175/BAMS-D-11-00233.1

Moncrieff MW, Shapiro M, Slingo J, Molteni F. (2007) Collaborative research at the intersection of weather and climate. *WMO Bulletin*, **56**:204–11.

Moncrieff MW, Klinker E. (1997) Organized convective systems in the tropical western Pacific as a process in general circulation models. *Quarterly Journal of the Royal Meteorological Society* **123**:805–28.

Rossow WB, Schiffer RA. (1999) Advances in understanding clouds from ISCCP. *Bulletin of the American Meteorological Society* **80**:2261–87.

Satoh M, Stevens B, Judt F, *et al.* (2019) Global cloud-resolving models. *Current Climate Change Reports*. https://doi.org/10.1007/s40641-019-00131-0

Thorpe AJ, Miller MJ, Moncrieff MW. (1982) Two-dimensional convection in non-constant shear: A model of mid-latitude squall lines. *Quarterly Journal of the Royal Meteorological Society* **108**:739–62.

Waliser DE, Moncrieff MW, Burridge D, *et al.* (2012) The "Year" of Tropical Convection (May 2008 to April 2010): Climate variability and weather highlights. *Bulletin of the American Meteorological Society* **93**:1189–218. doi:10.1175/2011BAMS3095.1

Slide 1

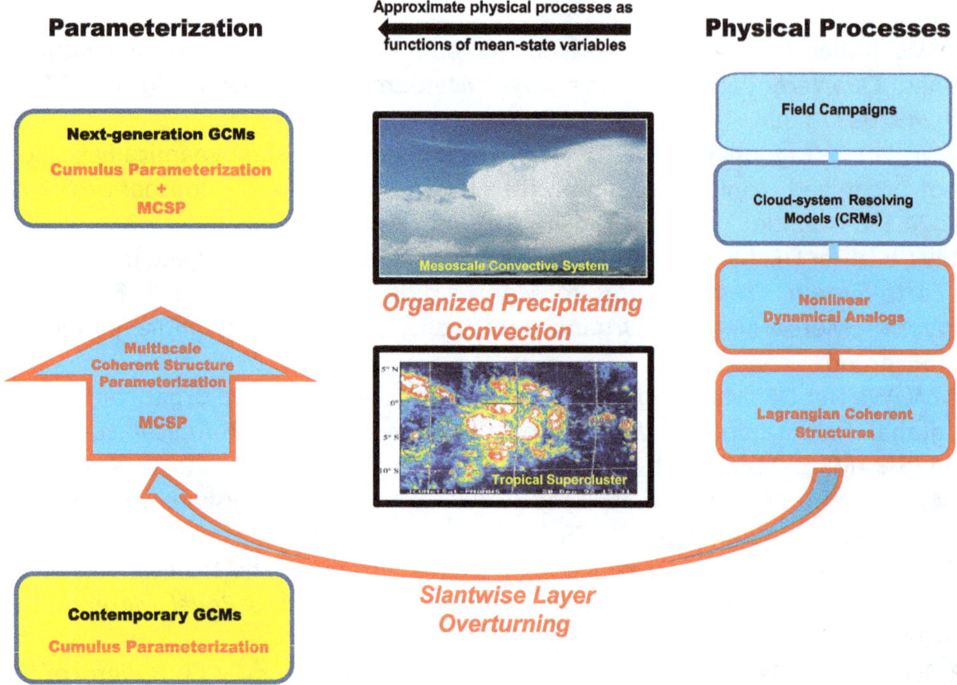
Organized Convection Parameterization for GCMs
Parameterization
Approximate physical processes as functions of mean-state variables
Physical Processes
Next-generation GCMs
Cumulus Parameterization
+
MCSP
Mesoscale Convective System
Organized Precipitating Convection
Field Campaigns
Cloud-system Resolving Models (CRMs)
Nonlinear Dynamical Analogs
Lagrangian Coherent Structures
Multiscale Coherent Structure Parameterization
MCSP
Tropical Supercluster
Slantwise Layer Overturning
Contemporary GCMs
Cumulus Parameterization

Slide 2

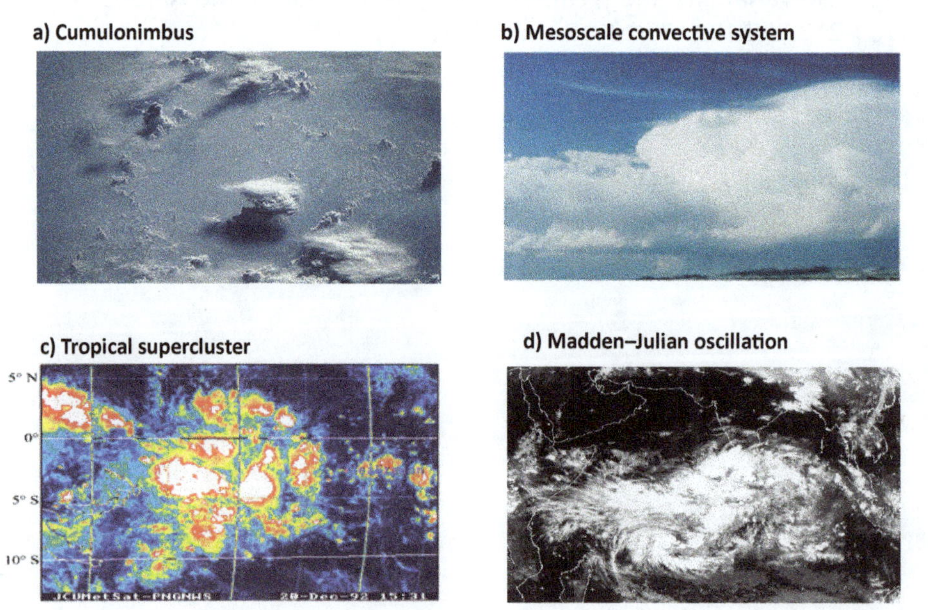
Coherent Multiscale Character of Convective Organization
a) Cumulonimbus
b) Mesoscale convective system
c) Tropical supercluster
d) Madden–Julian oscillation
5° N
0°
5° S
10° S

Slide 3

Global Ubiquity of Mesoscale Convective Organization

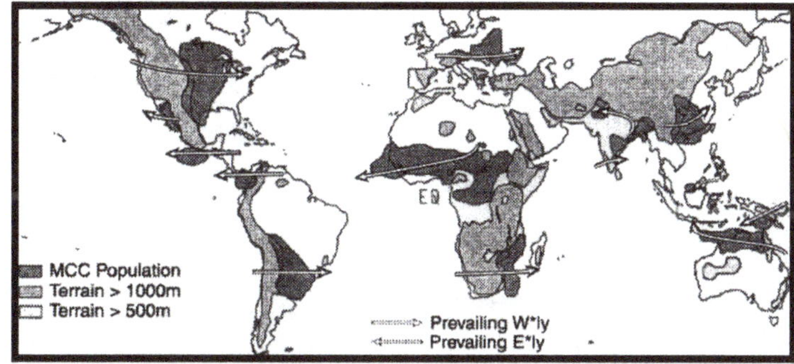

a) Global distribution of mesoscale convective complexes downstream of elevated terrain moving in the prevailing wind direction. From (Laing and Fritsch, 1997)

b) Mesoscale convective systems downstream of U.S. Continental Divide

Slide 4

Mesoscale Convective Systems in the Tropics

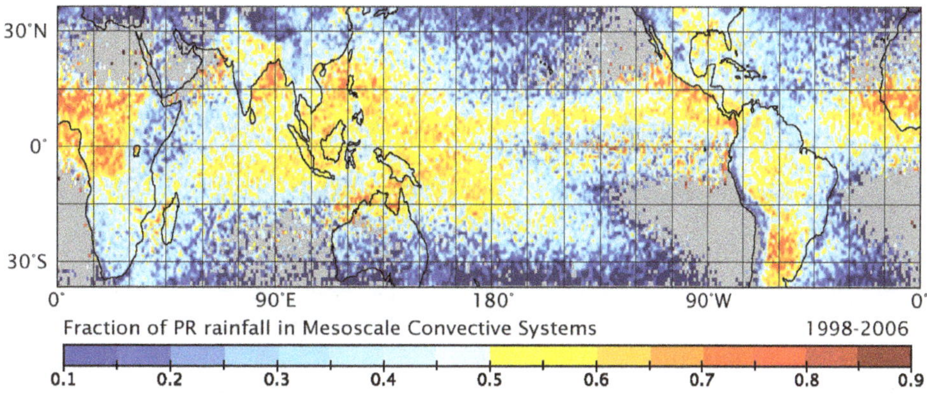

Tropical Rainfall Measuring Mission (TRMM) Precipitation Radar show MCSs:

- **Provide up to 75% of tropical rainfall**
- **Populate regions where GCMs have significant bias**
- **Are directly associated with large-scale meteorological features**

MCSs are missing from GCMs: not resolved, not parameterized

Slide 5

Zonal Propagation of Systems by Latitude and Season

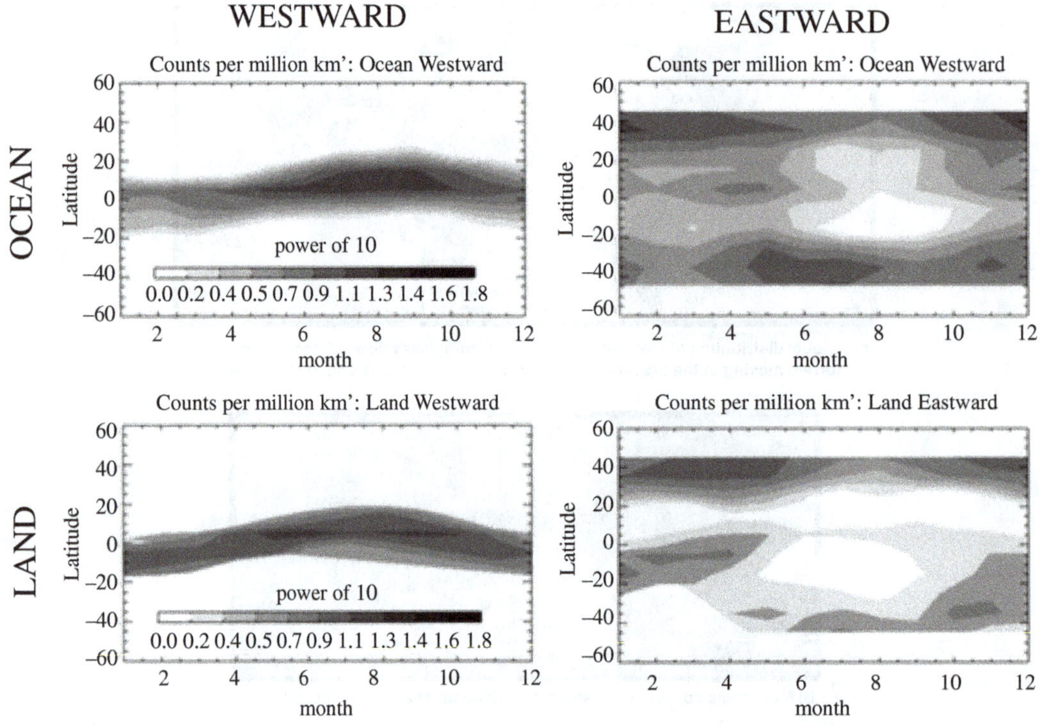

Slide 6

Hierarchical Convective Organization of MJO

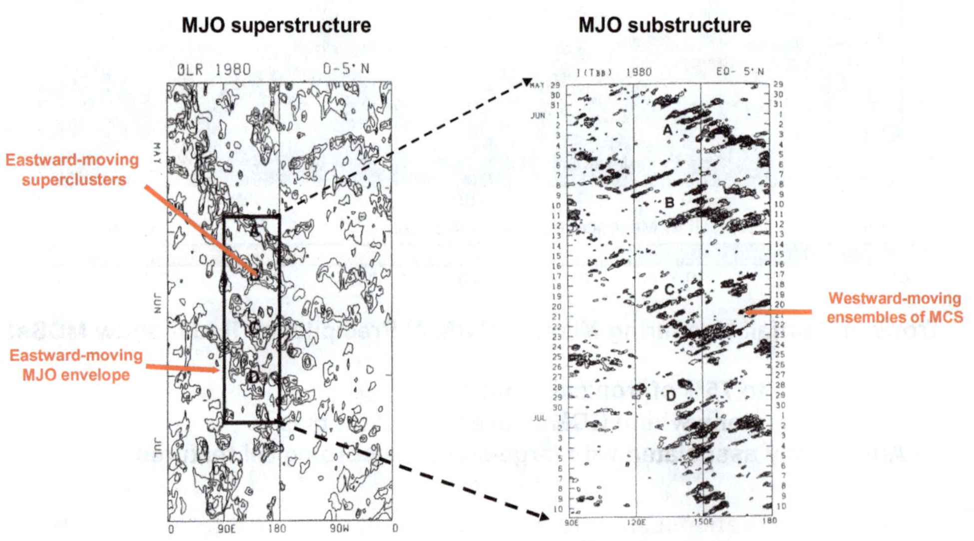

Slide 7

Self-similarity of MCSs and Tropical Superclusters

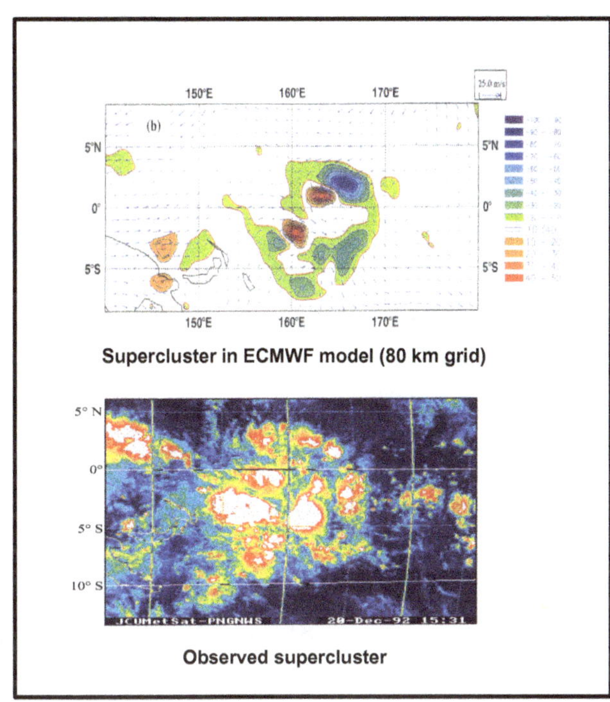

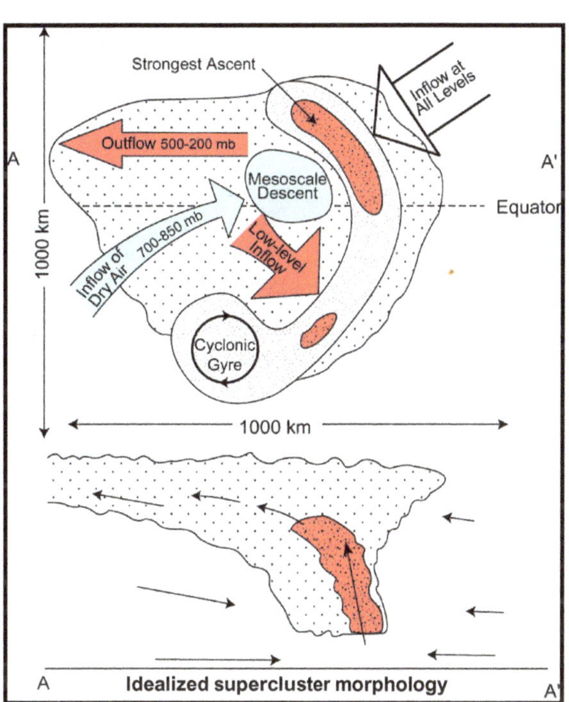

Slide 8

Multiscale Coherent Structure and Slantwise Layer Overturning Paradigms

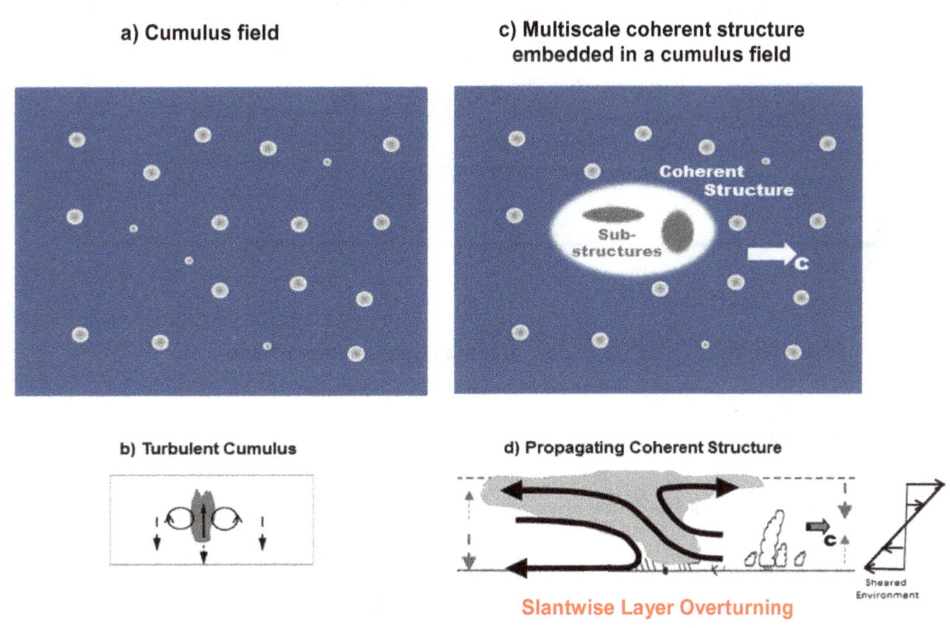

Slide 9

Lagrangian dynamics

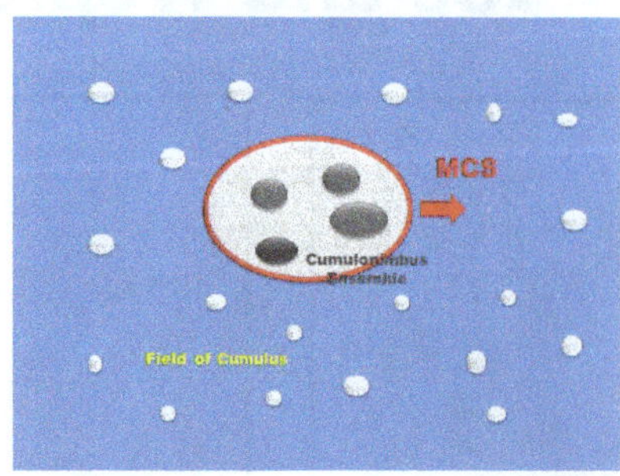

- **Lagrangian coherent structures: Nonlinear propagating systems steady in the system-relative frame-of-reference, i.e.,** $D/_{Dt} = \partial/_{\partial x} + \partial/_{\partial y} + \partial/_{\partial z}$
- **Transform Eulerian equations of momentum, thermodynamics, energy and vorticity into Lagrangian form, i.e.,** $\frac{D}{Dt} F_i = 0$
- **Integrate the transformed equations along trajectories (ψ) to give Lagrangian conserved quantities, i.e.,** $F_i = C_i(\psi)$
- **Conserved quantities define nonlinear dynamical models of organized convection**

Slide 10

Slantwise layer overturning model

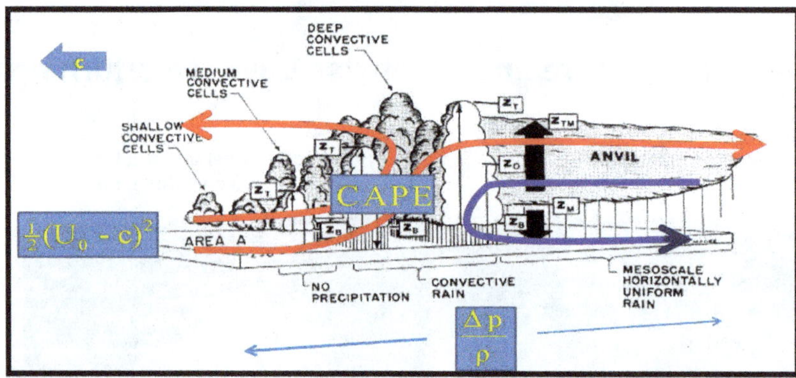

- **Dimensionless variables that control the regime of organization:**

Bernoulli Number $E = \frac{\Delta p/\rho}{\frac{1}{2}(U_0 - c)^2}$; **Richardson Number** $R = \frac{CAPE}{\frac{1}{2}(U_0 - c)^2}$

- **Two-dimensional slantwise layer overturning models satisfy the vorticity equation**

$$\nabla^2 \psi = G(\psi) + \int_{z_0}^{z} \left(\frac{\partial F}{\partial \psi}\right)_{z'} dz'$$

where G (ψ) and F(ψ, z) are environmental shear and parcel buoyancy, respectively, $z_0(\psi)$ is the inflow height, and (z - z_0) the vertical displacement of trajectory ψ

Slide 11

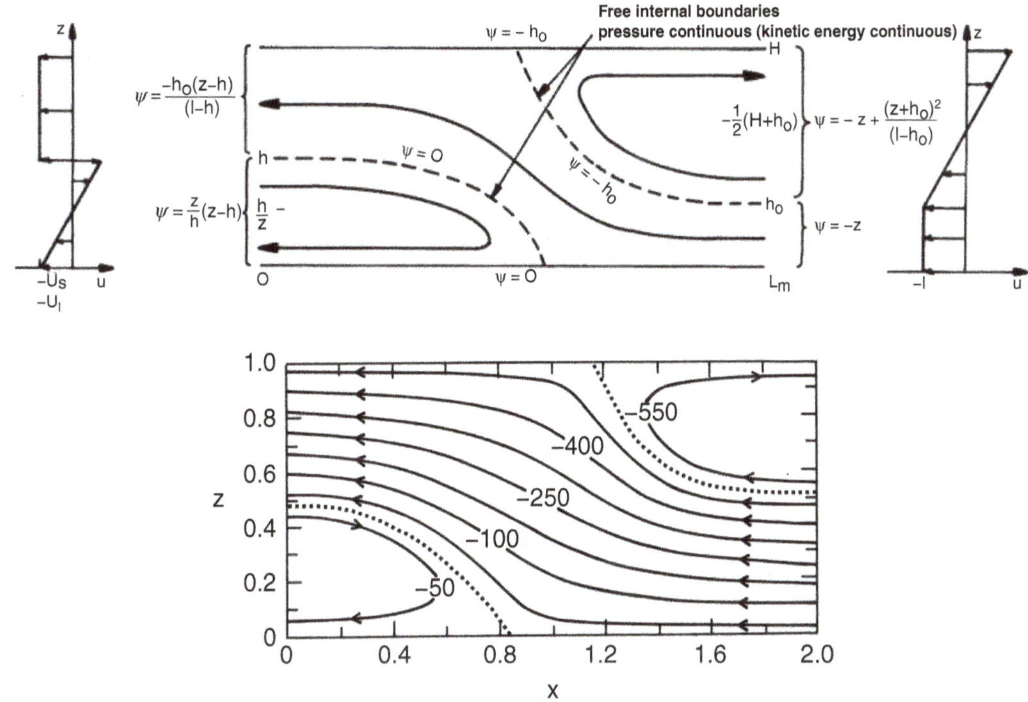

Slide 12

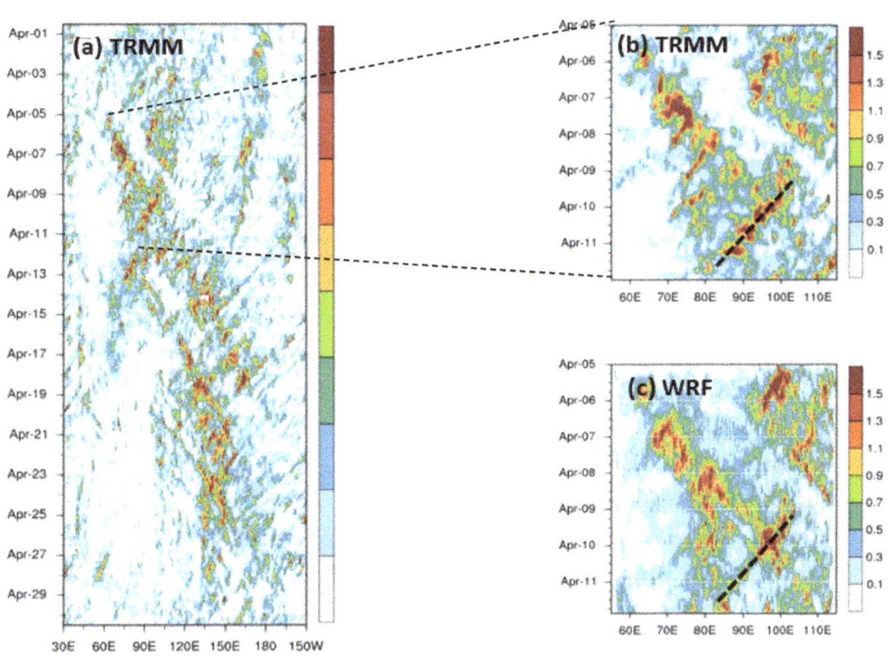

Slide 13

Simulated Slantwise Layer Overturning

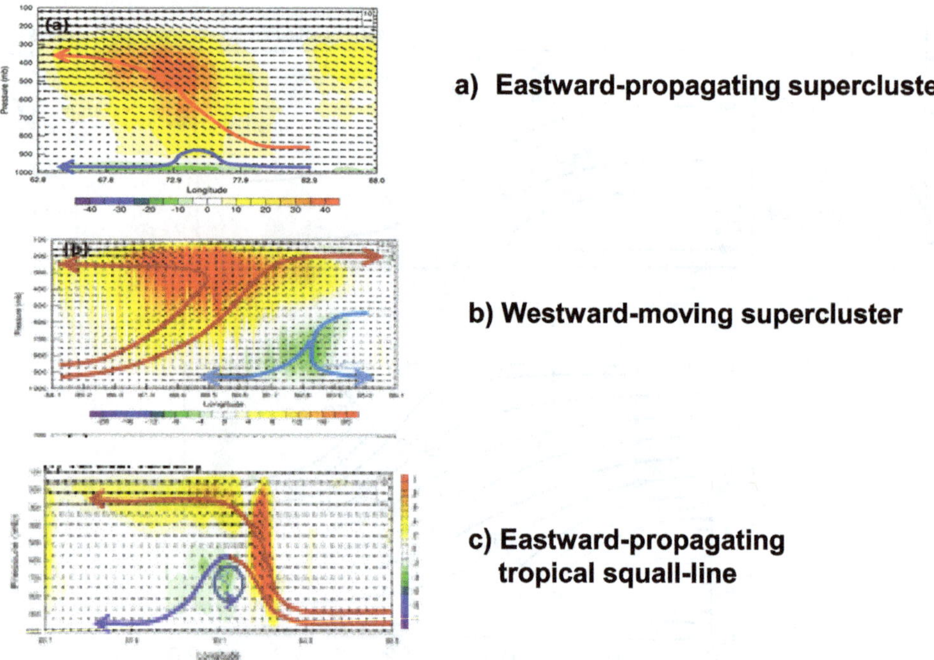

Slide 14

Prototype MCSP

- **Proof-of-concept for MCSP is the principal objective**
- **Represent the mesoscale rates-of-change in canonical form:**
  - **1st and/or 2nd baroclinic (top-heavy) mesoscale heating rate**
  - **1st baroclinic acceleration by mesoscale momentum transport**
- **Complementary strategy: Add mesoscale slantwise overturning to cumulus transport**

$$\left[\frac{\delta}{\delta t}\right]_{total} = \left[\frac{\delta}{\delta t}\right]_{cumulus} + \left[\frac{\delta}{\delta t}\right]_{mesoscale}$$

- **Large-scale effects of organized convection are unambiguously measured as differences between GCM runs with and without the mesoscale rates-of-change**
- **Minimal computational overhead**

Slide 15

Minimalist Heat and Momentum Tendencies

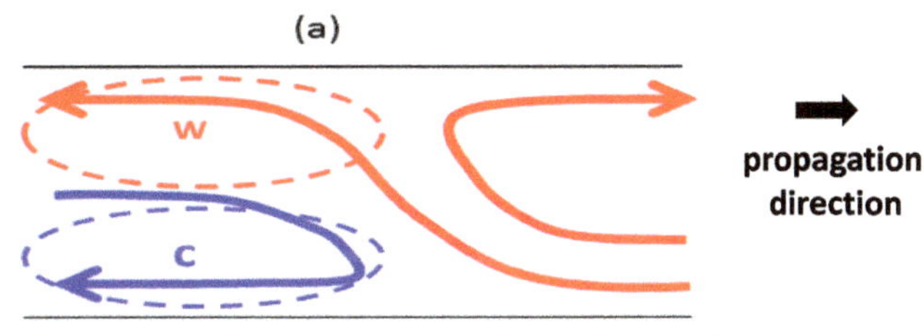

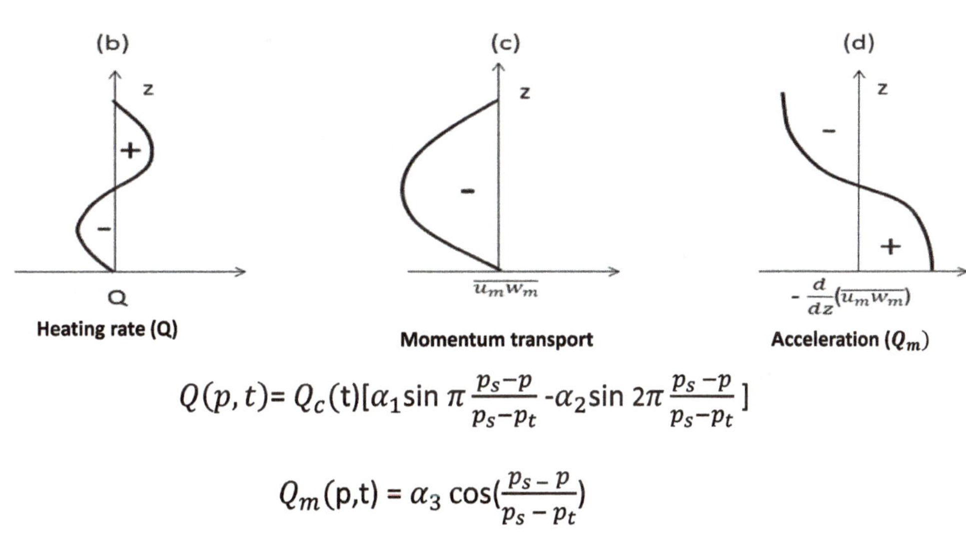

$$Q(p,t) = Q_c(t)[\alpha_1 \sin \pi \frac{p_s - p}{p_s - p_t} - \alpha_2 \sin 2\pi \frac{p_s - p}{p_s - p_t}]$$

$$Q_m(\mathrm{p,t}) = \alpha_3 \cos(\frac{p_s - p}{p_s - p_t})$$

Slide 16

MCSP Effects on the Large-scale Distribution of Tropical Convection
(8-year Averages)

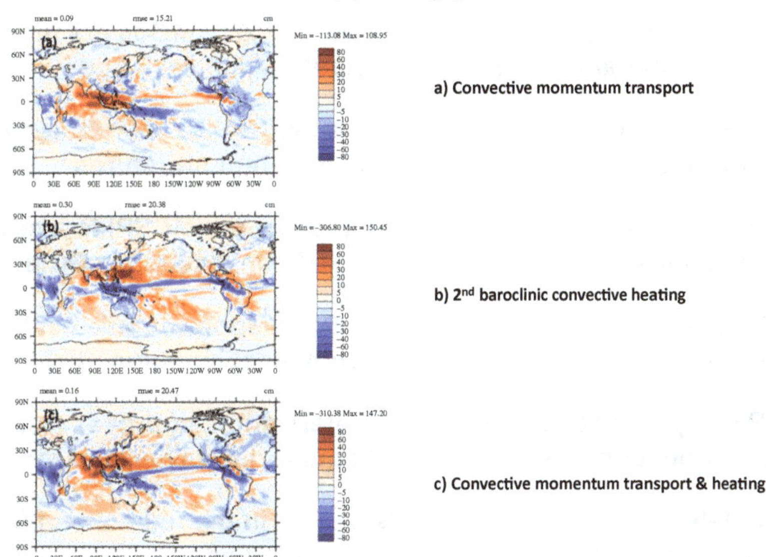

Slide 17

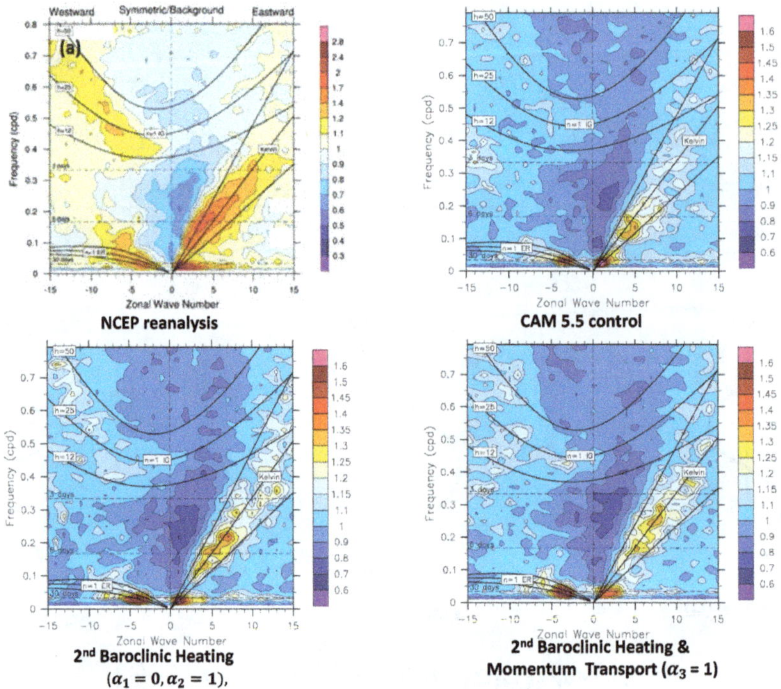

Slide 18

Take-away Messages

- **Mesoscale convective systems (MCSs) provide over half of the total precipitation in the Tropics**
- **Neither adequately resolved nor parameterized, MCSs are missing from contemporary GCMs, an important dynamical issue**
- **MCSs and tropical superclusters have self-similar morphology, collectively considered as multiscale coherent structures**
- **Multiscale coherent structure parameterization (MCSP) is a new dynamical-based treatment of organized convection for GCMs**
- **Slantwise layer overturning, the MCSP transport module exchanges entire tropospheric layers, distinct from local cumulus mixing**
- **MCSP complements cumulus parameterization, improves tropical precipitation, convectively coupled equatorial waves, and the MJO**
- **Computationally efficient, MCSP is usable in the full suite of GCMs, including Earth System Models and Ensembles**

Slide Captions

Slide 1 Organized convection parameterization for global climate models

This slide presents the anatomy of a new dynamically based paradigm for parameterizing convective organization in global models. The specific mechanisms are described in more detail in the following slides. The fact that organized convection is neither resolved nor parameterized, and hence missing from contemporary GCMs, has wide-ranging implications, including the distribution, type and intensity of precipitation, effects on tropical waves, interaction with the land and ocean, the representation of severe weather, and various effects on climate variability and change. Heat and momentum tendencies for organized convection parameterization are provided by nonlinear dynamical models of slantwise overturning. A minimalist form of this new paradigm implemented in a leading GCM acts as a proof-of-concept. Note that cumulus parameterization and planetary boundary layer parameterization are retained partly because they have positive feedback to the mean-state, including its recovery from the organized convection, such as evaporatively cooled downdrafts and surface exchange. Overall, the addition of organized convection processes completes contemporary convective parameterizations.

Slide 2 Coherent multiscale character of convective organization

The multiscale character of moist convective organization is illustrated by the following images: (a) O(10 km) cumulonimbus in sheared environments are structurally distinct and longer-lived than unorganized transient cumulus, (b) O(100 km) MCSs are ensembles of cumulonimbus that evolve into larger and longer-lived mesoscale convective complexes (MCCs) in sheared environments, notably over continents, (c) O(1,000 km) tropical superclusters as an embedded hierarchy of cumulonimbus, and MCSs occur extensively over the Indian Ocean, tropical western Pacific, and the Indonesian maritime continent regions, notably in the convectively active phase of the MJO, and (d) as the leading mode of tropical intraseasonal variability, the O(10,000 km) MJO is a remarkable multiscale envelope of cumulus, MCSs, superclusters, and convectively coupled waves with a lifetime of months. MJO initiation can either be internal to the tropics or a product of extratropical excitation by global Rossby waves and wintertime cold surges from the Asian continent. Because propagation is a ubiquitous feature, organized convection affects a significant fraction of the Earth's atmosphere and its interaction with the land and ocean.

Slide 3 Global ubiquity of mesoscale convective systems

Geostationary satellite observations show a high percentage of rainfall from propagating precipitation systems on scales upward from O(100 km).

(a) Using brightness temperatures obtained from satellite-based measurements as a proxy for deep convection, Laing and Fritsch (1997) revealed the ubiquitous relationship between MCCs, orography, and the vertical wind shear associated with the mid-latitude and subtropical jet streams. Over continents, MCCs initiate as MCSs in the lee of the US Continental Divide, Ethiopian Highlands, Andes, Tibetan plateau, and India's Ghats.

(b) MCSs and MCCs over the US develop in the lee of the Continental Divide in sheared environments during the US's warm season (April–September) and commonly travel eastward for O(1,000 km). Fritsch *et al.* (1986) concluded that such events are "very likely the most prolific precipitation producers in the United States" and "may be a crucial precipitation-producing deterrent to drought".

Slide 4 Mesoscale convective systems in the tropics

Radar data from the Tropical Rainfall Measuring Mission (TRMM) satellite distinguishes mesoscale convective organization from ordinary convection. A significant fraction of tropical rain falls from the stratiform cloud decks of the MCS. It is more than a coincidence that MCSs are co-located with meteorological phenomena that seriously challenge GCMs, e.g., the ITCZ, equatorial Africa, intraseasonal variability, and the world's monsoons. Up to 75% of the precipitation in those regions originates from ensembles of heavily precipitating cumulonimbus embedded in MCSs and their stratiform regions (Tao and Moncrieff, 2009). Consequently, MCSs significantly affect the tropical circulation and evince a crucial parameterization challenge. Based on analytic models of organized convection in shear-flow over the US continent and by using a regional mesoscale model, Moncrieff and Liu (2006) devised a hybrid predictor-corrector parameterization for the MCS whereby cumulus parameterization provides convective heating (predictor), which is then adjusted by top-heavy convective heating (corrector) to represent the stratiform region of the MCS. This work motivated the Cao and Zhang (2017) parameterization study.

Slide 5 Zonal propagation of systems by latitude and season

Despite being infrequent by comparison to ordinary convection, precipitating convection organized into MCSs and larger-scale convective systems dominate the atmospheric heating profiles and provide a large fraction of tropical precipitation. The dynamical character of MCSs directly affects the thermodynamic and kinetic energy spectrum and the horizontal pressure gradient of the tropical atmosphere across space and time. Being important precipitation systems in their own right, it is unreasonable to expect that organized convection can be satisfactorily understood in terms of long-term averages that combine different categories of convection (Moncrieff, 2019). The tracking of MCSs identifies about an equal distribution of eastward and westward-moving systems. Applied to high-resolution satellite data, the tracking procedure usefully quantifies the internal structure of mesoscale systems (see [Vant-Hull *et al.*, 2016] for more information).

The analysis of satellite measurements shows that westward-moving systems are closely coupled to eastward-moving Kelvin waves and the MJO, as follows.

Slide 6 Hierarchical convective organization of the MJO

The Nakazawa (1988) analyses of geostationary satellite images, an original depiction of the multiscale character of the convectively active phase of the MJO, reveal intriguing structural characteristics. The eastward-moving convective envelope of the MJO is a family of eastward-moving superclusters that, in turn, contain westward-moving MCSs. Note that the resolution of these images is insufficient to show the cumulonimbus substructure of the MCSs. The presence of multiscale propagating features in the MJO, at odds with contemporary cumulus parameterization with entraining plumes as the transport module, motivated the development of the MCSP dynamical-based parameterization of convective organization.

Slide 7 Self-similarity of MCSs and tropical superclusters

Documentation of the multiscale convective organization in the MJO was a highlight of the internationally coordinated 1992–1993 Tropical Ocean Global Atmosphere Coupled Ocean Atmosphere Response Experiment (TOGA COARE). Numerical experiments with the ECMWF global medium-range weather forecasting system at a grid spacing of 80 km quantify the interaction between tropical superclusters and the December 1992 MJO during TOGA COARE (Moncrieff and Klinker, 1997). The simulated MJO consists of three interacting scales of organization: cumulonimbus, MCSs and superclusters. It is remarkable that the 80-km-grid IFS permitted explicit superclusters and embedded MCSs (top-left diagram). The effects of the supercluster on the synoptic and large-scale environment dominated the convective parameterization in the IFS. The mesoscale momentum transport by the supercluster is consistent with the Moncrieff (1992) archetypal model (Slide 11). In other words, MCSs and superclusters have self-similar properties and hence underpin the generalized form of MCSP.

Slide 8 Multiscale coherent structure and slantwise layer overturning paradigms

The multiscale coherent structure concept is central to the dynamical-based parameterization of organized convection.

(a) Contemporary cumulus parameterization treats a field of cumulus as an ensemble of transient elements.

(b) An entraining plume with a vertical updraft and downdraft is the local transport module, and compensating descent maintains mass-balance in the grid-volume.

(c) The propagating coherent structure represents either an MCS with embedded cumulonimbus (upper picture, Slide 1) or a tropical supercluster having embedded MCSs (lower picture, Slide 1), evincing the self-similarity of these systems (Slide 7).

(d) The coherent structure features three interactive flow branches: (1) a rearward-slanting jump-like updraft, (2) an overturning updraft, and (3) an overturning updraft that approximates the MCS and supercluster transports as slantwise layer overturning. Slides 9–11 set the coherent structure and slantwise layer overturning concepts into parameterization context.

Slide 9 Lagrangian convective dynamics

Transformation of coordinates from an Eulerian to a system-relative Lagrangian frame-of-reference has the fundamental advantage that organized convective transports are derivable from far-field inflow and outflow variables without detailed knowledge of near-field complexity. The rearward slant means the warm up-branch overlies the cool down-branch and enables precipitation to efficiently fall into, evaporate, and sustain mesoscale descent. The long-lived character of slantwise layer overturning is consistent with a steadiness assumption ($\partial/\partial t = 0$), thereby simplifying the Lagrangian derivative to $D/Dt = \partial/\partial x + \partial/\partial y + \partial/\partial z$. The convective heating rate (Q) is a separable function of vertical velocity, $Q = \mathrm{w}\Gamma$, where w is the vertical velocity of air-parcels within the system and $\Gamma(z)$ is the moist adiabatic lapse rate. The transformed equations can then be written in analytically integrable form, $DF_i/Dt = 0$, where F_i represents mass, thermodynamics, total energy, vorticity, and horizontal momentum plus the horizontal pressure gradient. Lagrangian integration along trajectories yields a complete set of conserved quantities: the mathematical foundation of the slantwise layer overturning models. Vertical exchange by organized slantwise overturning is clearly distinguished from local lateral mixing by cumulus in contemporary convective parameterization and a key missing process in contemporary cumulus parameterizations. See Moncrieff and Green (1972) for a detailed derivation of the Lagrangian formulation.

Slide 10 Slantwise layer overturning model

The concept of slantwise layer overturning, the transport module for MCSP, has long been observed around the world. Here is a quotation from Hamilton and Archbold (1945) regarding tropical squall lines over Nigeria, Africa: "The principle of an induced circulation which moves the air initially at the surface to near the top of the [convective] cell implies that the air initially near the top of the cell moves down to the surface and stays there. In fact, the air between the two levels is simply turned upside down". Mesoscale-organized convection was a highlight of the 1960–1970 era of field campaigns and cloud-resolving numerical simulations. Observations of the US squall lines led Newton and Newton (1959) to interlink vertical shear, pressure gradient, and tilted updrafts. The Browning and Ludlam (1962) radar analysis of a convective storm over southern England revealed a "well-organized and steady flow pattern". Analytic dynamical models set these observations onto a firm theoretical foundation and identified distinct regimes of organization (Moncrieff, 1981). The building blocks of MCS are ensembles of propagating cumulonimbus in different stages of their life cycle, and the mesoscale slantwise overturning is driven by the attendant horizontal pressure gradients (Lafore and Moncrieff, 1989). Although most intense in strongly sheared environments over mid-latitude continents, similar circulations occur in weaker shear over the tropical oceans. The colored trajectories overlying the standard MCS model depict slantwise layer overturning generated by the cumulonimbus ensemble. The analytic models involve three categories of energy: (i) convective available potential energy (CAPE), (ii) inflow kinetic energy AKE$\frac{1}{2}(U_0 - c)^2$ where and c are inflow and propagation speeds, respectively, and (iii) work done by the cross-system pressure gradient, WPG $= \frac{\Delta p}{\rho}$. These quantities define the convective Richardson number, R = CAPE/AKE, and Bernoulli number, E = WPG/AKE, which strongly control the regime of convective organization. In two dimensions, slantwise layer overturning is governed by a vorticity equation (Moncrieff and Green, 1972) whose solution provides near and far-field structure and transport properties. This approach also addresses the far-field properties of 3-D convective organization (Moncrieff and Miller, 1976; Lane and Moncrieff, 2015; Moncrieff and Lane, 2015; Liu and Moncrieff, 2017). Note that Houze (2018) recorded impressive progress in mesoscale convective system research over the past hundred years.

Slide 11 Archetypal dynamical model

A likely reason why organized convection parameterization has received unduly little attention in parameterization is its perceived calls for simplicity. The Lagrangian formulation retains nonlinearity yet minimalizes complexity in the form of the Moncrieff (1992) archetypal model. There are three regimes of slantwise overturning, two of which are present in Slide 13. The upper-limit E = 1 is a basically hydraulic system where a jump-like updraft replaces the mesoscale downdraft in near-saturated environments (Liu and Moncrieff, 2017). E = 0 with a jump updraft, an overturning updraft, and a mesoscale downdraft is the canonical classical MCS. The lower-limit E = –8/9 is a propagating wave-like structure with an inflow entirely from the front of the system.

Slide 12 MJO simulation

The convectively active phase of the MJO features three cloud categories: lower-tropospheric cumulus congestus, deep convection, and upper-tropospheric stratiform cloud (Johnson *et al.*, 1999).

(a) The 8–11 April 2009 MJO in the YOTC-ECMWF database was simulated using the NCAR Weather and Forecasting Model (WRF) with a nested computational domain. The MJO initiated over the western Indian Ocean partly disintegrated as it traversed the Indonesian maritime continent and briefly reorganized in the western Pacific prior to a rapid demise.

(b) TRMM observations show the MJO is an embedded family of eastward-propagating Kelvin-wave-like systems that, in turn, contain embedded westward-propagating MCS-like features. This multiscale organization resembles the Nakazawa (1988) satellite analysis of an MJO (Slide 6). The WRF with a very large outer domain and 4-km grid captured the full MJO.

(c) The inner domain with a 1.3-km grid simulated the mesoscale organization described below.

Slide 13 Simulated slantwise layer overturning

The three archetypal forms of slantwise overturning represented by the analytic models have the rearward-slant associated with observed tropical systems.

(a) The westward relative-inflow at all levels shows the eastward-moving supercluster as a solitary-wave-like system.

(b) The westward-moving supercluster also slants rearward, so the warm ascent overlies the cool descent resulting in a second-baroclinic profile of convective heating.

(c) Four eastward propagating tropical squall lines have inflow at all levels, indicating solitary-wave-like behavior similar to category (a).

Slide 14 Minimalist form of MCSP

Extensively observed over land and ocean around the world, MCSs have been simulated by cloud-resolving models. Note that MCSs populate tropical regions where GCMs display serious bias. Slides 8–10 show Lagrangian nonlinear dynamical models for MCSP that provide organized heat and momentum transports as a tendency-forcing for the resolved-scale equations. Recall that in contrast to superparameterization, MCSP complements rather than replaces cumulus parameterization and boundary-layer turbulence parameterization. These processes are retained in recognition of their beneficial effects on the mean-state of GCMs. Consequently, for the first time, the differences between GCM experiments with and without MCSP unambiguously measure the large-scale effects of organized convection. Field campaigns, satellite measurements, and numerical simulations confirm the relevance of slantwise layer overturning to MCSP.

Slide 15 Minimalist heat and momentum tendencies

The heating and momentum tendencies for organized convection are mathematically complicated solutions of nonlinear dynamical models (Slide 10), whereas these tendencies are simplified in the minimalist MCSP to sine and cosine functions (Moncrieff *et al.*, 2017).

(a) The system-relative airflow indicates the stratiform warm slantwise overturning branch that overlies the cool mesoscale descent sustained by the evaporation of rain.

(b) The vertical profile of the heating rate tendency is top-heavy with the maximum/minimum values in the upper/lower troposphere.

(c) The rearward slant of the airflow means that the horizontal (u_m) and vertical (w_m) velocity perturbations are positively correlated, so their product, the vertical transport of zonal momentum, is positive.

(d) It follows that the zonal acceleration, the negative of the vertical gradient of momentum transport, has a cosine profile. The sign of the momentum transport is opposite to the propagation vector. For example, an eastward-propagating system provides eastward momentum transport with maximum value in the middle of the convective layer, in agreement with field measurements (LeMone and Moncrieff, 1993). The kinetic energy generation of organized convection can be comparable to the convective available potential energy (Wu and Moncrieff, 1996). Furthermore, the scale-invariance of slantwise overturning is evinced by the similar airflow patterns of O(10 km) cumulonimbus, O(100 km) MCS, and O(1,000 km) superclusters consistent with Moncrieff and Klinker (1997), see Slide 7. The principal effects of the minimalist MCSP on the large-scale distribution of tropical precipitation and tropical waves are summarized below.

Slide 16 MCSP effects on the large-scale distribution of tropical precipitation

(a) Large-scale effects of momentum transport and (b) second-baroclinic heating by MCSP in CAM occur over the Indian Ocean, tropical western Pacific, Indonesian maritime continent, South Pacific Convergence Zone, and equatorial Africa, which demonstrates the upscale effects of organized convection. Precipitation over the maritime continent is reduced by momentum transport but increased by top-heavy heating. The dipole-like pattern of precipitation tends to be oriented east-west for the convective momentum transport parameterization and north-south for the convective heating. In most GCMs, ITCZ precipitation tends to be excessive and spatially continuous. The top-heavy heating extends the South Pacific Convergence Zone. The effects of convective heating are bigger than the convective momentum transport, although that depends on the relative values of the α parameters (Slide 15). The decreased precipitation by top-heavy heating in the ITCZ region is consistent with the existence of new meso-synoptic dynamical instability mechanisms (Khouider and Moncrieff, 2015). The possibility that the spontaneous generation of upscale evolution and large-scale coherence by MCSP involves undiscovered scale-selection principles has been suggested (Moncrieff, 2004, Secs. 8(b) and (c); Moncrieff and Waliser, 2015, Secs. 15.4.1 and 15.4.3; Khouider and Moncrieff, 2015), and calls for further study.

Slide 17 MCSP effects on convectively coupled equatorial waves and the MJO

The wavenumber-frequency diagram (Wheeler and Kiladis, 1999) of precipitation across the 15°S–15°N equatorial belt shows MCSP beneficially affects equatorial waves.

(a) The National Centers for Environmental Prediction (NCEP) reanalysis features the MJO, Kelvin waves, westward Rossby waves, and inertio-gravity waves.

(b) While the CAM control simulation has a reasonable MJO-like system, the Kelvin waves are weak. The improved MJO, compared to earlier versions of CAM, may stem from the moister lower-troposphere due to the higher entrainment rate. The reason why wavenumber 1 of the MJO differs from wavenumbers 1–5 in the NCEP reanalysis suggests it is a moisture mode that resembles the Majda and Stechmann (2009) "skeleton model" of intraseasonal oscillations.

(c) and (d) The MCSP CAM runs have positive values of α_3 to focus on eastward-moving meso-convective systems, and the MJO is consistent with (Biello *et al.*, 2007) and (Khouider and Han, 2013). Momentum transport and top-heavy convective heating both add power in the wavenumber 2–4 range, consistent with the representation of MCSs and superclusters by MCSP. While 8-year-old CAM runs are too short to provide meaningful statistical information, they are a reasonable proof-of-concept of the upscale effects of organized convection on tropical precipitation and convectively coupled waves.

Slide 18 Take-away messages

These messages summarize the main results of the minimalist form of MCSP implemented in a state-of-the-art GCM, which have beneficial effects on equatorial-wave modes and MJO, and the global precipitation distribution is reasonably consistent with TRMM data.

References

Biello J, Majda A, Moncrieff MW. (2007) Meridional momentum flux and superrotation in the multiscale IPESD MJO model. *Journal of the Atmospheric Sciences* **64**:1636–51.

Browning KA, Ludlam FH. (1962) Airflow in convective storms. *Quarterly Journal of the Royal Meteorological Society* **88**:117–35.

Cao G, Zhang GJ. (2017) Role of vertical structure of convective heating in MJO simulation in NCAR CAM5.3. *Journal of Climate* **30**:7423–39.

Fritsch JM, Kane RJ, Chelius CR. (1986) The contribution of mesoscale convective weather systems to the warm-season precipitation in the United States. *Journal of Applied Meteorology and Climatology* **25**:1333–45.

Hamilton RA, Archbold JW. (1945) Meteorology of Nigeria and adjacent territory. *Quarterly Journal of the Royal Meteorological Society* **71**:231–62.

Houze RA. (2018) 100 years of research on mesoscale convective systems. *Meteorological Monographs* **59**:17.1–17.54. https://doi.org/10.1175/AMSMONOGRAPHS-D-18-0001.1

Johnson RH, Rickenbach TM, Rutledge SA, *et al.* (1999) Trimodal characteristics of tropical convection. *Journal of Climate* **12**:2397–418.

Khouider B, Han Y. (2013) Simulation of convectively coupled waves using WRF: A framework for assessing the effects of mesoscales on synoptic scales. *Theoretical and Computational Fluid Dynamics* **27**:437–89.

Khouider B, Moncrieff MW. (2015) Organized convection parameterization for the ITCZ. *Journal of the Atmospheric Sciences* **72**:3073–96. http://dx.doi.org/10.1175/JAS-D-15-0006.1

Lafore J-P, Moncrieff MW. (1989) A numerical investigation of the organization and interaction of the convective and stratiform regions of tropical squall lines. *Journal of the Atmospheric Sciences* **46**:521–44.

Laing AG, Fritsch JM. (1997) The global population of mesoscale convective complexes. *Quarterly Journal of the Royal Meteorological Society* **123**:389–485.

Lane TP, Moncrieff MW. (2015) Long-lived convective systems in a low-convective inhibition environment: Part I: Upshear propagation. *Journal of the Atmospheric Sciences* **72**:4297–318. doi:10.1175/2010JAS3418.1

LeMone MA, Moncrieff MW. (1993) Momentum and mass transport by convective bands: Comparisons vof highly idealized dynamical models to observations. *Journal of the Atmospheric Sciences* **51**:281–305.

Liu C, Moncrieff MW. (2017) Shear-parallel mesoscale convective systems in a low-inhibition mei-yu front environment. *Journal of the Atmospheric Sciences* **74**:4213–28. doi:10.1175/JAS-D-17-0121.1

Majda, AJ, Stechmann SN. (2009) The skeleton of tropical intraseasonal oscillations. *Proceedings of the National Academy of Sciences* **106**:8417–22.

Moncrieff MW. (1981) A theory of organized steady convection and its transport properties. *Quarterly Journal of the Royal Meteorological Society* **107**:29–50.

Moncrieff MW. (1992) Organized convective systems: Archetypal models, mass and momentum flux theory, and parameterization. *Quarterly Journal of the Royal Meteorological Society* **118**:819–50.

Moncrieff MW. (2004) Analytic representation of the large-scale organization of tropical convection. *Journal of the Atmospheric Sciences* **6**:1521–38.

Moncrieff MW. (2019) Toward a dynamical foundation for organized convection parameterization in GCMs. *Geophysical Research Letters* **46**. https://doi.org/10.1029/2019GL085316

Moncrieff MW, Green JSA. (1972) The propagation and transfer properties of steady convective overturning in shear. *Quarterly Journal of the Royal Meteorological Society* **98**:336–52.

Moncrieff MW, Klinker E. (1997) Organized convective systems in the tropical western Pacific as a process in general circulation models. *Quarterly Journal of the Royal Meteorological Society* **123**:805–828.

Moncrieff MW, Lane TP. (2015) Long-lived convective systems in a low-convective inhibition environment. Part II: Downshear propagation. *Journal of the Atmospheric Sciences* **72**:4319–36. doi:10.1175/JAS-D-15-0074.1

Moncrieff MW, Liu C. (2006) Representing convective organization in prediction models by a hybrid strategy. *Journal of the Atmospheric Sciences* **63**:3404–20.

Moncrieff MW, Miller MJ. (1976) The dynamics and simulation of tropical cumulonimbus and squall-lines. *Quarterly Journal of the Royal Meteorological Society* **102**:373–94. doi:10.1002/qj.49710243208373-394

Moncrieff MW, Waliser DE. (2015) Organized convection and the YOTC project. In *Seamless Prediction of the Earth-System: From Minutes to Months*. Eds. Brunet G, Jones S, Ruti, PM. WMO-No. 1156, World Meteorological Organization, pp. 283–309. http://library.wmo.int/pmb_ged/wmo_1156_en.pdf

Moncrieff MW, Liu C, Bogenschutz P. (2017) Simulation, modeling and dynamically based parameterization of organized tropical convection for global climate models. *Journal of the Atmospheric Sciences* **74**:1363–80. doi:10.1175/JAS-D-16-0166.1

Nakazawa T. (1988) Tropical superclusters within intraseasonal variations over the western Pacific. *Journal of the Meteorological Society of Japan* **66**:823–39.

Newton CW, Newton HR. (1959) Dynamical interactions between large convective clouds and the environment with vertical shear. *Journal of Meteorology* **16**:483–96.

Tao W-K, Moncrieff MW. (2009) Multiscale cloud system modeling, *Reviews of Geophysics* **47**:RG4002. doi:10.1029/2008RG000276

Vant-Hull B, Rossow W, Pearl C. (2016) Global comparisons of regional life cycle properties and motion of multiday convective systems: Tropical and midlatitude land and ocean. *Journal of Climate* **29**:5837–58.

Wheeler M, Kiladis GN. (1999) Convectively coupled equatorial waves: Analysis of clouds and temperature in the wave-number-frequency domain. *Journal of the Atmospheric Sciences* **56**:874–99.

Wu X, Moncrieff MW. (1996) Collective effects of organized convection and their approximation in general circulation models. *Journal of the Atmospheric Sciences* **53**:1477–95.

LECTURE 12

Cloud Updrafts, Climate Forcing and Climate Sensitivity

Leo Donner

Leo Donner GFDL/NOAA, Princeton University, USA

Leo Donner is a physical scientist at the NOAA Geophysical Fluid Dynamics Laboratory and a lecturer at Princeton University. His research focuses on the development of climate models, with emphasis on clouds and convection in these models.

A defining characteristic of Bill Rossow's leadership in using satellite observations has been its emphasis on tackling the most challenging questions on clouds in the climate system. Two fundamental issues in climate change — cloud feedbacks with associated implications for climate sensitivity and climate forcing through cloud-aerosol interactions — have stubbornly resisted resolution over a period of decades. In this chapter, I will examine cloud updrafts, largely not considered in relation to these problems, for possible clues they may provide. The chapter will culminate by looking, as Bill has often done, at the role that satellite observations could play in providing a critical observational basis.

Uncertainties in both anthropogenic climate forcing and the sensitivity of climate to these forcings limit the extent to which climate projections can meet critical societal needs. The observed climate transition from pre-industrial to present times depends simultaneously on climate forcing, sensitivity, and natural variability, precluding the determination of any of these three elements based on the historical record alone. In the future, greenhouse gases will likely continue to accumulate, while this will be less so for aerosols, with their shorter lifetimes and probable limitation through controls on air pollution. Climate sensitivity will increasingly shape projections of future climate change, as it applies to net forcing dominated increasingly by greenhouse gases with relatively less masking by aerosols.

The Intergovernmental Panel on Climate Change Fifth Assessment Report estimates anthropogenic radiative forcing (for 2011 relative to 1750) associated with aerosols to lie between -1.9 and -0.1 Wm^{-2} with medium confidence. Cloud "adjustments" due to anthropogenic changes in aerosols are estimated with low confidence to be in a range from -1.3 to -0.1 Wm^{-2} and are the largest single source of uncertainty in radiative forcing. The corresponding radiative forcing for well-mixed greenhouse gases (carbon dioxide, methane, nitrous oxide and halocarbons) is between 2.2 and 3.8 Wm^{-2} with high or very high confidence (IPCC, 2013). Forster *et al.* (2013) and Kiehl (2007) have indeed shown that climate models participating in Climate Model Intercomparison Project Phases 3 and 5 have an adjusted radiative forcing ranging from just under 1 Wm^{-2} to just over 2.5 Wm^{-2}. The subset of these climate models that successfully simulated pre-industrial to present-day global-mean temperature increase did so with offsets between net adjusted forcings and equilibrium climate sensitivities. Sensitivities in this subset ranged down from about 4.8 to 2.0 K for the doubling of carbon dioxide with inversely varying net forcing ranging up from about 1.0 to 2.0 Wm^{-2}. These results suggest that the understanding of pre-industrial to present-day climate change can be explained by a range of anthropogenic forcings with a corresponding (inversely related) range of climate sensitivities. The historical record of global-mean temperature change can only constrain forcing and sensitivity within these ranges, and even this interpretation requires the assumption that climate models capture natural variability reasonably.

A possible path toward reducing uncertainty is (1) to identify processes in climate models that control anthropogenic forcing and sensitivity and (2) to identify observable characteristics of these processes as subsets of what are often referred to as "emergent constraints" (Klein and Hall, 2015). This chapter suggests that cloud updraft speeds comprise such constraints for both forcing and sensitivity. Until recently, updraft speeds have received scant observational attention and their possible relevance for climate forcing and sensitivity has not been emphasized. The connection between anthropogenic climate forcing by aerosols, through their interactions with clouds, emerges directly from the well-established theory of activation of cloud particles. Updraft speeds, through their control on supersaturation, are among the major factors determining cloud drop number concentration. The cloud drop number concentration, for a given quantity of condensate, is directly related to the sizes of cloud particles, which, in turn, are central to cloud radiation and microphysics. Feingold (2003) and McFiggans *et al.* (2006) show that aerosol number concentration, the width of the aerosol size distribution, and updraft speed are the three most important properties regulating effective drop sizes and number concentrations in liquid clouds for a large range of aerosol and atmospheric conditions. For homogeneous aerosol freezing, Kay and Wood (2008) show that in some regimes updraft speed is the primary control on ice crystal number concentrations. The implication for modeling aerosol forcing in climate models is clear, i.e., physically realistic representation of cloud-aerosol processes requires realistic updraft speeds at the scale of activation. In climate models, these updraft speeds are often not resolved and must be parameterized, e.g., in convective and stratocumulus cloud systems. Even resolved vertical velocities require careful consideration regarding their consistency with aerosol and microphysical processes. As Donner *et al.* (2016) point out, vertical velocities resolved by dynamical cores in climate models may scale nonlinearly with resolution. Observed vertical velocities, averaged over large scales, also exhibit nonlinear dependence on the horizontal averaging scale.

Updraft speeds may also provide an observable constraint related to climate sensitivity, but the connection is more nuanced than for aerosol climate forcing. A number of climate models exhibit a strong dependence of their climate sensitivity on the value chosen for convective entrainment, e.g., (Zhao, 2014; Stainforth *et al.*, 2005; Sanderson *et al.*, 2010; Klocke *et al.*, 2011). Direct measurement of entrainment is challenging, but, using the plume assumptions for convection commonly employed in climate models, updraft speeds and entrainment are closely related (Masunaga and Luo, 2016).

Although cumulus parameterizations in earlier generations of climate models often simulated convective mass fluxes without decomposing those fluxes into vertical velocities and areas, some parameterizations do provide subgrid distributions of convective updraft speeds, e.g., (Donner *et al.*, 2011). These parameterized updraft speeds can be compared with retrievals from radars and profilers (Donner *et al.*, 2016).

Extending the domain of observed updraft speeds from localized field campaigns is a major challenge. Consistent with the theme of Bill Rossow's expansive use of satellites to view clouds on a global scale, I conclude by noting innovative efforts to observe cloud updrafts using a suite of satellite observations (Masunaga and Luo, 2016). Compositing observations from CloudSat, CALIPSO and TRMM allow estimating changes in cloud-top properties, from which plume ascent rates and vertical velocities can be inferred.

In summary, vertical velocities at both resolved and unresolved scales have received little attention in the development of climate models. Accurately simulated vertical velocities in climate models and appropriate treatment of their scaling properties could narrow uncertainty in climate forcing and sensitivity. An appropriate celebration of Bill Rossow's contributions to the innovative use of satellite observations is the prospect of continued innovation in using these observations and interfacing them with model development.

References

Donner LJ, O'Brien TA, Rieger D, *et al.* (2016) Are atmospheric updrafts a key to unlocking climate forcing and sensitivity? *Atmospheric Chemistry and Physics* **16**:12983–192. doi:10.5194/acp–16–12983–2016

Donner LJ, Wyman BL, Hemler RS, *et al.* (2011) The dynamical core, physical parameterizations, and basic simulation characteristics of the atmospheric component of the GFDL global coupled model CM3. *Journal of Climate* **24**:3484–519. doi:10.1175/2011JCLI3955.1

Feingold G. (2003) Modeling of the first indirect effect: Analysis of measurement requirements. *Geophysical Research Letters* **30**:1997. doi:10.1029/2003GL017967

Forster PM, Andrews T, Good P, *et al.* (2013) Evaluating adjusted forcing and model spread for historical and future scenarios in the CMIP5 generation of climate models. *Journal of Geophysical Research: Atmospheres Research Letters* **118**:1139–50. https://doi.org/10.1002/jgrd.50174

IPPC. (2013) Summary for policymakers. In *Climate Change 2013: The Physical Science Basis. Contribution of Working Group I to the Fifth Assessment Report of the Intergovernmental Panel on Climate Change*. Eds. Stocker TF, Qin D, Platter G-K, *et al.* Cambridge University Press, Cambridge/New York.

Kay J, Wood R. (2008) Timescale analysis of aerosol sensitivity during homogeneous freezing and implications for upper tropospheric water vapor budgets. *Geophysical Research Letters* **35**:L10809. doi:10.1029/2007GL032628

Kiehl JT. (2007) Twentieth century climate model response and climate sensitivity. *Geophysical Research Letters* **34**:L22 710. doi:10.1029/2007GL031383

Klein S, Hall A. (2015) Emergent constraints for cloud feedbacks. *Current Climate Change Reports* **1**, 276–87. https://doi.org/10.1007/s40641–015–0027–1.

Klocke D, Pincus R, Quaas J. (2011) On constraining estimates of climate sensitivity with present-day observations through model weighting. *Journal of Climate* **24**:6092–99. doi:10.1175/2011JCLI4193.1

Masunaga H, Luo Z. (2016) Convective and large-scale mass flux profiles over tropical oceans determined from synergistic analysis of a suite of satellite observations. *Journal of Geophysical Research: Atmospheres* **121**. doi:10.1002/2016JD024753

McFiggans, G., Artaxo P, Baltensperger U, *et al.* (2006) The effect of physical and chemical aerosol properties on warm cloud droplet activation. *Atmospheric Chemistry and Physics* **6**:2593–649. doi:10.5194/acp–6–2593–2006

Sanderson B, Shell K, Ingram W. (2010) Climate feedbacks determined using radiative kernels in a multi-thousand member ensemble of AOGCMs. *Climate Dynamics* **35**:1219–36.

Stainforth, D, Aina T, Christensen C, *et al.* (2005) Uncertainty in predictions of the climate response to rising levels of greenhouse gases. *Nature* **433**:403–06.

Zhao M. (2014) An investigation of the connections among convection, clouds, and climate sensitivity in a global climate model. *Journal of Climate* **27**:1845–62. doi:10.1175/JCLI–D–13–00145.1

Slide 1

Cloud Updrafts, Climate Forcing, and Climate Sensitivity

Leo Donner
GFDL/NOAA, Princeton University

Study of Cloud and Water Processes in Weather and Climate through Satellite Observations

Slide 2

IPCC AR5 Estimates Total Aerosol Forcing to Be -0.9 [-1.9 to -0.1] Wm^{-2}

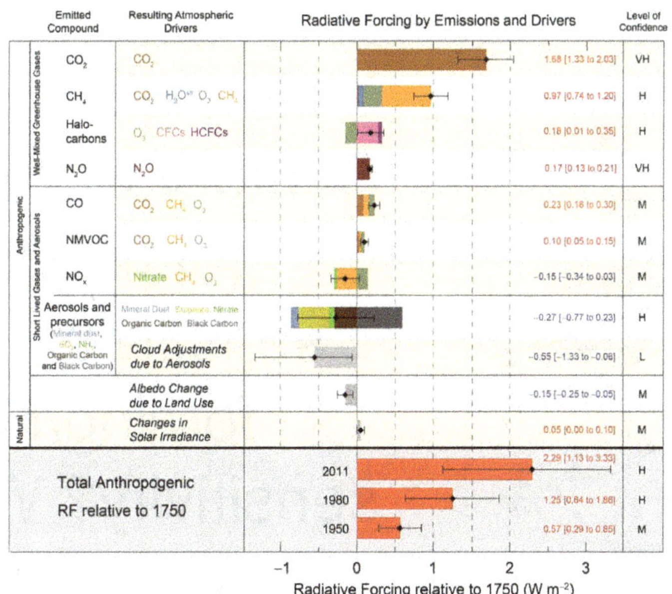

Figure SPM.5, p. 14 in IPCC, 2013: Climate Change 2013: *The Physical Science Basis. Contribution of Working Group I to the Fifth Assessment Report of the Intergovernmental Panel on Climate Change* [Stocker, T.F., D. Qin, G.-K. Plattner, M. Tignor, S.K. Allen, J. Boschung, A. Nauels, Y. Xia, V. Bex, and P.M. Midgley (eds.)]. Cambridge University Press, Cambridge, United Kingdom, and New York, NY, USA, 1535 pp.

Slide 3

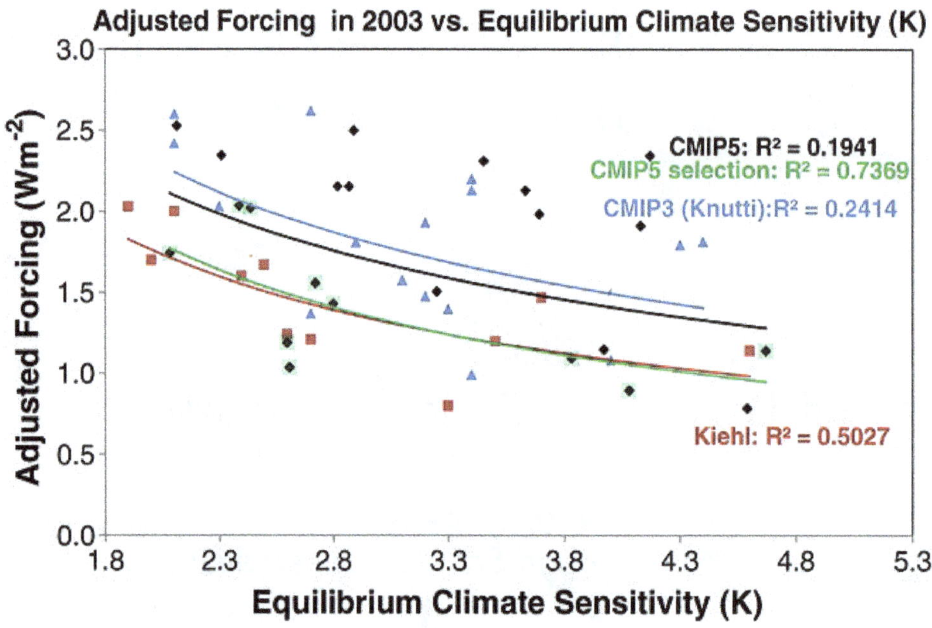

Forster, *et al.* (2013) *Journal of Geophysical Research* **118**:1139–50.
doi:10.1002/jgrd.50174
http://onlinelibrary.wiley.com/doi/10.1002/jgrd.50174/full#jgrd50174-fig-0007

Slide 4

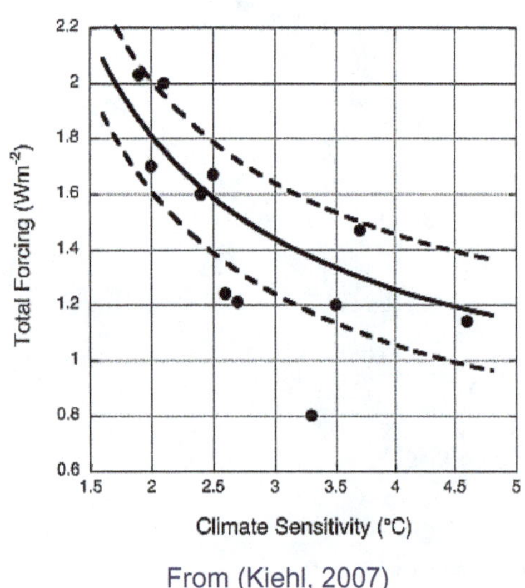

From (Kiehl, 2007)

How did the 20th Century warm? High forcing/low sensitivity or low forcing/high sensitivity? Why is it important?

Slide 5

Climate Forcing and Cloud Updrafts

Forcing from interactions between clouds and human-produced aerosols is a key uncertainty in current climate models. Cloud dynamics, cloud-scale updraft speeds, in particular, are a major control on this forcing.

Slide 6

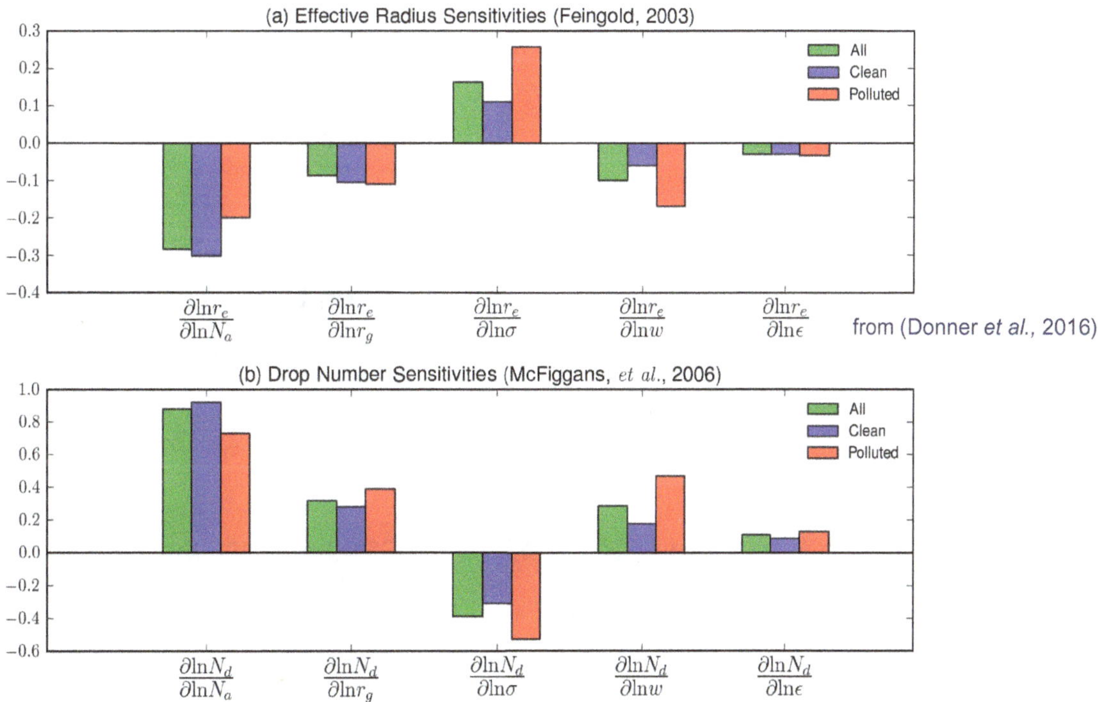

Droplet effective radius and drop number concentration are important controls on radiative and microphysical properties of clouds. Variations in aerosol concentrations, aerosol median size, breadth of aerosol size distribution σ, updraft speed, and ammonium sulfate fraction ϵ produce variations in drop size and number.

Slide 7

Aerosol Sensitivity

$$\eta_a \equiv \frac{d(\ln N_i)}{d\ \ln(N_a)}$$

N_i (cm^{-3}) contoured as a function of updraft velocity (V) and aerosol concentration (N_a).
Colors indicate the aerosol sensitivity parameter (η_α).

r_{a_dry}= 0.2 µm (mono-disperse)
α_i (deposition coefficient) = 0.1;
T_0 = -50°C, P_0 = 250 hPa;

Courtesy of Xiaohong Liu, U. Wyoming

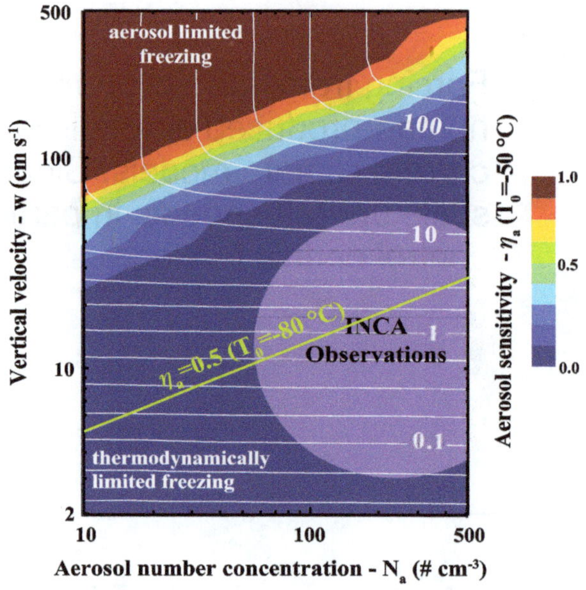

from (Kay and Wood. 2008)

Slide 8

Dependence of Climate Sensitivity on Convective Entrainment (Zhao, 2014)

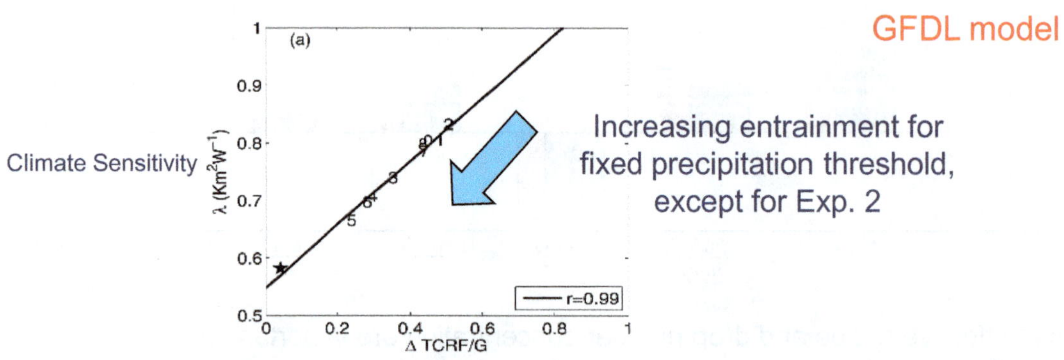

Change in Cloud-Radiative Effect, Normalized by Energy-Balance Change

Climate sensitivity dependence on entrainment also shown by Stainforth *et al.* (2005), Sanderson *et al.* (2010) and for shallow but not deep convection (Klocke *et al.*, 2011).

Figure from Zhao, M., 2014: An investigation of the connections among convection, clouds, and climate sensitivity in a global climate model. *Journal of Climate*, **27** doi:10.1175JCLI-D-13-00145.1.

Slide 9

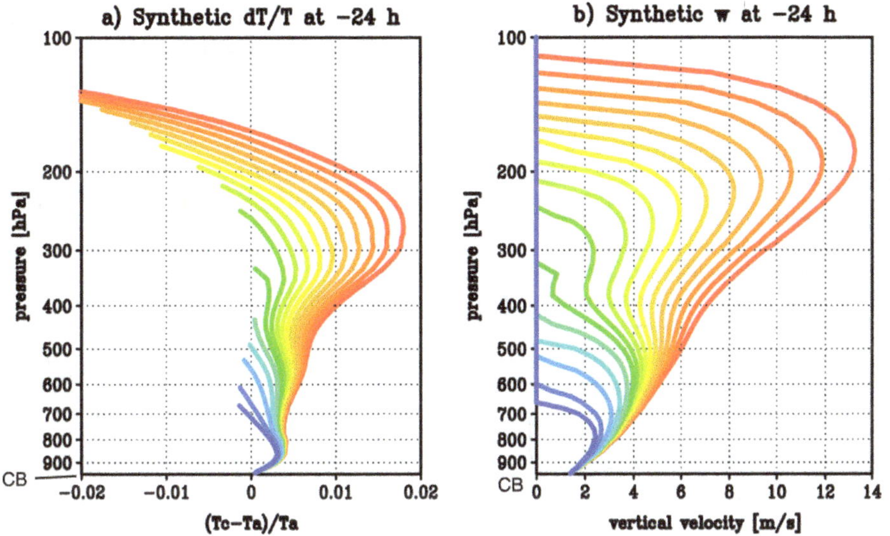

Plume entrainment rates from 0 to 0.4 km^{-1} as red goes to blue.

Journal of Geophysical Research: Atmospheres
Volume 121, Issue 13, pages 7958–74 doi:10.1002/2016JD024753
http://onlinelibrary.wiley.com/doi/10.1002/2016JD024753/full#jgrd53104-fig-0002

from (Masunaga and Luo, 2016)

Slide 10

MC3E PDFs of Cumulus Vertical Velocity in GFDL AM3 and Radar Observations

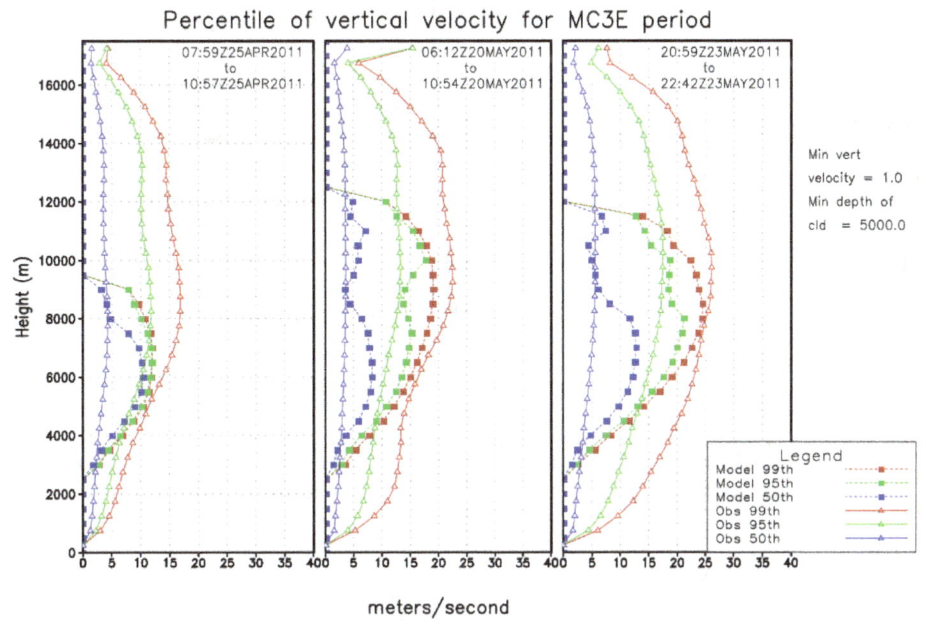

Slide 11

Convective and large-scale mass flux profiles over tropical oceans determined from synergistic analysis of a suite of satellite observations

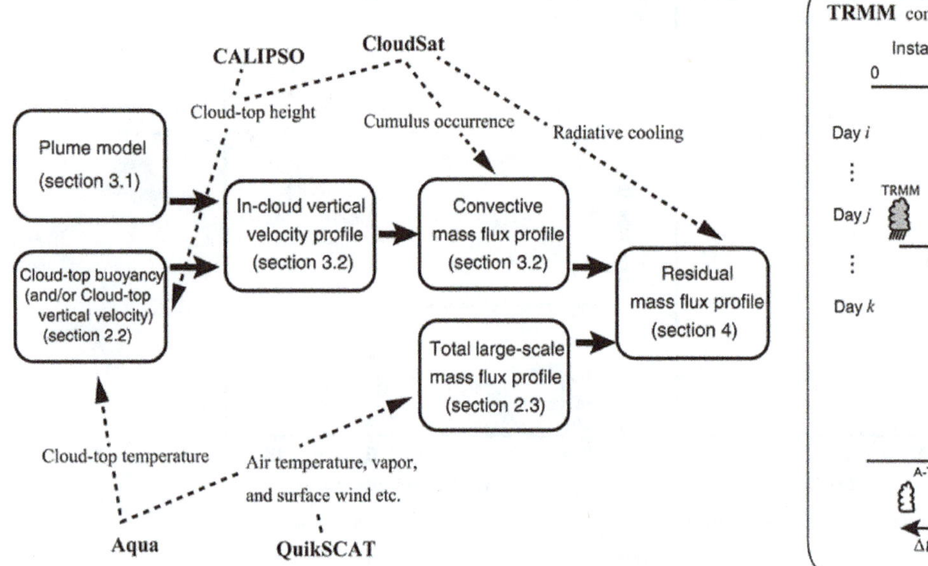

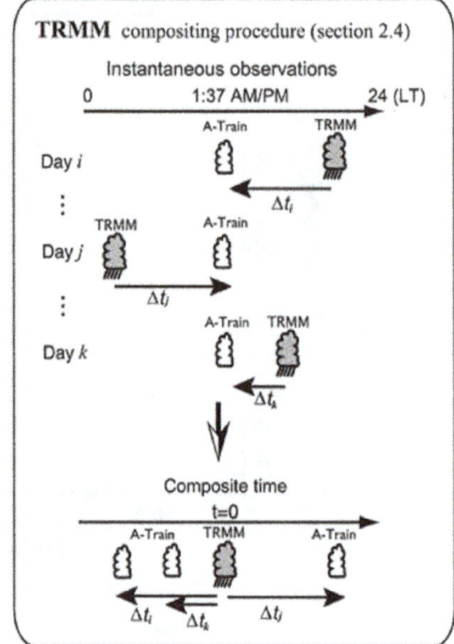

Journal of Geophysical Research: Atmospheres
Volume 121, Issue 13, pages 7958–74. doi:10.1002/2016JD024753
http://onlinelibrary.wiley.com/doi/10.1002/2016JD024753/full#jgrd53104-fig-0001

from (Masunaga and Luo, 2016)

Slide 12

Convective and large-scale mass flux profiles over tropical oceans determined from synergistic analysis of a suite of satellite observations

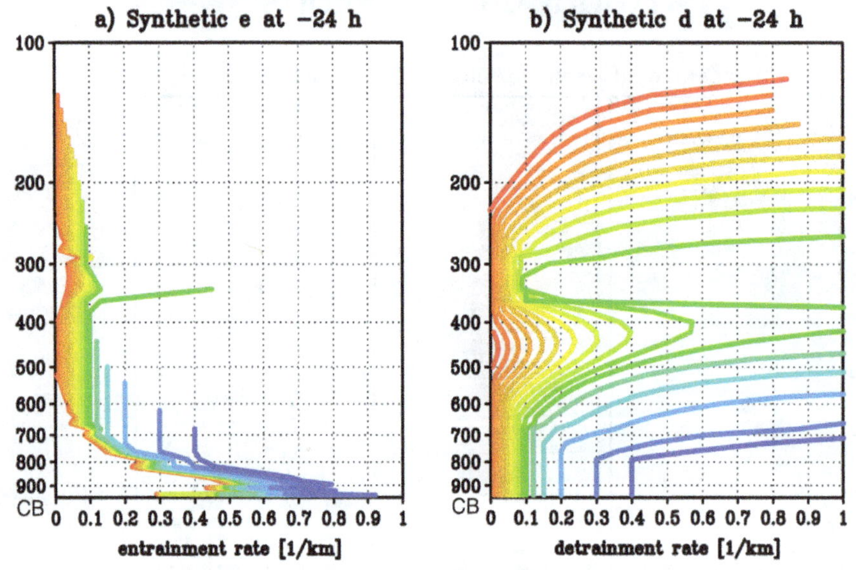

Plume entrainment rates from 0 to 0.4 km^{-1} as red goes to blue.

from Masunaga and Luo (2016, *JGR*)

Journal of Geophysical Research: Atmospheres
Volume 121, Issue 13, pages 7958-7974, 12 July 2016 doi: 10.1002/2016JD024753
http://onlinelibrary.wiley.com/doi/10.1002/2016JD024753/full#jgrd53104-fig-0003

Slide 13

Conclusions

Cloud updrafts have received limited attention in climate models and observations.

Compensating uncertainties in climate sensitivity and climate forcing limit understanding of pre-industrial to present-day climate change and the ability to project future changes.

Cloud updrafts are a major control on how clouds respond to aerosols.

Entrainment, which updraft speeds constrain, is an important control on sensitivity in some climate models.

Physically based, observationally constrained modeling of updrafts may provide an opportunity to reduce uncertainty in climate sensitivity and forcing.

Slide Captions

Slide 1 Title

Slide 2 "Cloud Adjustments due to Aerosols" are the most uncertain anthropogenic forcing of the climate system. Low-confidence estimates from the Intergovernmental Panel on Climate Change range from nearly zero to −1.3 Wm^{-2}.

Slide 3 Climate models (highlighted in green and red) able to successfully simulate observed pre-industrial to present-day warming exhibit compensation between forcing and sensitivity. Overall, CMIP models do not display this compensation, and many do not successfully simulate observed warming (not highlighted). These models suggest that observed global warming can be explained by a range of inversely related anthropogenic forcings and sensitivities.

Slide 4 Future climate change will be driven more by greenhouse gases than aerosols, as aerosols have shorter lifetimes than dominant anthropogenic greenhouse gases, and aerosols are likely to be regulated by air pollution policy. "Masking" by aerosols will be less. Projecting warming requires knowledge of sensitivity. Assuming models are reasonably correct in their estimates of variability from pre-industrial times to the present day, progress on understanding forcing can narrow uncertainty in climate sensitivity.

Slide 5 Climate Forcing and Cloud Updrafts

Slide 6 Droplet effective radius r_e and drop number concentration N_d are important controls on radiative and microphysical properties of clouds. Variations in aerosol concentrations N_a, aerosol median size r_g, breadth of aerosol size distribution σ, updraft speed *w*, and ammonium sulfate fraction ε produce variations in the drop size and number.

Slide 7 In order to study the sensitivity of ice number density N_i resulting from homogeneous freezing to aerosol concentration N_a, Kay and Wood (2008) defined an aerosol sensitivity parameter η_a. Note there are regimes where the ice number density is controlled primarily by vertical velocity.

Slide 8 Convective entrainment rates are related to climate sensitivity in some models. Using observations to constrain entrainment could constrain sensitivity. Updraft speeds are among the constraints on entrainment.

Slide 9 Entrainment rates are closely related to observable cloud properties like temperature and updraft speeds.

Slide 10 Parameterized PDFs of cumulus vertical velocities at MC3E from GFDL AM3 (Donner *et al.*, 2011; Donner *et al.*, 2016) and dual-Doppler radar (Collis *et al.*, 2013) can be compared to evaluate entrainment in this climate model.

Slide 11 Satellite observations enable widespread observations of convective mass fluxes and scaling up geographically the results from field campaigns.

Slide 12 Satellite-inferred mass fluxes can be used to estimate entrainment over large areas and varying types of convective clouds and systems.

Slide 13 Conclusions

References

Collis S, Protat A, May PT, Williams C. (2013) Statistics of storm updraft velocities from TWP-ICE including verification with profiling measurements, *Journal of Applied Meteorology and Climatology* **52**:1909–1922. https://doi.org/10.1175/JAMC-D-12-0230.1

Donner LJ, O'Brien TA, Rieger D, *et al.* (2016) Are atmospheric updrafts a key to unlocking climate forcing and sensitivity? *Atmospheric Chemistry and Physics* **16**:12983–192. doi:10.5194/acp–16–12983–2016

Donner LJ, Wyman BL, Hemler RS, *et al.* (2011) The dynamical core, physical parameterizations, and basic simulation characteristics of the atmospheric component of the GFDL global coupled model CM3. *Journal of Climate* **24**:3484–519. doi:10.1175/2011JCLI3955.1

Feingold G. (2003) Modeling of the first indirect effect: Analysis of measurement requirements. *Geophysical Research Letters* **30**:1997. doi:10.1029/2003GL017967

Forster PM, Andrews T, Good P, *et al.* (2013) Evaluating adjusted forcing and model spread for historical and future scenarios in the CMIP5 generation of climate models. *Journal of Geophysical Research: Atmospheres Research Letters* **118**:1139–50. https://doi.org/10.1002/jgrd.50174

Kay J, Wood R. (2008) Timescale analysis of aerosol sensitivity during homogeneous freezing and implications for upper tropospheric water vapor budgets. *Geophysical Research Letters* **35**:L10809. doi:10.1029/2007GL032628

Kiehl JT. (2007) Twentieth century climate model response and climate sensitivity. *Geophysical Research Letters* **34**:L22710. doi:10.1029/2007GL031383

Klocke D, Pincus R, Quaas J. (2011) On constraining estimates of climate sensitivity with present-day observations through model weighting. *Journal of Climate* **24**:6092–9. doi:10.1175/2011JCLI4193.1

Masunaga H, Luo Z. (2016) Convective and large-scale mass flux profiles over tropical oceans determined from synergistic analysis of a suite of satellite observations. *Journal of Geophysical Research: Atmospheres* **121**. doi:10.1002/2016JD024753

McFiggans, G., Artaxo P, Baltensperger U, *et al.* (2006) The effect of physical and chemical aerosol properties on warm cloud droplet activation. *Atmospheric Chemistry and Physics* **6**:2593–649. doi:10.5194/acp–6–2593–2006

Sanderson B, Shell K, Ingram W. (2010) Climate feedbacks determined using radiative kernels in a multi-thousand member ensemble of AOGCMs. *Climate Dynamics* **35**:1219–36.

Stainforth, D, Aina T, Christensen C, *et al.* (2005) Uncertainty in predictions of the climate response to rising levels of greenhouse gases. *Nature* **433**:403–6.

Zhao M. (2014) An investigation of the connections among convection, clouds, and climate sensitivity in a global climate model. *Journal of Climate* **27**:1845–62. doi:10.1175/JCLI–D–13–00145.1

Lecture 13

Use of Satellite Observations in Climate Model Evaluation of Clouds: A Time History Based on the NASA/GISS MODEL

George Tselioudis
Research Physical Scientist, NASA Goddard Institute for Space Studies, New York, NY, USA

George Tselioudis (PhD in Earth Sciences from Columbia University, 1992) is a Research Scientist at the NASA Goddard Institute for Space Studies and an Adjunct Professor at the Department of Applied Physics and Applied Mathematics at Columbia University. His research focuses primarily on the use of satellite observations and general circulation models to understand physical processes and their interactions in the Earth's atmosphere and to quantify the resulting feedbacks on climate change, with emphasis on cloud, radiation, and precipitation processes and feedbacks.

The importance of clouds in determining the Earth's radiative budget was well known to the Earth and planetary scientists who, back in the late 1960s and early 1970s, built the first climate models. Cloud cover was simulated in early climate models as a simple function of the grid box relative humidity, while cloud properties like optical depth were assigned as a function of cloud height. Convective clouds were simulated using the convective instability theory. Right from the start of climate modeling efforts, the need to evaluate the clouds simulated by the models was urgent, as cloud amount and property variability played a large role in determining the models' simulated climate. In this lesson, we will follow the efforts of the NASA Goddard Institute for Space Studies (GISS) climate modeling group to evaluate the clouds simulated by the GISS climate model against the available observations of each different time period, and thus improve the representation of clouds in the model.

The first version of the GISS climate model was documented in a paper by Somerville *et al.* (1974), which was published in the *Journal of the Atmospheric Sciences*. The model included a stratiform cloud parameterization that produced clouds whenever supersaturation occurred in a grid box, which at the time was 4 by 5 degrees wide, and the mixing ratio was reduced until the relative humidity reached 100%. The excess condensation was moved to the lower layer where it evaporated if it was unsaturated or precipitated if it was saturated, and a cloud was formed that was assumed to be black in the infrared and had a visible optical depth that depended on the cloud height and layer thickness. There was also a moist convection parameterization that produced shallow or deep convective clouds depending on the vertical gradients of static energy and saturation static energy. An instantaneous array of total (convective plus supersaturation) clouds is shown in Slide 2. Note that due to the lack of graphing capabilities in the mainframe computers of the time, cloud occurrence is indicated by a printed "0" sign. The pattern shows the expected cloud-free areas over major dessert regions and clouds occurring in major storm tracks, but the authors needed observational validation of their cloud simulations. At the time, available observations came either from surface observations that were collected by the national meteorological services (but covered only a small portion of the Earth's continents) or by the very first satellite retrievals coming from the series of Television InfraRed Observation Satellites (TIROS) launched by NASA in the 1960s. As GISS is a NASA organization, the authors appropriately compared their model output with TIROS nephanalysis data. The bottom-right panel in Slide 2 shows a comparison of zonal mean cloud cover between the model output and the TIROS retrievals. It can be seen that the model is doing well in the tropical convective zones and mid-latitude storm tracks but severely lacks cloud cover in the subtropical zones. Keep this feature in mind as we will return to it in the next few figures. The authors also compared their global mean cloud cover number of 52% to that derived from surface observer data (53% at the time) and noted the good agreement between the two. Later satellite figures, however, showed that this number was closer to 70%. In this first version of the climate model, the evaluation of the cloud field is restricted to the zonal mean cloud distribution and global mean cloud amount, which was for the first time calculated from the TIROS satellite retrievals.

The next incarnations of the GISS climate model, named Models I and II, were described in a paper by Hansen *et al.* (1983). The cloud scheme was again based on the saturation mixing ratio of the grid box, but this time, a subgrid-scale variability of temperature was introduced, which allowed for varying relative humidity and, therefore, cloud cover. The simulated cloud cover in the model is compared to observations from surface observer cloud climatology, in a sense a step backward since the original model was compared to satellite retrievals. This time around, global maps of cloud cover are produced, and the cloud distribution patterns around the globe are used for the evaluation (Slide 3). The authors emphasize that the model reproduces the major cloud features, such as the Intertropical Convergence Zone (ITCZ) and the storm tracks, but also note that there are regions around the globe where the model either overpredicts or underpredicts the cloud amount. It is noted at the end of the paper that the crude treatment of clouds is a major issue in model climate simulations.

In the late 1980s and early 1990s, there were major developments with respect to cloud research, both in climate models and in observations. In the modeling field, groups started including prognostic cloud schemes that predicted the water/ice content of the cloud and thus interactively calculated cloud optical properties. In the observational arena, the first global, multi-year climatology of cloud properties (the International Satellite Cloud Climatology Project [ISCCP]) was produced. The ISCCP satellite climatology, in addition to cloud cover, included retrievals of cloud optical depth and cloud top pressure (CTP), thus allowing for the complete calculation of cloud effects on the radiative budget components. It was only natural that the ISCCP dataset was used widely to evaluate the new and improved model cloud simulations.

A prognostic cloud scheme was introduced in the GISS Model II in 1996, which is described in a paper by Del Genio *et al.* (1996). The scheme allowed for cloud life cycle effects and permitted cloud optical properties to be determined interactively. This interactive determination of optical properties, together with the already existing vertical layering of the model clouds, allowed for comparisons with the cloud optical depth and CTP retrievals included in the ISCCP dataset. This paper was the first one that also attempted to take into account the top-down view of the satellite when comparing model and satellite fields. It calculated cloud top in a grid box using only the cloud of the highest vertical layer and took provisions to account for satellite under-detection of optically thin clouds. This attempt to produce top-down model views for comparison with satellite observations would later evolve to the production of satellite simulators, first for the ISCCP and later for other satellite datasets. This time around, in addition to cloud cover, global maps of cloud optical properties like cloud optical thickness (COT) and CTP are produced and compared to the satellite retrievals. In addition to the ISCCP, other cloud and radiation datasets were first constructed in the 1980s, such as the radiative flux dataset from the broadband radiometer of the Earth Radiation Budget Experiment (ERBE) and the cloud water path dataset from the microwave radiometer of Special Sensor Microwave Imager (SSMI), and those datasets are also used for model evaluation in (Del Genio *et al.*, 1996). The use of multiple satellite datasets allowed the evaluation of the model clouds in more detail and the examination of cloud deficiencies as a function of cloud type. The comparison with the ISCCP and SSMI showed that the model tends to produce too few and too bright low clouds. This compensation between cloud cover and cloud albedo accounted for the fact that the model does not simulate an excessively unrealistic shortwave radiative budget when compared to the ERBE. In addition, the model was found to underpredict thin cirrus clouds, and this deficiency is more pronounced in tropical latitudes.

We can see that the introduction of prognostic cloud water in the model and the retrieval of COT in the observations allowed for much more direct attribution of model cloud deficiencies to particular cloud types in specific dynamic regimes. This kind of attribution is crucial for the attempts of the modelers to improve model cloud representation, as it points to specific cloud types and regimes, and therefore, to specific atmospheric processes that may be responsible for the cloud simulation deficiencies. The next version of the GISS climate model, named ModelE and detailed by Schmidt *et al.* (2006), employed the stratiform cloud scheme introduced by Del Genio *et al.* (1996), with relatively small modifications. The existence of the too-few-too-bright low clouds in the model is demonstrated in this paper through comparisons of COD-CTP (TAU-PC) histograms from the ISCCP and the ISCCP simulator applied to model observations. The comparison also showed a lack of middle level and high thin (cirrus clouds).

Following the evaluation of the model using mean fields from satellite retrievals, a more detailed mapping of model cloud deficiencies related to CTP and COT was obtained. In the case of the GISS model, for example, the main deficiencies that were identified were the too-few-and-too-bright low clouds and the lack of mid-level and high thin clouds. The next level of understanding that needed to be obtained involved the atmospheric conditions under which those deficient clouds were formed. That information would allow modelers to know which dynamic or thermodynamic processes were not properly simulated in the model and hence responsible for the biases in the cloud properties. One method to achieve this goal is to composite cloud properties over different dynamic or thermodynamic regimes. Tselioudis and Jakob (2002) used mid-tropospheric vertical velocity in the northern mid-latitude region as a separator and composed TAU-PC histograms in uplift and subsidence regions. They found that the biases in the low cloud model simulation were occurring primarily in the subsidence regime, while the lack of mid-level and high thin clouds occurred primarily in the uplift regime. A more detailed picture of model cloud deficiencies in the vicinity of Southern Hemisphere (SH) mid-latitude storms was obtained by Bodas-Salcedo *et al.* (2014), who composited and evaluated cloud and radiation properties around low-pressure storm centers. They found that the largest model biases occurred in the cold air outbreak region behind the storm center.

Another method to obtain regimes over to evaluate the model cloud simulations is to apply data mining techniques on the cloud properties themselves. This was done by Tselioudis and Jakob (2002) and Tselioudis *et al.* (2021), who applied a clustering algorithm to ISCCP TAU-PC histograms to obtain cloud regimes. The global application of the clustering algorithm produced cloud regimes, named Weather States (WSs), that range from deep convective and mid-latitude storm clouds, to cirrus, arctic, mid-level, and fair-weather clouds, to shallow cumulus and stratocumulus clouds. On the global scale, the distribution of the WSs places them in expected circulation patterns, like the deep convection in the ITCZ and the mid-latitude storm clouds in the storm tracks. When examined over dynamical features like mid-latitude storms, those WSs are found in the expected circulation belts, like the storm and deep convective clouds in the frontal conveyor belts and the mid-level and shallow cumulus clouds in cold air outbreaks behind the fronts. Those WSs have been recently used in model evaluation studies. The first evaluation of CMIP3 models using regional WSs by Williams and Webb (2009) showed a large model spread and deficiencies in cumulus congestus (mid-level) and transition (shallow cumulus) clouds in the extra-tropics and stratocumulus clouds in the tropics. More recently, Tselioudis *et al.* (2021) evaluated CMIP6 models using global WSs, and found improvements in the representation of stratocumulus clouds compared to CMIP3 and CMIP5 ensembles, but there were still deficiencies in the simulation of shallow cumulus and mid-level clouds. To better understand the dynamic conditions under which the deficient cloud regimes occur, they plotted an 850 mb wind roses histogram for the global WSs. It showed that the shallow cumulus WS occurs both in a mid-altitude regime of westerly-northwesterly winds and in a tropical trade wind regime, while the mid-level cloud WS occurs almost exclusively in a mid-latitude westerly-northwesterly wind regime. This implies that further regime separation is needed to map the atmospheric mechanisms related to the formation of the shallow cumulus clouds.

The evaluation of climate model cloud simulations has progressed from comparisons of space and time-averaged quantities to the evaluation of cloud-defined regimes defined from compositing over dynamical quantities or from clustering of cloud property histograms. This has helped identify the weather regimes where model cloud deficiencies occur and thus provide information on the processes that may be responsible for those deficiencies. In the case of the GISS model, this evaluation has focused attention on the simulation of shallow cumulus clouds that occur in the tropical trade winds and cold air outbreaks behind frontal systems and to mid-level clouds that occur predominantly in cold air outbreaks. However, more detailed regime separation analysis is needed to dig deeper into cloud formation or dissipation mechanisms that must be improved in the models in order to improve the model cloud biases.

References

Bodas-Salcedo A, Williams KD, Ringer MA, *et al.* (2014) Origins of the solar radiation biases over the southern ocean in CFMIP2 models. *Journal of Climate* **27**:41–56. https://doi.org/10.1175/JCLI-D-13-00169.1

Del Genio AD, Yao M-S, Kovari W, Lo KK-W. (1996) A prognostic cloud water parameterization for global climate models. *Journal of Climate* **9**:270–304. doi:10.1175/1520-0442(1996)009<0270:APCWPF>2.0.CO;2

Hansen J, Russell G, Rind D, *et al.* (1983) Efficient three-dimensional global models for climate studies: Models I and II. *Monthly Weather Review* **111**:609–62. doi:10.1175/1520-0493(1983)111<0609:ETDGMF>2.0.CO;2

Schmidt GA, Ruedy R, Hansen JE. (2006) Present day atmospheric simulations using GISS ModelE: Comparison to in-situ, satellite and reanalysis data. *Journal of Climate* **19**:153–92. doi:10.1175/JCLI3612

Somerville RCJ, Stone PH, Halem M, *et al.* (1974) The GISS model of the global atmosphere. *Journal of the Atmospheric Sciences* **31**:84–117. doi:10.1175/1520-0469(1974)031<0084:TGMOTG>2.0.CO;2

Tselioudis G, Jakob C. (2002) Evaluation of midlatitude cloud properties in a weather and a climate model: Dependence on dynamic regime and spatial resolution. *Journal of Geophysical Research* **107**(D24):4781. doi:10.1029/2002JD002259

Tselioudis G, Remillard J, Tropf D, *et al.* (2021) Evaluation of clouds, radiation, and precipitation in CMIP6 models using global Weather States derived from ISCCP-H cloud property data. *Journal of Climate* **34**:7311–24. doi:10.1175/JCLI-D-21-0076.1

Williams KD, Webb MJ. (2009) A quantitative performance assessment of cloud regimes in climate models. *Climate Dynamics* **33:**141–57. https://doi.org/10.1007/s00382-008-0443-1

Slide 1

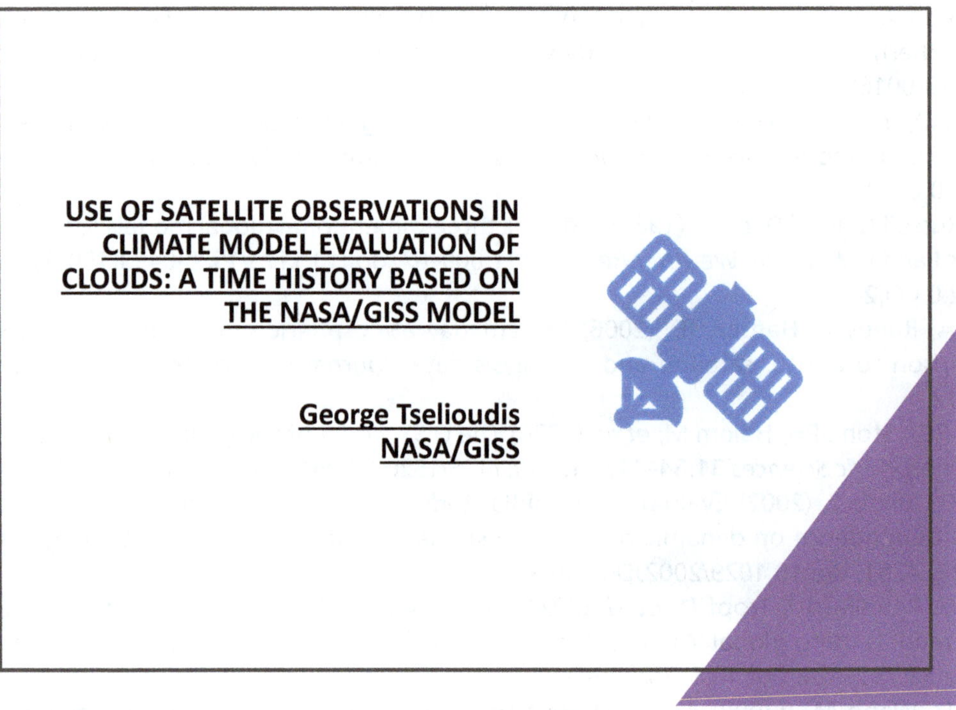

USE OF SATELLITE OBSERVATIONS IN
CLIMATE MODEL EVALUATION OF
CLOUDS: A TIME HISTORY BASED ON
THE NASA/GISS MODEL
George Tselioudis
NASA/GISS

Slide 2

Early GISS model evaluation using data from the very first meteorological satellite!

January 1974

Model output

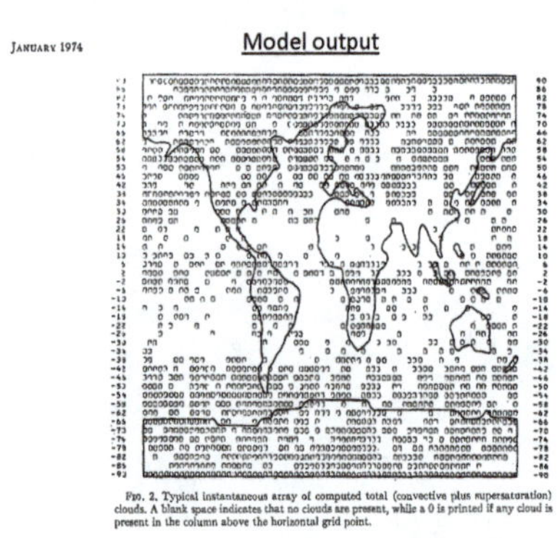

Fig. 2. Typical instantaneous array of computed total (convective plus supersaturation) clouds. A blank space indicates that no clouds are present, while a 0 is printed if any cloud is present in the column above the horizontal grid point.

Comparisons of cloud cover in the original GISS model with retrievals from the TIROS satellite. The model instantaneous field (left) shows cloud presence with a "0" sign, while the model zonal mean cloud cover (below) shows a lack of clouds in the subtropical regions compared to the satellite retrievals.

(Somerville *et al.*, 1974)

January 1974

Model/satellite cloud fraction

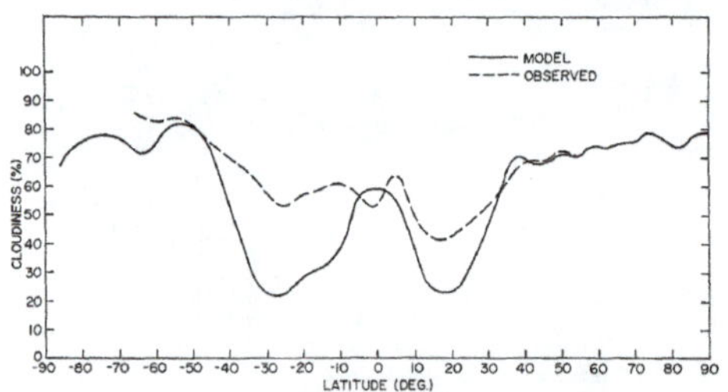

Fig. 23. Latitudinal variation of zonally averaged cloudiness from the GISS model (January mean) and from Clapp's (1964) analyses of TIROS observations (December–February).

Slide 3

Next GISS model evaluated against surface observations

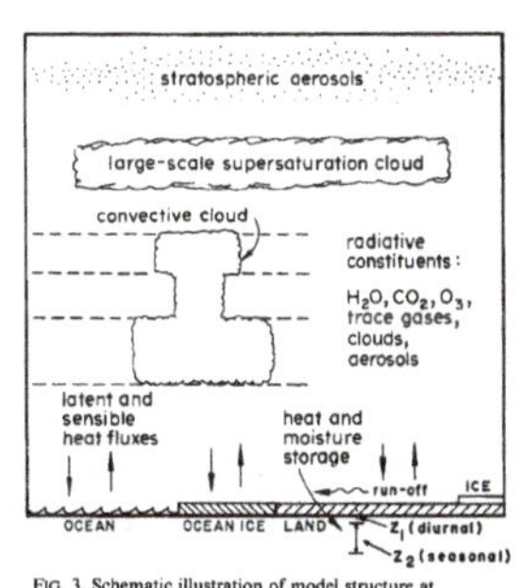

FIG. 3. Schematic illustration of model structure at a single gridbox.

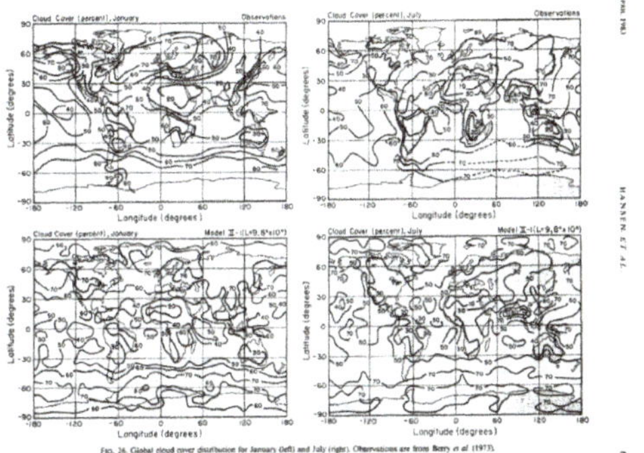

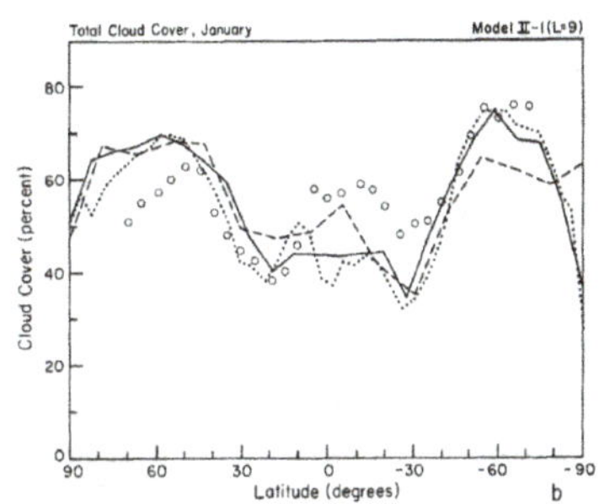

Comparison of the global (top) and zonal mean (left) cloud cover in the GISS Model I–II with a surface observer climatology, a step backward compared to the satellite evaluation of the original GISS model.

(Hansen *et al.*, 1983)

Slide 4

The first global satellite cloud climatology (ISCCP) and new GISS model with the prognostic cloud property scheme.

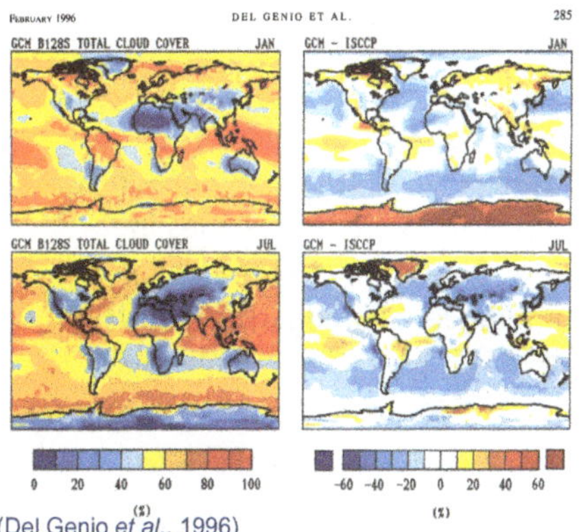

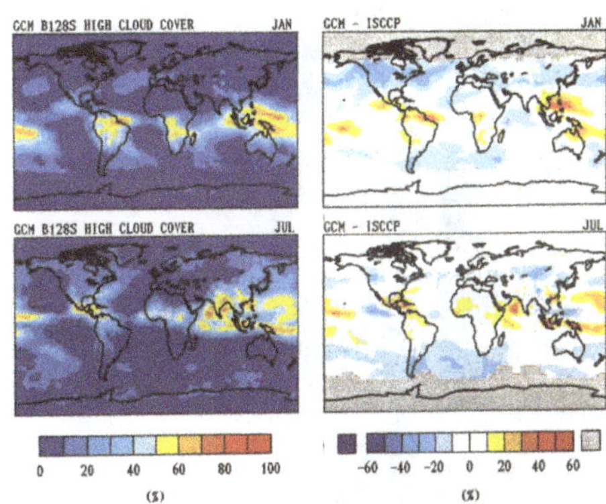

(Del Genio *et al.*, 1996)

Slide 5

Additional satellite climatologies make possible the evaluation of model radiative components (ERBE – top) and cloud liquid water path (SSMI – bottom).

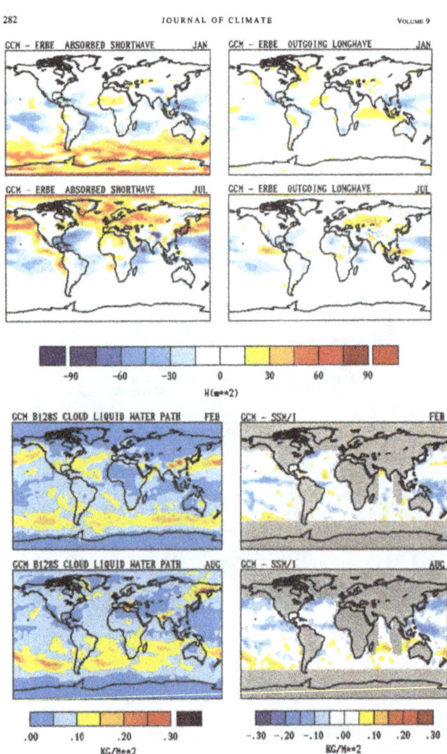

(Del Genio *et al.*, 1996)

Slide 6

Every 3 hours for each grid box around the globe, the ISCCP-D1 dataset provides cloud frequencies at 42 cloud optical thickness – cloud top pressure categories. These ISCCP retrievals of cloud top pressure and cloud optical thickness allow the definition of cloud types based on the reflectivity of the cloud and the location of its top.

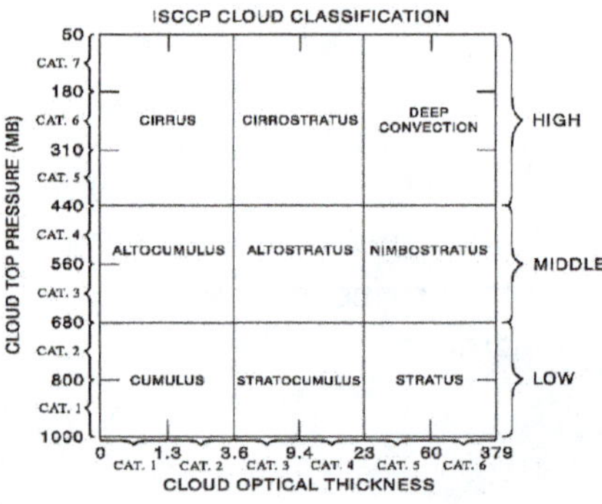

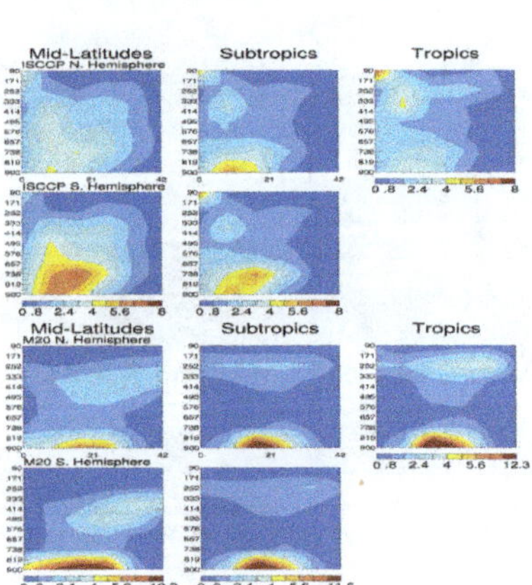

Evaluation of the model cloud types showed deficiencies in the simulation of cumulus, middle level, and cirrus clouds.

(Schmidt *et al.*, 2006)

Slide 7

When the model cloud evaluation was broken into uplift and subsidence regimes in the mid-latitude storm tracks, it was found that the model deficiencies in cumulus and mid-level clouds were mostly present in the subsidence behind the fronts while the deficient representation of cirrus clouds was present in the frontal uplift regime.

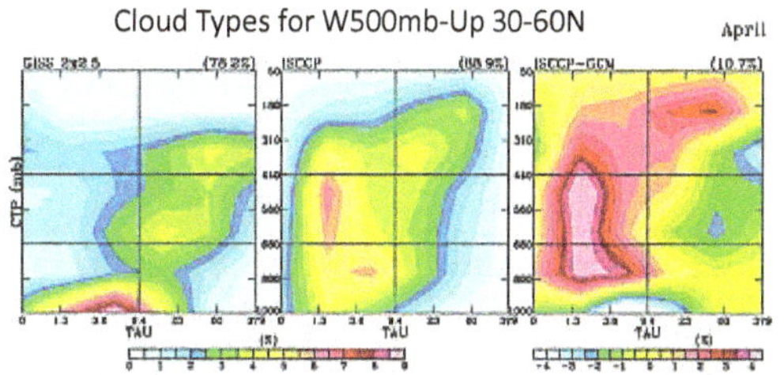

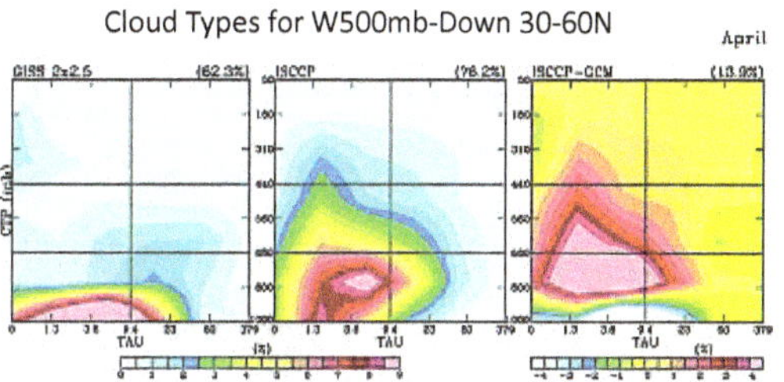

(Tselioudis and Jakob, 2002)

Slide 8

Evaluation of the CMIP5 model cloud simulations around mid-latitude storms in the Southern Ocean also showed that the models did not produce enough clouds in the cold air outbreak regime behind the storm center. As a result, the models simulated excessive solar heating in that regime, which resulted in excessive solar heating of the Southern mid-latitude region.

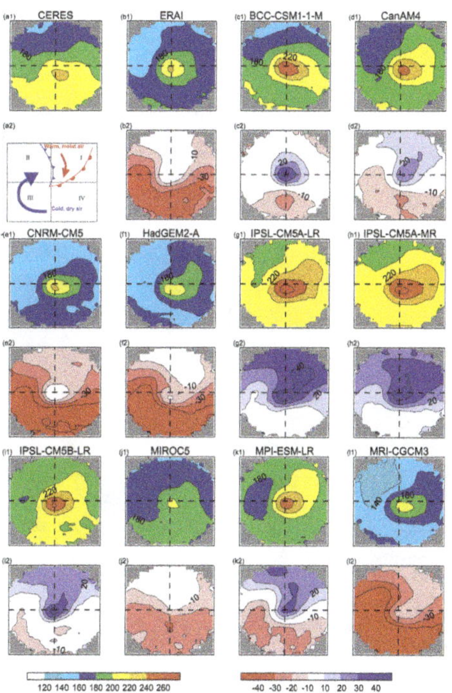

(Bodas-Salcedo et al., 2014)

Slide 9

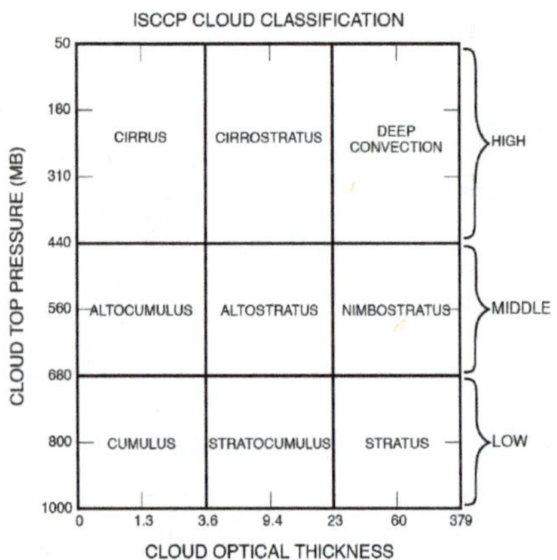

Cloud types according to ISCCP

The ISCCP cloud type definitions are based on fixed and somewhat subjective boundaries of cloud top pressure and cloud optical thickness.

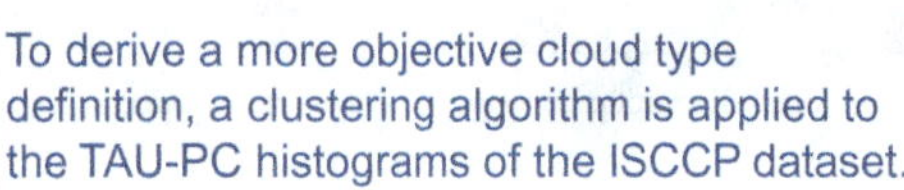

To derive a more objective cloud type definition, a clustering algorithm is applied to the TAU-PC histograms of the ISCCP dataset.

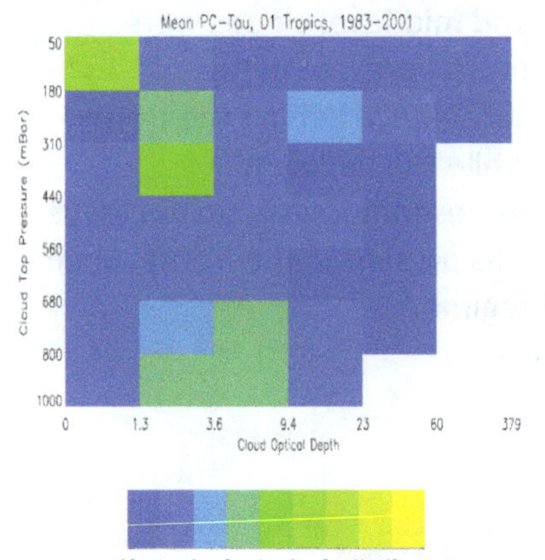

Slide 10

The global application of the clustering algorithm produces Cloud Regimes named Weather States, which range from deep convective and mid-latitude storm clouds to cirrus, arctic, mid-level, and fair-weather clouds, to shallow cumulus and stratocumulus clouds.

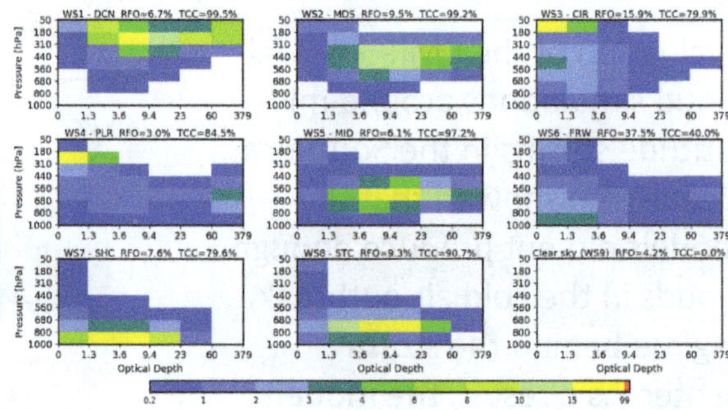

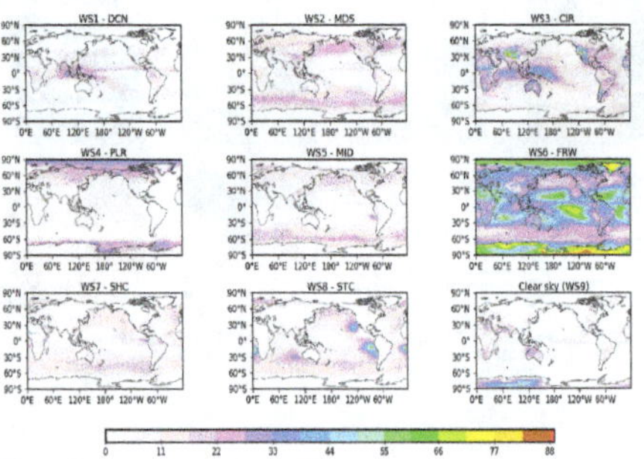

On the global scale, the distribution of the Weather States places them in expected circulation patterns, like the deep convection in the ITCZ and the mid-latitude storm clouds in the storm tracks.

(Tselioudis *et al.*, 2021)

Slide 11

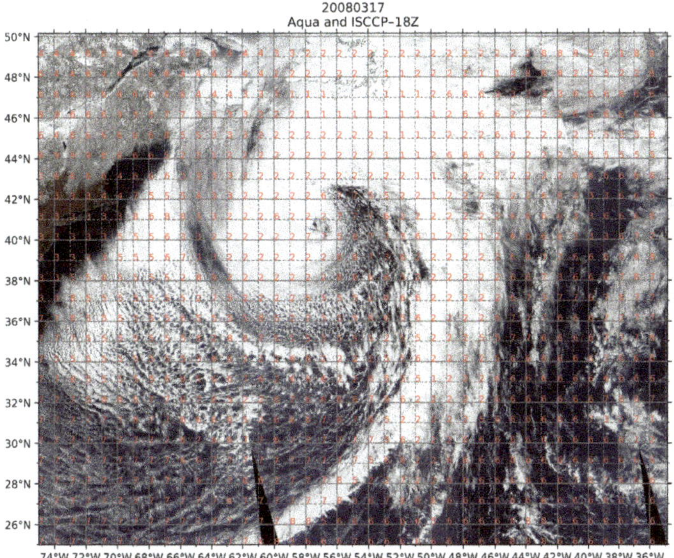

On the synoptic scale, the Weather scales are again found in the expected circulation belts like the storm and deep convective cloud in the frontal conveyor belts and the mid-level and shallow cumulus cloud in the cold air outbreaks behind the fronts.

(Tselioudis *et al.*, 2021)

Slide 12

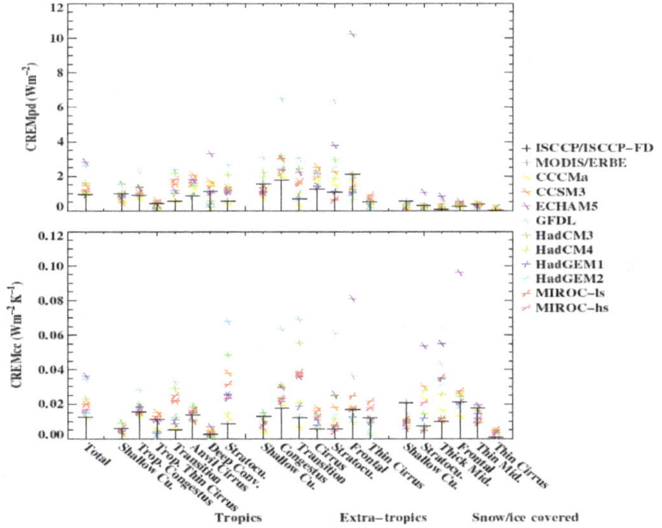

The first evaluation of CMIP3 models using regional Weather States showed large model spread and deficiencies in cumulus congestus (mid-level) and transition (shallow cumulus) clouds in the extra-tropics and stratocumulus clouds in the tropics.

(Williams and Webb, 2009)

Slide 13

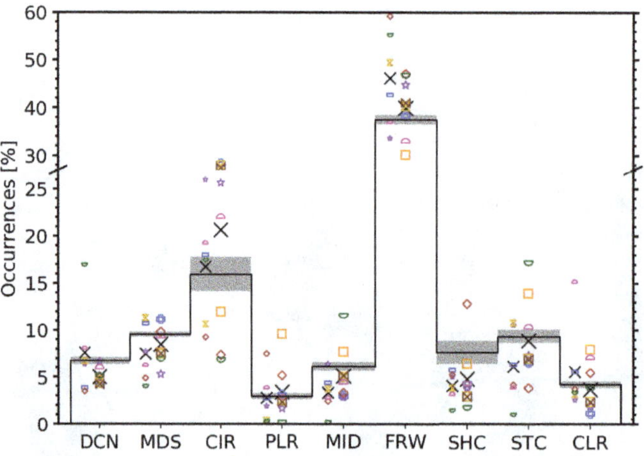

A recent evaluation of CMIP6 models using global Weather States showed improvements in the representation of stratocumulus clouds, but there are still deficiencies in the simulation of shallow cumulus and mid-level clouds.

(Tselioudis *et al.*, 2021)

Slide 14

An 850 mb wind roses histogram for the global Weather States (WSs) shows that the shallow cumulus WS occurs both in a mid-latitude regime of westerly-northwesterly winds and in a tropical trade wind regime, while the mid-level cloud WS occurs almost exclusively in a mid-latitude westerly-northwesterly wind regime. This implies that further regime separation is needed to map the atmospheric mechanisms related to the formation of the shallow cumulus clouds.

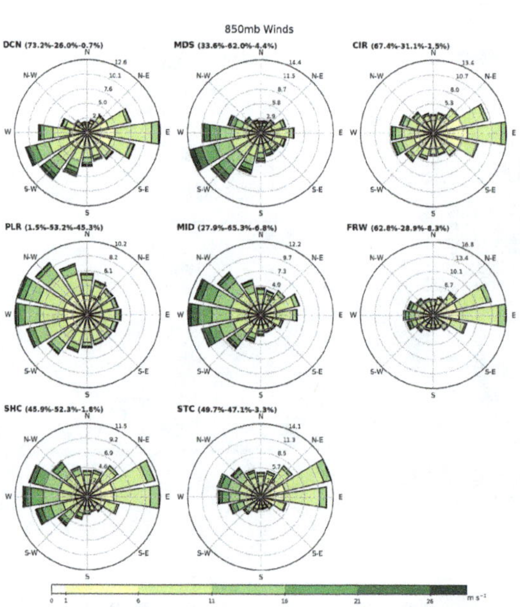

(Tselioudis *et al.*, 2021)

Slide 15

SUMMARY

- Satellite observations have been used since the early days of climate modeling to evaluate the simulation of clouds in climate models.

- The cloud properties retrieved from satellites have progressed from simple cloud spatial cover to cloud optical depth, water path, and top pressure over the past 40 years. In the same period, climate models have progressed to the simulation of prognostic cloud water content and thus cloud radiative properties.

- The evaluation of climate model cloud simulations has progressed from comparisons of space- and time-averaged quantities to the evaluation of cloud-defined regimes defined from compositing over dynamical quantities or from the clustering of cloud property histograms. This has helped identify the weather regimes where model cloud deficiencies occur and thus provide information on the processes that may be responsible for those deficiencies. However, more detailed analysis is needed to better isolate cloud formation or dissipation mechanisms that need to be better simulated in the models in order to improve the model cloud simulations.

Slide Captions

Slide 1 Title slide

Slide 2 Comparisons of cloud cover in the original Goddard Institute for Space Studies (GISS) model with retrievals from the Television InfraRed Observation Satellite (TIROS) (pictured, bottom left). The top-left panel shows a model instantaneous cloud field (left) where the cloud presence is noted by a "0" sign. In the bottom-right panel, the model and satellite zonal mean cloud cover curves and reveals a lack of clouds in the subtropical regions compared to the satellite retrievals.

Slide 3 Comparison of global (top right) and zonal mean (bottom-left) cloud cover in the GISS Model I–II with surface observer climatology, a step backward compared to the satellite evaluation of the original GISS model.

Slide 4 Comparisons of the total cloud cover (left) and high cloud cover (right) between the International Satellite Cloud Climatology Project (ISCCP) and GISS Model II. The comparisons show an excess of cloud in the model in the tropical regions and a deficit of cloud in the subtropical and mid-latitude regions. The cloud excess in the tropics is due to an excess of high-level clouds, while the deficit in the subtropics/mid-latitudes is due more to a deficit in the mid-level and low clouds.

Slide 5 Comparisons between the GISS Model II and ERBE-absorbed shortwave and outgoing longwave radiation (top), and the GISS Model II and SSMI cloud liquid water path (bottom). The comparisons show excess amount of shortwave radiation absorbed in the model in the summer middle and high latitudes, which can be attributed to the lack of low and mid-level clouds discussed in the previous slide.

Slide 6 Cloud classification histograms derived from 3-hourly retrievals of cloud optical thickness (COT) and cloud top pressure (CTP) (top right). Those ISCCP TAU-PC histograms allow the definition of cloud types based on set limits of the reflectivity of the cloud and the location of its top. The bottom left shows a comparison of TAU-PC histograms from the ISCCP and its simulator applied on output from the GISS Model E. The evaluation of the model cloud types shows deficiencies in the simulation of cumulus, middle level, and cirrus clouds.

Slide 7 Evaluation of GISS Model E TAU-PC histograms against the ISCCP in the mid-latitude storm regions. The comparison is separated into uplift (top) and subsidence (bottom) regimes, and the difference between the mode and observations is shown in the third column. It can be seen that the model deficiencies in cumulus and middle-level clouds are mostly present in the subsidence regime that usually follows mid-latitude storms, while the deficient representation of cirrus clouds is present in the frontal uplift regime.

Slide 8 Evaluation of an ensemble of CMIP5 model cloud simulations around mid-latitude storms in the Southern Ocean. The plot shows absorbed shortwave radiation in a composite center on the low-pressure center of mid-latitude storms for CERES observations (top left), the ERA-Interim reanalysis (top, second from left), and 10 CMIP5 models. The box underneath each model shows the difference between the models and the Clouds and the Earth's Radiant Energy System (CERES). It can be seen that most models did not produce enough clouds in the cold air outbreak regime behind the storm center. As a result, the models simulated excessive solar heating in that regime, which resulted in excessive solar heating of the Southern mid-latitude region.

Slide 9 The ISCCP TAU-PC histogram with the cloud type definitions. The definitions are based on fixed and somewhat subjective boundaries of CTP and COT. To obtain a more objective cloud type definition, a clustering algorithm is applied to the TAU-PC histograms of the ISCCP dataset.

Slide 10 The global application of the clustering algorithm produces eight cloud regimes, named Weather States (WSs), and their TAU-PS histograms are shown at the top-right panel. They range from deep convective and mid-latitude storm clouds to cirrus, arctic, mid-level, and fair-weather clouds to shallow cumulus and stratocumulus clouds. The ninth WS is reserved for clear sky conditions. The global distribution of the nine WSs and clear sky are shown at the bottom-left panel. On the global scale, the distribution of the WSs places them in the expected circulation patterns, like the deep convection in the ITCZ and the mid-latitude storm clouds in the storm tracks.

Slide 11 Visible satellite picture of a mid-latitude storm in the North Atlantic, with a superimposed grid with the number representing the corresponding WS. It can be seen that in the synoptic scale, the weather scales are again found in the expected circulation belts, like the storm and deep convective cloud in the frontal conveyor belts and the mid-level and shallow cumulus cloud in the cold air outbreaks behind the fronts.

Slide 12 The Cloud Radiative Effect (top) and the Cloud Radiative Feedback (bottom) of CMIP3 models (colored signs), broken into regional WSs and compared to ISCCP-FD and MODIS-ERBE (black and gray signs). This first evaluation of CMIP3 models showed large model spread and deficiencies in cumulus congestus (mid-level) and transition (shallow cumulus) clouds in the extra-tropics and stratocumulus clouds in the tropics.

Slide 13 Frequency of occurrence of the nine WSs in the ISCCP (black line) and CMIP5 (left column color signs) and CMIP6 models (right column color signs). This evaluation shows improvements in the representation of stratocumulus clouds between CMIP5 and CMIP6, but there are still deficiencies in the simulation of shallow cumulus and mid-level clouds.

Slide 14 Wind roses histograms of the 850 mb wind for the global WSs showing the predominant wind direction and speed for each WS. It shows that the shallow cumulus WS occurs both in a mid-latitude regime of westerly-northwesterly winds and tropical trade wind regime, while the mid-level cloud WS occurs almost exclusively in a mid-latitude westerly-northwesterly wind regime. This implies that further regime separation is needed to map the atmospheric mechanisms related to the formation of the shallow cumulus clouds.

Slide 15 Summary

References

Bodas-Salcedo A, Williams KD, Ringer MA, *et al.* (2014) Origins of the solar radiation biases over the southern ocean in CFMIP2 models. *Journal of Climate* **27**:41–56. https://doi.org/10.1175/JCLI-D-13-00169.1

Del Genio AD, Yao M-S, Kovari W, Lo KK-W. (1996) A prognostic cloud water parameterization for global climate models. *Journal of Climate* **9**:270–304. doi:10.1175/1520-0442(1996)009<0270:APCWPF>2.0.CO;2

Schmidt GA, Ruedy R, Hansen JE. (2006) Present day atmospheric simulations using GISS ModelE: Comparison to in-situ, satellite and reanalysis data. *Journal of Climate* **19**:153–92. doi:10.1175/JCLI3612

Somerville RCJ, Stone PH, Halem M, *et al.* (1974) The GISS model of the global atmosphere. *Journal of the Atmospheric Sciences* **31**:84–117. doi:10.1175/1520-0469(1974)031<0084:TGMOTG>2.0.CO;2

Tselioudis G, Jakob C. (2002) Evaluation of midlatitude cloud properties in a weather and a climate model: Dependence on dynamic regime and spatial resolution. *Journal of Geophysical Research* **107**(D24):4781. doi:10.1029/2002JD002259

Tselioudis G, Remillard J, Tropf D, *et al.* (2021) Evaluation of clouds, radiation, and precipitation in CMIP6 models using global Weather States derived from ISCCP-H cloud property data. *Journal of Climate* **34**:7311–24. doi:10.1175/JCLI-D-21-0076.1

Williams KD, Webb MJ. (2009) A quantitative performance assessment of cloud regimes in climate models. *Climate Dynamics* **33**:141–57. https://doi.org/10.1007/s00382-008-0443-1

Lecture 14

The Role of Upper Tropospheric Cloud Systems in Climate: Building Observational Metrics for Process Evaluation Studies

Claudia J. Stubenrauch

Laboratoire de Météorologie Dynamique, IPSL, Sorbonne University, Paris, France

Claudia Stubenrauch is a Senior Research Scientist at Laboratoire de Météorologie Dynamique, which is part of Institut Pierre Simon Laplace. She has developed long-term global cloud climate databases from spaceborne IR sounder observations and has led the GEWEX Cloud Assessment. Her research interest focuses on UT clouds, their microphysics, and their effect on climate.

Introduction

With the development of the International Satellite Cloud Climatology Project (ISCCP) database (Schiffer and Rossow, 1983; Rossow and Schiffer, 1999), William B. Rossow laid the foundation for a deeper comprehension of cloud properties and cloud processes on a global scale. His vision of how these data are to be used is clearly noticeable in the way the statistics of the different cloud and atmospheric variables were brought together, leading to hundreds of scientific publications.

Clouds play a dominant role in the energy and water cycles of our planet. Hence, it is of immense interest to evaluate their feedback on climate change; that is, how they change in a warming climate and how these changes, in turn, influence the climate system. As the cloud particles strongly interact with both incoming sunlight and the Earth's emitted thermal radiation, small changes in the clouds already have a potent effect on their feedback. For reliable climate change predictions, it is therefore of utmost importance to improve our understanding and representation of the cloud processes that control their feedback (e.g., [Stephens, 2005]).

In particular, the role of upper tropospheric (UT) cloud feedback is still highly uncertain, as reported by the Intergovernmental Panel on Climate Change (IPCC) (Boucher *et al.*, 2013). Yet UT clouds represent 40% of the Earth's cloud cover and even 60% of the tropical total cloud cover (e.g., [Stubenrauch *et al.*, 2013; 2017]) and exert a strong greenhouse effect. In the tropics, they form either as isolated cirrus by *in situ* freezing or as cirrus anvils from convective outflow, which rapidly transports water into the upper troposphere and detrains ice crystals. By modulating the Earth's energy budget and heat transport, they also affect the large-scale atmospheric circulation (Slide 3). As cirrus heating tends to stabilize the atmosphere, it may regulate convection itself (e.g., [Stephens *et al.*, 2004; Lebsock *et al.*, 2010]) and thus influence the response of the Earth's water cycle to climate forcing.

Living much longer than the convective towers, the large stratiform cirrus anvils produce radiative heating that is expected to be as important for large-scale circulation as the released latent heat in the initial stage of convection. In tropical convective regions, more than 50% of the total heating is attributed to cirrus radiative heating (e.g., [Sohn, 1999]). It is the gradient between this heating and the cooling of the surrounding clear sky and low-level cloud fields that impacts the atmospheric circulation and thus influences precipitation patterns.

Current climate models predict an increase in global precipitation of 1 to 3% per degree Celsius warming. However, society is most impacted by the regional distribution, frequency, duration and intensity of precipitation events in a warming climate. These features are driven by moisture and its convergence, which increase with warming (e.g., [Trenberth, 1999]), but regional projections of the occurrence of flooding and drought remain a substantial challenge (Allan *et al.*, 2014). Meanwhile, observations reveal that in the moistest regions of the tropics, where UT clouds are most abundant, heavy rainfall increases much above the expected thermodynamic response of 7% per degree Celsius warming (Wodzicki and Rapp, 2016; Stephens *et al.*, 2018). Moreover, a link between the change in tropical rainfall and the frequency of organized convection has been brought to attention (Tan *et al.*, 2015). This suggests a dynamical change in the climate system.

Organized convection, defined as convection that aggregates and grows upscale, creates very large, long-lived UT cloud systems with cirrus anvils extending over several thousand square kilometers (e.g., [Houze, 2004]). Still, how the properties of these anvils relate to the convective intensity is not well understood and poorly represented in climate models. Progress has been hampered by the difficulty to represent ice cloud processes and the organization of convection itself, as well as by a lack of key measurements that directly connect the smaller deep convective towers to the much larger cirrus anvils.

Deep convective cloud systems are involved in a series of positive and negative feedback processes, mostly through their anvils. To identify the most influential feedback mechanisms, large-scale modeling is necessary, in general, by confronting simulations of a present and future climate. However, the outcome of such feedback studies may rely upon the representation of convection and detrainment in these models (e.g., [Del Genio *et al.*, 2005; Zhao, 2014]).

Advancing our understanding of these convective storm systems is considered one of the major challenges of atmospheric sciences in the coming decade (Bony *et al.*, 2015). Therefore, one of the World Climate Research Programme's (WCRP) core projects, the Global Energy and Water Exchanges (GEWEX) has initiated the international working group of "Process evaluation studies on upper tropospheric clouds and convection" (UTCC PROES, https://gewex-utcc-proes.aeris-data.fr/), uniting experts from observations, cloud and climate modeling to foster collaborations. The GEWEX UTCC PROES initiative specifically aims to develop new diagnostic methods using existing observations to examine the processes that detrain UT clouds from convection and the interconnection between convection and the radiative heating induced by outflowing cirrus anvils.

So far, observational studies of tropical mesoscale convective systems have mainly concentrated on the convective towers and the thick cirrus anvils (e.g., [Liu *et al.*, 2007; Yuan and Houze, 2010; Roca *et al.*, 2014). Their structure and radiative effects have been first studied over the whole tropics by Machado and Rossow (1993), using a clustering technique on cold infrared (IR) brightness temperatures. Other studies focused on their vertical structure along spaceborne radar-lidar nadir tracks (e.g., [Takahashi and Luo, 2014; Igel and van der Heever, 2015; Deng *et al.*, 2016; Takahashi *et al.*, 2017]). The organization of convection was studied by statistical analysis of "cloud regimes", defined by similar cloud property distributions within grid cells (e.g., [Rossow *et al.*, 2005; Jakob and Schumacher, 2008; Rossow *et al.*, 2013; Tan *et al.*, 2015]). This analysis technique revealed that these regimes have distinct mesoscale structures that vary according to atmospheric circulation, thus suggesting a connection between radiative effects and dynamics. Recent observation analysis techniques of convection provide insight into processes (e.g., [Masunaga and Luo, 2016]). In general, the synergy between different datasets is crucial, and all these studies have considerably increased our knowledge of tropical convection.

IR sounder cloud observations, including cirrus with a visible optical depth as low as 0.2, have been recently used to reconstruct UT cloud systems (Stubenrauch *et al.*, 2017; Protopapadaki *et al.*, 2017). These data revealed that thin cirrus (with a visible optical depth of less than about 1.4) corresponds to about 30% of/around the area of mature deep convective systems and that deeper convection leads to relatively more thin cirrus within/around larger anvils. As the thinner cirrus heat the upper troposphere (Slide 17), this may have a far-reaching impact on the feedback to climate warming. At present, this UT cloud system database is being complemented by a cloud vertical structure, in particular, radiative heating rates (Henderson *et al.*, 2013) from spaceborne lidar and radar.

By relating the observed anvils to the processes shaping them, this Cloud System Concept creates a bridge between the observations and paradigms that guide our understanding and modeling of clouds. Similarly applied to observations and model simulations at different scales, it will enable new insights into the processes linked to convection, detrainment and microphysics, and guide improvements of their representation in climate models, necessary prerequisites for assessing their feedback to a warming climate.

These processes are well described by cloud-resolving models (CRMs) because they explicitly resolve convective updrafts (at a scale of 1 km). By suppressing and including specific interactions (e.g., radiation, detrainment, microphysics), the sensitivity of anvil characteristics to these processes can then be quantified (i.e., [Stephens *et al.*, 2008]).

First collaborations within GEWEX UTCC PROES are being formed, and the following slides present a short overview and some highlights from the workshop of March 2017 in New York. Meanwhile, another workshop was held in October 2018 in Paris. All presentations are available at https://gewex-utcc-proes.aeris-data.fr/. The workshop planned for September 2020 is postponed to 2022 due to the pandemic.

References

Allan RP, Liu C, Zahn M, *et al.* (2014) Physically Consistent responses of the global atmospheric hydrological cycle in models and observations. *Surveys in Geophysics* **35**:533–52. doi:10.1007/s10712-012-9213-z

Bony S, Stevens B, Frierson DMW, *et al.* (2015), Clouds, circulation and climate sensitivity. *Nature Geoscience*, **8**:261–68.

Boucher O, Randall D, Artaxo P, *et al.* (2013) Clouds and aerosols. In *Climate Change 2013: The Physical Science Basis. Contribution of Working Group I to the Fifth Assessment Report of the Intergovernmental Panel on Climate Change*. Eds. Stocker TF, Qin D, Plattner G-K, *et al.* Cambridge University Press, pp. 571–657. doi:10.1017/CBO9781107415324.016

Del Genio AD, Kovari W, Yao M, Jonas J. (2005) Cumulus microphysics and climate sensitivity. *Journal of Climate* **18**:2376–87. doi:10.1175/JCLI3413.1

Deng M, Mace GG, Wang Z. (2016) Anvil productivities of tropical deep convective clusters and their regional differences, *J. Atmos. Sci.* **73**:3467–87. doi:10.1175/JAS-D-15-0239.1

Henderson DS, L'Ecuyer T, Stephens G, *et al.* (2013) A multisensor perspective on the radiative impacts of clouds and aerosols. *Journal of Applied Meteorology and Climatology* **52**:853–71. doi:10.1175/JAMC-D-12-025.1

Houze RA Jr. (2004) Mesoscale convective systems. *Reviews of Geophysics* **42**:RG4003. doi:10.1029/2004RG000150

Igel MR, van den Heever SC. (2015) The relative influence of environmental characteristics on tropical deep convective storm morphology as observed by CloudSat. *Journal of Geophysical Research* **120**:4304–22.

Jakob C, Schumacher C. (2008) Precipitation and latent heating characteristics of the major tropical western Pacific cloud regimes. *Journal of Climate* **21**:4348–64.

Lebsock MD, Stephens GL, Kummerow C. (2010) An observed tropical oceanic radiative-convective cloud feedback. *Journal of Climate* **23**:2065–78. doi:10.1175/2009JCLI3091.1

Liu C, Zipser EJ, Nesbitt SW. (2007) Global distribution of tropical deep convection: Different perspectives from TRMM infrared and radar data. *Journal of Climate* **20**:489–503.

Machado LAT, Rossow WB. (1993) Structural characteristics and radiative properties of tropical cloud clusters. Mon. *Weather Rev.* **121**:3234–60. doi:10.1175/1520-0493(1993)121<3234:SCARPO>2.0.CO;2

Masunaga H, Luo ZJ. (2016) Convective and large-scale mass flux profiles determined from synergistic analysis of a suite of satellite observations. *Journal of Geophysical Research* **121**:7958–74. doi:10.1002/2016JD024753

Protopapadaki E-S, Stubenrauch CJ, Feofilov AG. (2017) Upper tropospheric cloud systems derived from IR sounders: Properties of cirrus anvils in the tropics. *Atmospheric Chemistry and Physics* **17**:3845–59. doi:10.5194/acp-17-3845-2017

Roca R, Aublanc J, Chambon P, *et al.* (2014) Robust observational quantification of the contribution of mesoscale convective systems to rainfall in the tropics. *Journal of Climate* **27**:4952–58.

Rossow WB, Schiffer RA. (1999) Advances in understanding clouds from ISCCP. *Bulletin of the American Meteorological Society* **80**:2261–88.

Rossow WB, Mekonnen A, Pearl C, Goncalves W. (2013) Tropical precipitation extremes. *Journal of Climate* **26**:1457–66. doi:10.1175/JCLI-D-11-00725.1

Rossow WB, Tselioudis G, Polak A, Jakob C. (2005) Tropical climate described as a distribution of weather states indicated by distinct mesoscale cloud property mixtures. *Geophysical Research Letters* **32**. doi:10.1029/2005GL024584

Schiffer RA, Rossow WB. (1983) The International Satellite Cloud Climatology Project (ISCCP): The first project of the World Climate Research Programme. *Bulletin of the American Meteorological Society* **64**:779–84.

Sohn B-J. (1999) Cloud-induced infrared radiative heating and its implications for large-scale tropical circulations. *Journal of the Atmospheric Sciences* **56**:2657–72.

Stephens GL. (2005) Cloud feedbacks in the climate system: A critical review. *Journal of Climate* **18**:237–73.

Stephens GL, Hakuba MZ, Webb MJ, *et al.* (2018) Regional intensification of the tropical hydrological cycle during ENSO. *Geophysical Research Letters* **45**:4361–70. doi:10.1029/2018GL077598

Stephens GL, Van den Heever S, Pakula LA. (2008) Radiative convective feedback in idealized states of radiative-convective equilibrium. *Journal of the Atmospheric Sciences* **65**:3899–916.

Stephens GL, Webster PJ, Johnson RH, *et al.* (2004) Observational evidence for the mutual regulation of the tropical hydrological cycle and tropical sea surface temperatures. *Journal of Climate* **17**:2213–24.

Storer R. https://www7.obs-mip.fr/wp-content-aeris/uploads/sites/20/2017/12/02_water_balance_Storer.pdf

Stubenrauch CJ, Feofilov AG, Protopapadaki SE, Armante R. (2017) Cloud climatologies from the infrared sounders AIRS and IASI: Strengths and applications. *Atmospheric Chemistry and Physics* **17**:13625–44. doi:10.5194/acp-17-13625-2017

Stubenrauch CJ, Rossow WB, Kinne S, *et al.* (2013) Assessment of global cloud datasets from satellites: Project and database initiated by the GEWEX Radiation Panel. *Bulletin of the American Meteorological Society* **94**:1031–49.

Takahashi H, Luo ZJ. (2014) Characterizing tropical overshooting convection from joint analysis of CloudSat and geostationary satellite observations. *Journal of Geophysical Research* **119**:112–21. doi:10.1002/2013JD020972

Takahashi H, Luo ZJ, Stephens GL. (2017) Level of neutral buoyancy, deep convective outflow, and convective core: New perspectives based on 5 years of CloudSat data. *Journal of Geophysical Research: Atmospheres* **122**:2958–69. doi:10.1002/2016JD025969

Tan J, Jakob C, Rossow WB, Tselioudis G. (2015) The role of organized deep convection in explaining observed tropical rainfall changes. *Nature* **519**:451–54. doi:10.1038/nature14339

Trenberth KE. (1999) Conceptual framework for changes of extremes of the hydrological cycle with climate change. *Climatic Change* **42**:327–39. doi:10.1007/978-94-015-9265-9\18

Wodzicki KR, Rapp AD. (2016) Long-term characterization of the Pacific ITCZ using TRMM, GPCP, and ERA-Interim. *Journal of Geophysical Research* **121**:3153–70. doi:10.1002/2015JD024458

Yuan J, Houze Jr RA. (2010) Global variability of mesoscale convective system anvil structure from A-train satellite data. *Journal of Climate* **23**:5864–88.

Zhao M. (2014) An investigation of the connections among convection, clouds, and climate sensitivity in a Global Climate Model. *Journal of Climate* **27**:1845–62. doi:10.1175/JCLI-D-13-00145.1

Slide 1

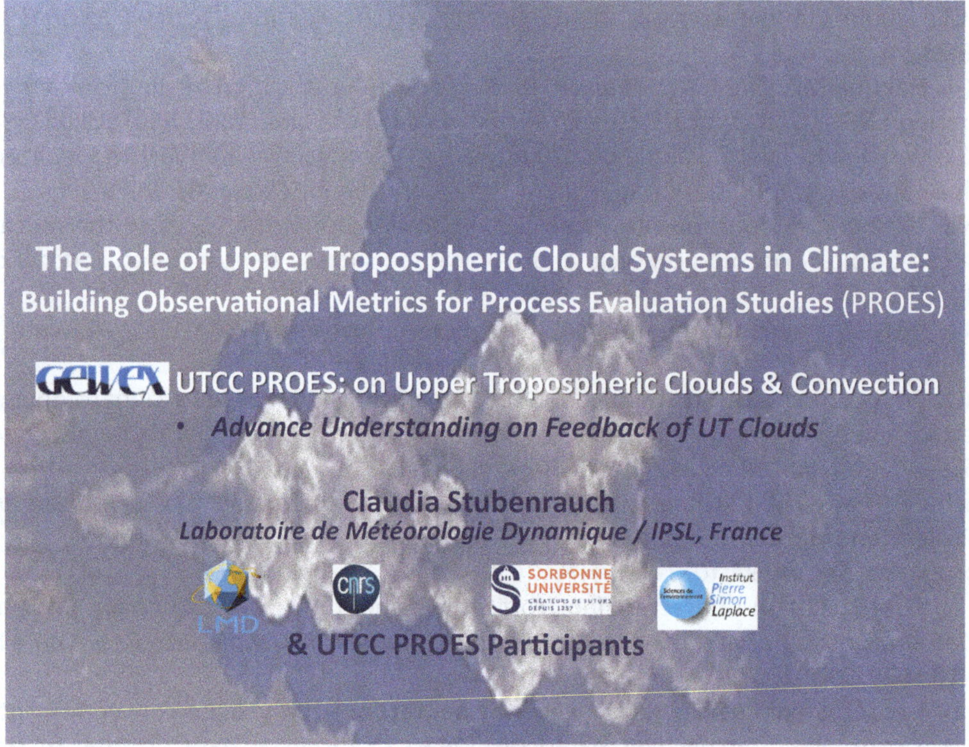

Slide 2

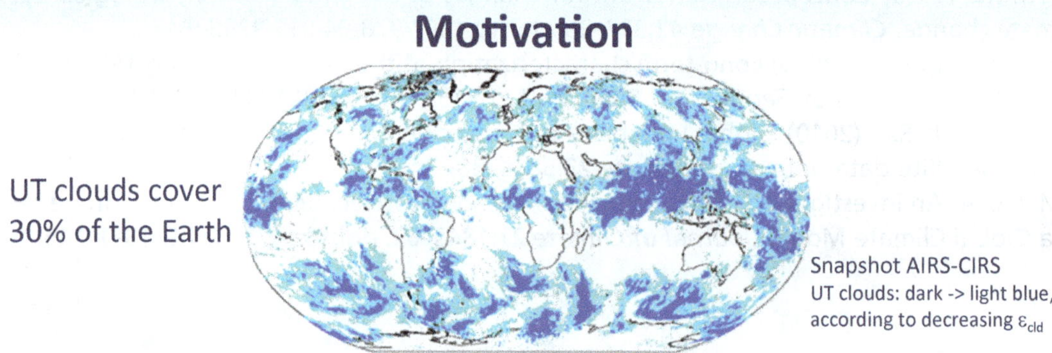

Motivation
UT clouds cover
30% of the Earth
Snapshot AIRS-CIRS
UT clouds: dark -> light blue,
according to decreasing ε_{cld}

UT clouds play a vital role in the climate system by modulating the Earth's energy budget and UT heat transport.

Convective tropical regions: > 50% radiative heating by cirrus (Sohn, 1999)

They often form mesoscale systems extending over 1,000 km, as outflow of convective / frontal systems or *in situ* by large-scale forcing.

How does convection affect UT clouds and vice versa?
Critical to feedbacks: cirrus radiative heating -> atmospheric circulation

Slide 3

Elements of UT Cloud Feedback

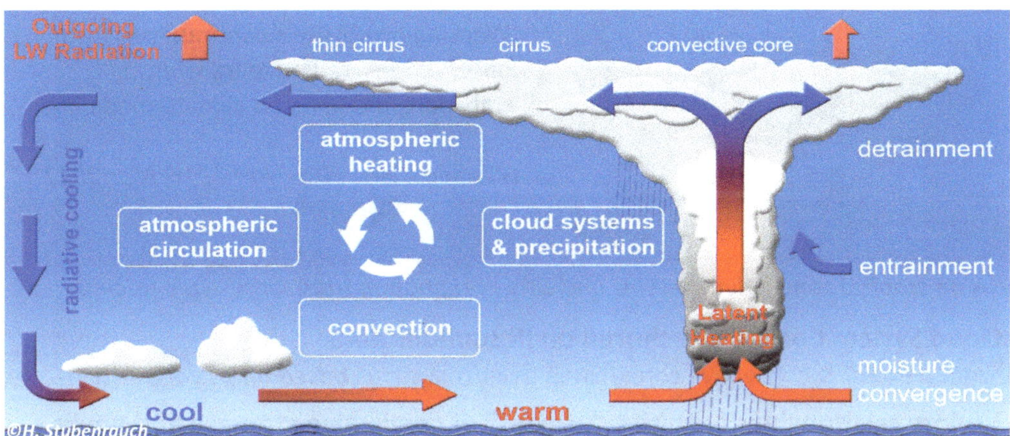

Height and amount of UT clouds constrained by clear sky mass and energy budget

Large-scale modeling is necessary to identify the most influential feedback mechanisms => models should be in agreement with observations

Goals:

- Understand the relation between convection, cirrus anvils and radiative heating.
- Provide observation-based metrics to evaluate detrainment processes in models.

Slide 4

Hypotheses on UT Cloud Feedback to Warmer Environment

➢ **Thermostat** (Ramanathan and Collins, 1991): Increased boundary layer H_2O -> increased convective condensate -> larger, thicker Ci anvils -> increased SW forcing
(negative cloud and positive H_2O feedback)

➢ **IRIS effect** (Lindzen *et al.*, 2001; Lin *et al.*, 2002): Increased convective condensate -> increased precip. efficiency -> reduced detrainment -> reduced anvil area
(slightly positive cloud & negative H_2O feedback (radiative properties accounted for)*)*

Observations intermediate between thermostat and IRIS extremes (Del Genio *et al.*, 2005): Detrainment (*ice formation at upper levels*) and precipitation (*condensate removal at lower levels*) offset each other in regulating TOA balance (*neutral cloud feedback*)

➢ **Stability IRIS effect** (Bony *et al.*, 2016): Warmer environment -> anvils rise
-> increase in static stability -> reduction of convective outflow
enhanced convective aggregation -> increased IRIS effect (Mauritsen and Stevens, 2015), including cloud-radiative effects -> narrowing of rainy areas

➢ **Humidistat** (Stephens *et al.*, 2004; Lebsock *et al.*, 2010):
Self-regulating radiative – convective feedback mechanism via cirrus anvil heating

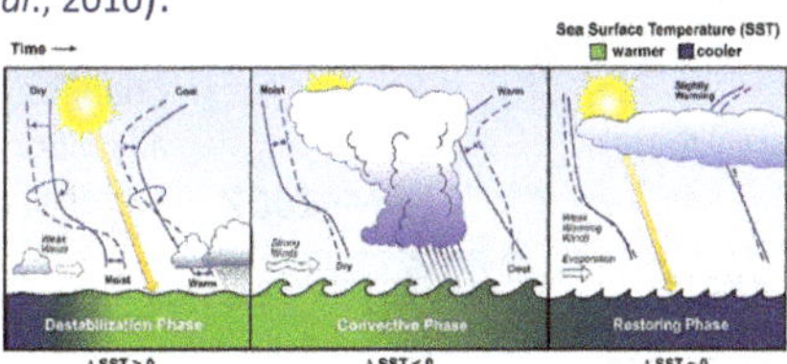

Slide 5

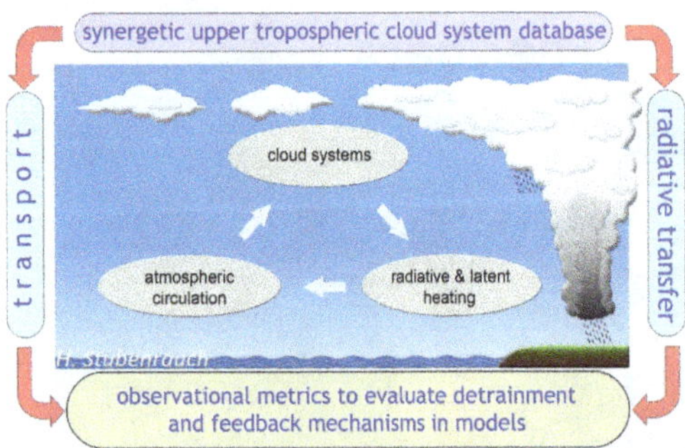

UTCC PROES Strategy

Working group **links communities** from observations, radiative transfer, transport, process and climate modeling

Meetings: Nov 2015, Apr 2016, Mar 2017 Oct 2018

Focus on tropical convective systems *and cirrus originating from large-scale forcing*

➢ **Cloud System Concept, anchored on IR sounder data**
horizontal extent & convective cores/cirrus anvil/thin cirrus ***based on*** p_{cld}, ε_{cld}

➢ **Explore relationships between "proxies" of convective strength and anvils**

➢ **Build synergetic data** (vert. dimension, atmosph. environment, temporal res.)

➢ **Determine heating rates** of different parts of UT cloud systems

➢ **Follow snapshots** by Lagrangian transport -> **evolution and feedbacks**

➢ **Investigate how cloud systems behave in CRM studies and in GCM simulations** *(under different parameterizations of convection/detrainment/microphysics)*

Slide 6

Why Use IR Sounders to Derive Cirrus Properties ?

TOVS, ATOVS
>1979 / ≥1995: 7:30 / 1:30 AM/PM

AIRS, CrIS
≥2002 / ≥2012 : 1:30 AM/PM

IASI (1,2,3), IASI-NG
≥2006 / ≥2012 / ≥2020 : 9:30 AM/PM

UT cloud amount *July*

AIRS-LMD ***2003–2015***

IASI-LMD ***2008–2015***

ISCCP ***1984–2007***

0 0.1 0.2 0.3 0.4 0.5 0.6 0.7 0.8 0.9

From GEWEX Cloud Assessment Database
Stubenrauch *et al.*. BAMS 2013

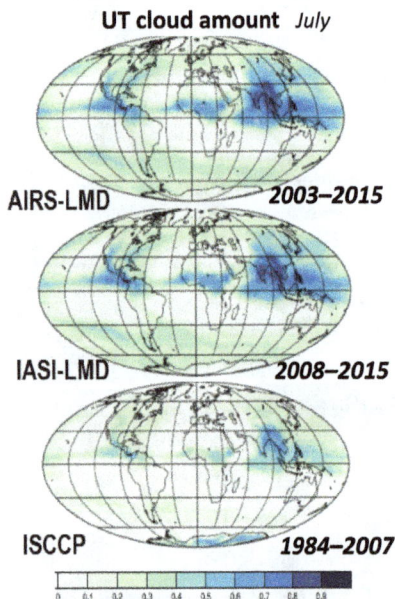

➢ Long time series and good areal coverage

➢ **Good IR spectral resolution -> sensitive to cirrus**
day & night, $COD_{vis} > 0.2$*, also above low clouds*

CIRS (Cloud retrieval from IR Sounders):
Stubenrauch *et al.*, *J. Clim. 1999, 2006; ACP 2010,* ***ACP 2017***
AIRS / IASI cloud climatologies -> French data centre AERIS
HIRS cloud climatology -> EUMETSAT CM-SAF (DWD)

Stubenrauch *et al.*, ACP 2017

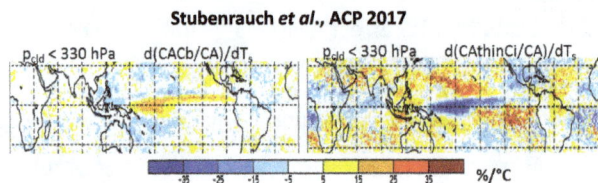

p_{cld} < 330 hPa d(CACb/CA)/dT_s

p_{cld} < 330 hPa d(CAthinCi/CA)/dT_s

-35 -25 -15 -5 5 15 25 35 %/°C

Changes in the occurrence of Cb and thin Ci clouds relative to all clouds per ° C warming show different geographical patterns and slight tropical increase in Ci, thCi rel to all clouds
-> change in heating gradients

Slide 7

From Cloud Retrieval to Cloud Systems

Clouds are **extended objects**, driven by dynamics -> **organized systems**

Method: (1) Group adjacent grid boxes with high clouds of similar height (p_{cld})

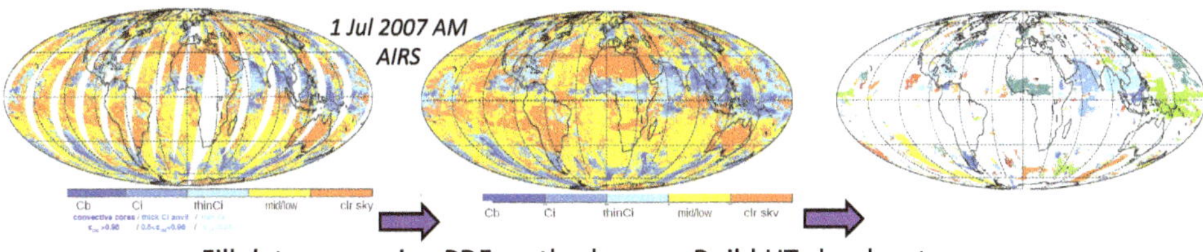

Protopapadaki *et al.*, ACP 2017

(2) Use ε_{cld} to distinguish convective core, thick cirrus, thin cirrus *(only IR sounder)*

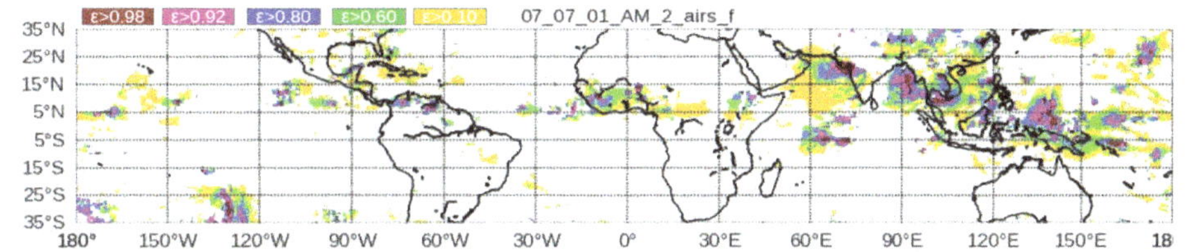

30° N–30° S: UT cloud systems cover 25%, those without convective core 5%
50% of these originate from convection (Luo and Rossow, 2004; Riihimaki *et al.*, 2012)

Slide 8

Observation Synergies

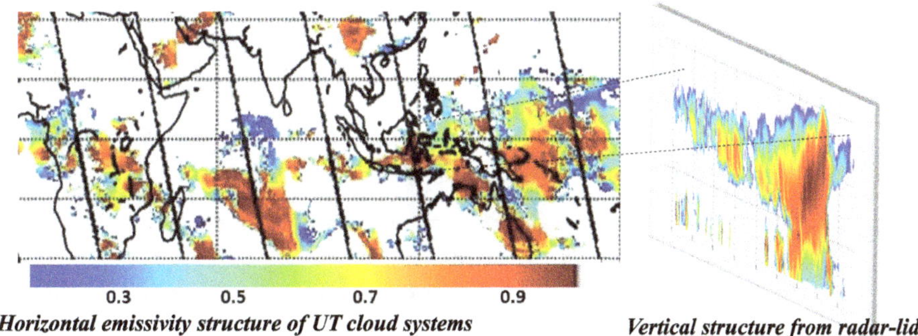

Horizontal emissivity structure of UT cloud systems
compared to other studies, IR sounders add thin cirrus (0.1 < ε_{cld} < 0.5)

Vertical structure from radar-lidar:
radar reflectivity of convective system

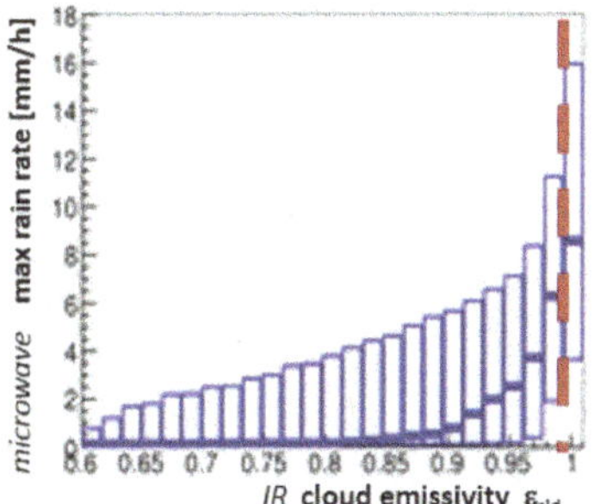

A-Train synergy (1h30 AM / PM)
AIRS – CALIPSO – CloudSat – AMSR-E

Microwave imager - IR sounder synergy:
definition of convective core : ε_{cld} > 0.98

Slide 9

Life Cycle of Deep Convective Cloud Systems

Max. of convection over land / ocean : 16–18 h / early morning

problem: most polar sunsynchronous observations do not catch this

-> Use good time resolution of geostationary satellite imagers

and track cold convective cores with T_B^{IR}

However T_B^{IR} depends on T_{cld} and on ε_{cld}

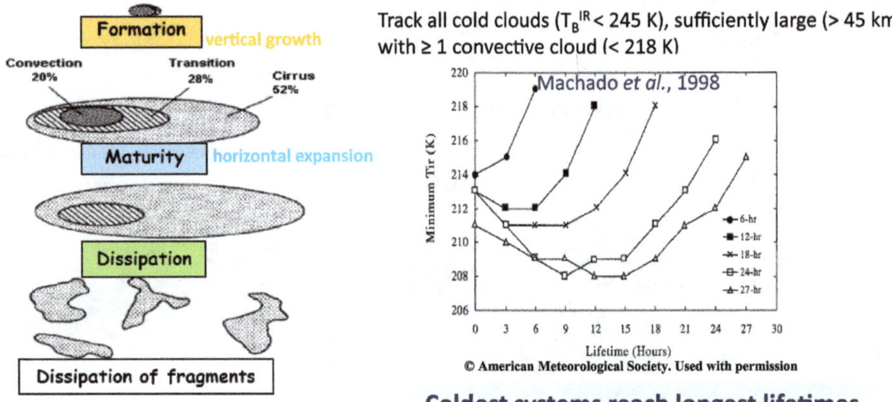

Coldest systems reach longest lifetimes

(Yuan and Houze, 2010) (<260 K + AMSR-E rain rate), (Fiolleau and Roca, 2013) (<233 K + TRMM rain rate)

Slide 10

Synergy with TRMM to analyze system life evolution

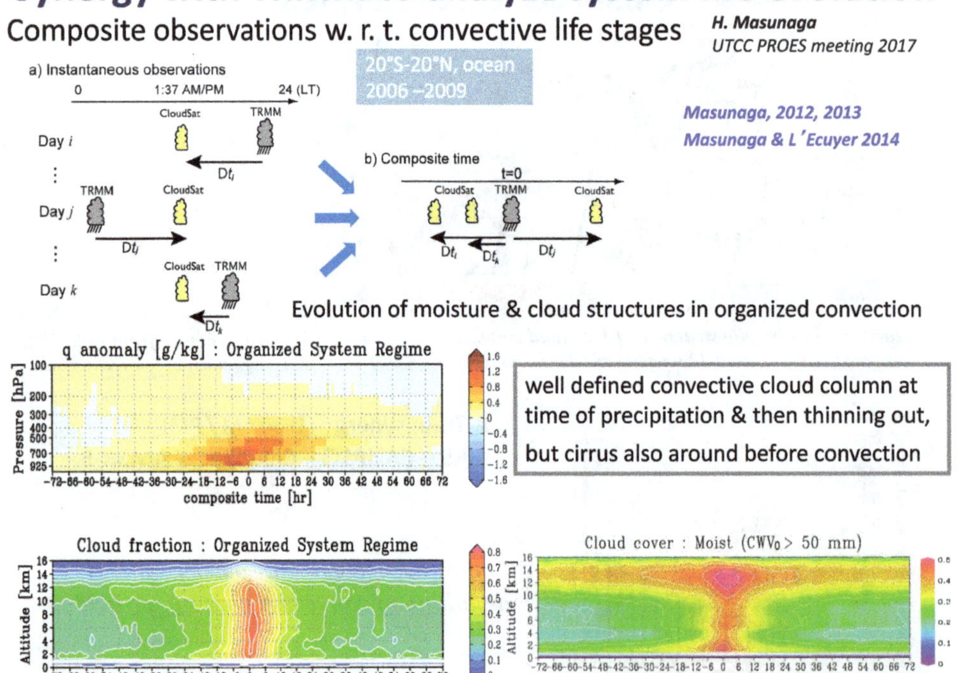

Slide 11

environment -> cloud system properties

CALIPSO-CloudSat **nadir track statistics**, ocean

S. van den Heever,
UTCC PROES meeting 2017

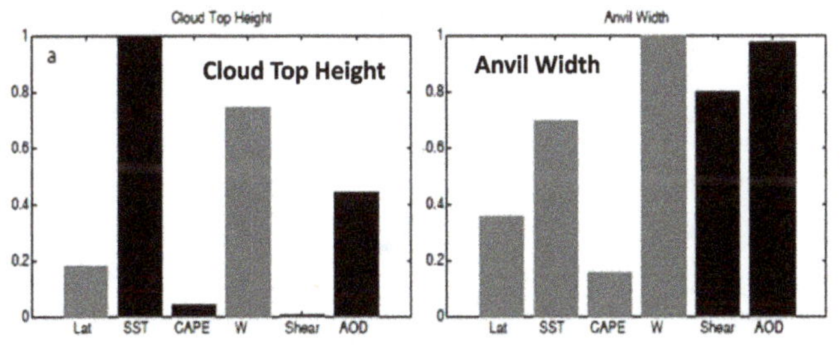

(Igel et al., 2014)

Normalized multiple linear regression coefficients for each predictor (latitude, SST, CAPE (ECMWF-CldSat), W_{500hPa}, shear (ERA-I), AOD (MODIS) for various cloud attributes. **Gray bars indicate negative values.**

vertical updraft & cld top height / anvil width well correlated

SST might be an important factor in convective strength

anvil width negatively related to SST

Slide 12

Goal: relate anvil properties to convective strength

Strategy: need proxies

- to identify convective cores

  $\varepsilon_{cld} > 0.98$ *(compared to AMSR-E rain rate)*

- to identify mature convective systems

  system core fraction : 0.1 – 0.3 *(reaching max core size)*

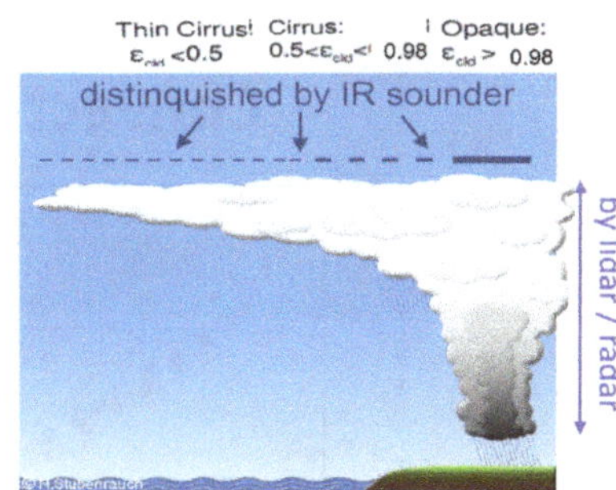

- to describe convective strength

  core temp. : T_{min}^{Cb} (Protopapadaki *et al.*, 2017)

  T_B^{IR} (Machado & Rossow, 1993)

  vertical updraft : ***CloudSat Echo Top Height*** / TRMM
  / conv mass transport (Takahashi & Luo, 2014 / Liu *et al.* (2007). Mullendore *et al.*, 2008)

  LNB : soundings / ***max mass flux outflow*** (Takahashi & Luo, 2012)

  heavy rain area: CloudSat-AMSR-E-MODIS (Yuan & Houze, 2010)

  core width : CloudSat (Igel *et al.*, 2014)

  mass flux : ERA-Interim + Lagrangian approach (Tissier *et al.*, 2016)
  A-Train + 1D cld model (Masunaga & Luo, 2016)

Slide 13

Convective Strength -> Cloud System Properties

(Protopapadaki *et al.*, 2017)

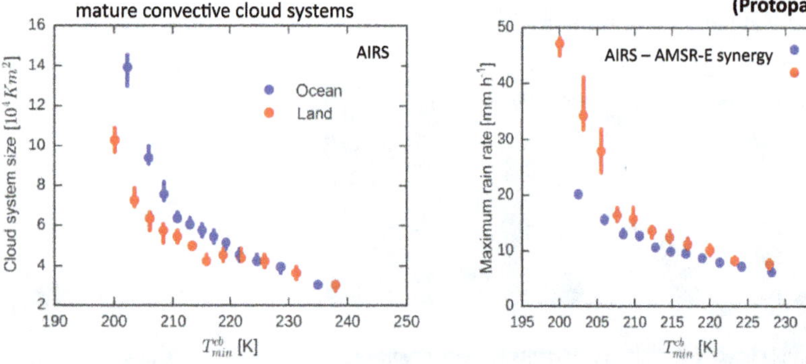

Cloud system size / max. rain rate increase with convective depth (colder cloud tops), but land–ocean differences:
at same height continental cloud systems stronger convective rain rate and smaller size
colder cores -> stronger max. RR => T^{cb}_{min} proxy for convective strength

TRMM study **(Liu *et al.*, 2007):**
larger updraft and convective cores, but smaller cloud systems
smaller updraft and convective cores, but larger cloud systems

CloudSat study (**Takahashi *et al.*, 2017**):
less entrainment - stronger entrainment

Slide 14

Convective Strength -> Anvil Properties

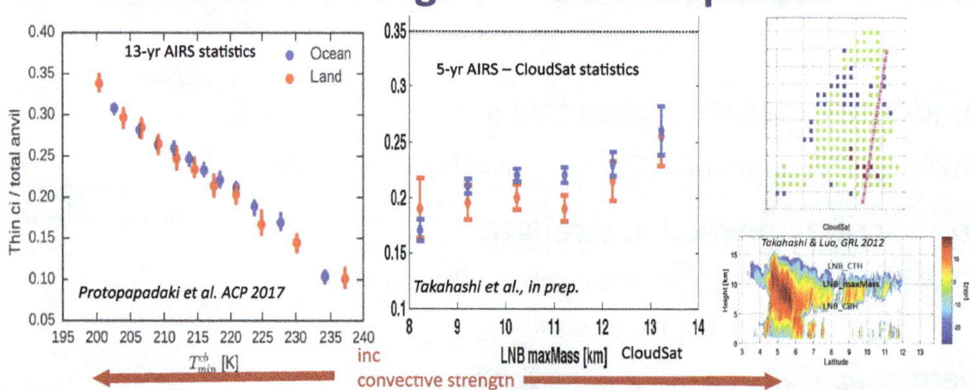

Mature convective systems: Increase of thin Ci with increasing convective strength!
similar land / ocean
Relation robust *using different proxies: T_{min}^{Cb} / LNB (max mass)*

Why?
H1: UT environmental predisposition (at higher altitude larger RH, T stratification)
H2: UT humidification from cirrus outflow
-> CRM studies

Slide 15

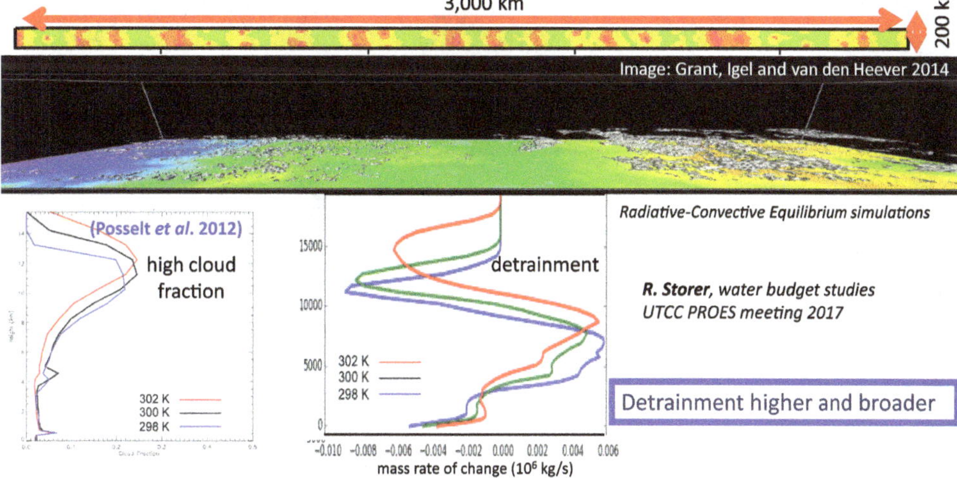

high cloud fraction

Slide 16

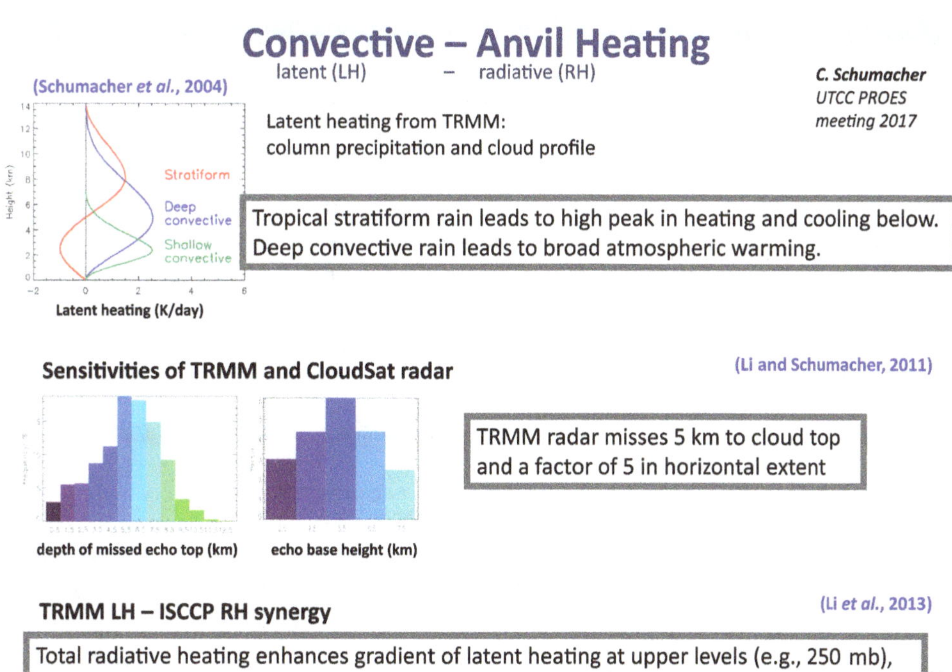

Convective – Anvil Heating
latent (LH) – radiative (RH)
C. Schumacher
mass rate of change (10⁶ kg/s)
R. Storer, water budget studies
UTCC PROES meeting 2017
Detrainment higher and broader
Increasing SST -> increased PW, convective intensity (w) and high cloud fraction, decrease in IR cooling -> slowing radiatively driven circulation
(Schumacher et al., 2004)
Stratiform
Deep convective
Shallow convective
Latent heating (K/day)
Latent heating from TRMM:
column precipitation and cloud profile
Tropical stratiform rain leads to high peak in heating and cooling below.
Deep convective rain leads to broad atmospheric warming.
Sensitivities of TRMM and CloudSat radar
(Li and Schumacher, 2011)
depth of missed echo top (km)
echo base height (km)
TRMM radar misses 5 km to cloud top and a factor of 5 in horizontal extent
TRMM LH – ISCCP RH synergy
(Li et al., 2013)
Total radiative heating enhances gradient of latent heating at upper levels (e.g., 250 mb), especially over Africa, Maritime Continent and South America, and enhances overall LH by ~20%

Slide 17

Heating Rates of UT Cloud Systems

UT heating due to cirrus -> impact on large-scale tropical atmospheric circulation

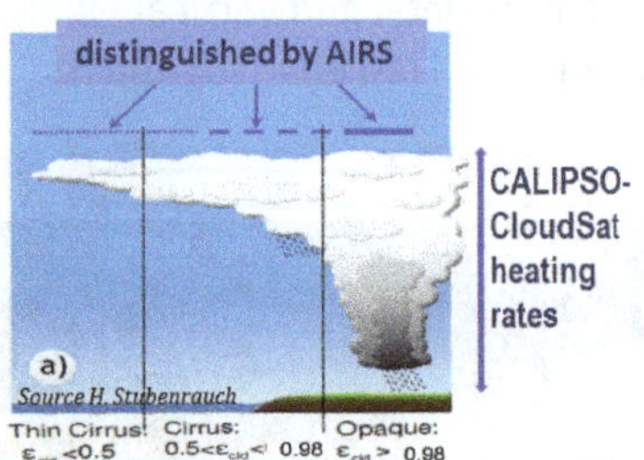

Heating will be affected by:

- areal coverage
- emissivity distribution
- vertical structure of cirrus anvils (layering and microphysics)

Propagate nadir track info on vertical structure across UT cloud systems

AIRS – CloudSat-CALIPSO synergy

Categorize NASA CloudSat FLXHR-LIDAR heating rates wrt to ε_{cld}, p_{cld}, vert. layering, thermodyn.

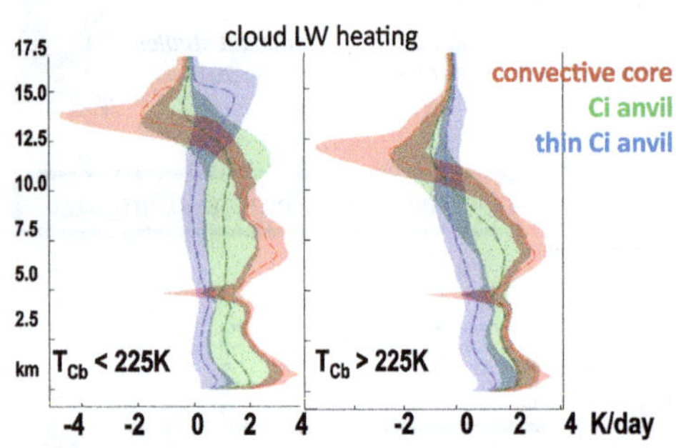

Clear distinction of heating associated with each category

Thin Ci heating increases with convective strength

Slide 18

UT Cloud System Concept to assess GCM parameterizations

analyze GCM clouds as seen from AIRS/IASI, via simulator *(M. Bonazzola, LMD)*

& construct UT cloud systems

-> evaluation of GCM convection schemes / detrainment / microphysics

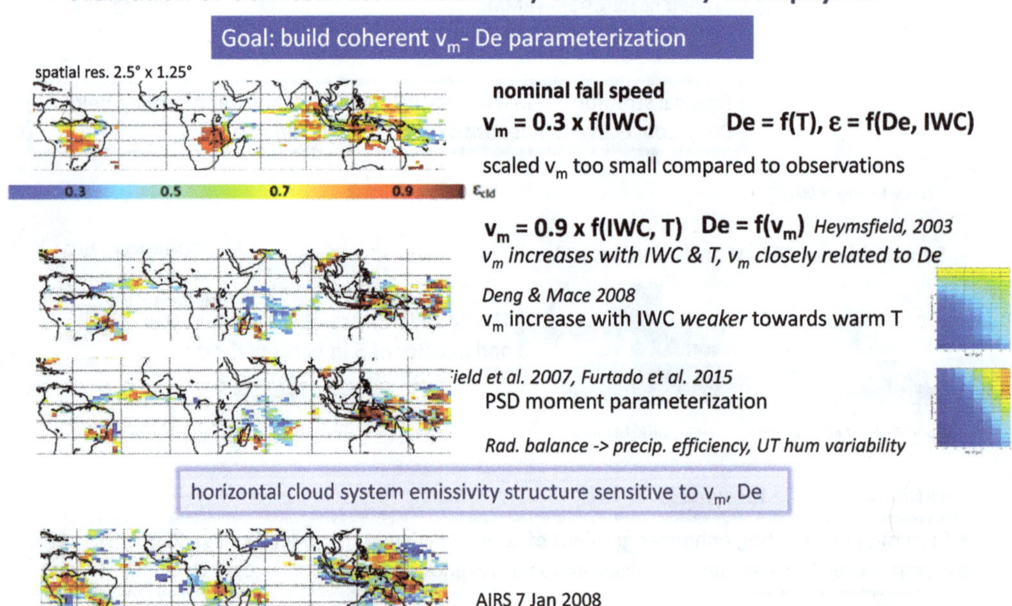

Goal: build coherent v_m- De parameterization
spatial res. 2.5° x 1.25°
nominal fall speed
v_m = 0.3 x f(IWC)
De = f(T), ε = f(De, IWC)
scaled v_m too small compared to observations
0.3 0.5 0.7 0.9 ε_{cld}
v_m = 0.9 x f(IWC, T) De = f(v_m) Heymsfield, 2003
v_m increases with IWC & T, v_m closely related to De
Deng & Mace 2008
v_m increase with IWC weaker towards warm T
ield et al. 2007, Furtado et al. 2015
PSD moment parameterization
Rad. balance -> precip. efficiency, UT hum variability
horizontal cloud system emissivity structure sensitive to v_m, De
AIRS 7 Jan 2008

Slide 19

process-oriented UT cloud system behaviour

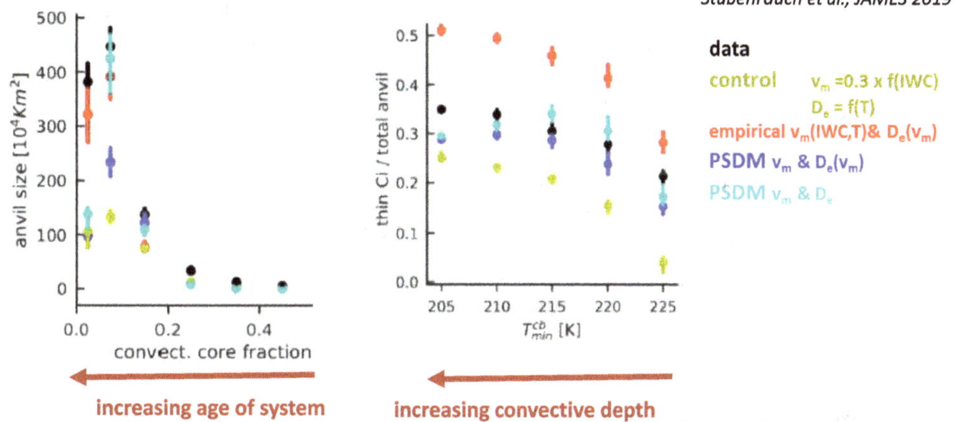

including T dependency of v_m -> larger spread in T
more realistic v_m –De very promising: leads to more realistic anvil size development and thin Ci increasing

Next steps: integrate single scattering properties developed by *Baran et al. 2016* from PSD's of F07
more realistic UT humidity variability threshold (AIRS climatology of Kahn and Teixeria, 2009)
precipitation – detrainment efficiency parameterization

Slide 20

Summary and Outlook

GEWEX UTCC PROES: Cooperations being formed, focusing on tropical convective systems
coord. C. Stubenrauch & G. Stephens *https://gewex-utcc-proes.aeris-data/fr*

➢ **AIRS and IASI cloud climatologies will be distributed by AERIS and be part of an updated GEWEX Cloud Assessment database** *(end 2018)*

➢ **UT cloud system concept based on IR sounder data powerful tool**

1) To study the relation between convection and anvil properties:

emissivity structure of mature systems changes with convective strength:
more surrounding thin cirrus

2) For process-oriented metrics to evaluate GCM parameterizations linked to convection/detrainment/microphysics *(fallspeed – De)*

➢ **Categorization of heating rates (A-Train synergy) wrt to ε_{cld}, p_{cld} shows clear distinction**
thin Ci heating larger for colder systems

❑ **Expand heating rates across UT cloud systems and integrate into feedback studies**
using Lagrangian transport and advanced analysis methods

❑ **Investigate mechanisms leading to emissivity structure in CRM RCE studies** *(large domain)*

Slide Captions

Slide 1

Slide 2 **Motivation:** Upper Tropospheric (UT) clouds cover ~30% of the Earth (which corresponds to ~40% of the Earth's total cloud cover) and exert a strong greenhouse effect. Therefore, they play a vital role in modulating the Earth's energy budget and heat transport. In tropical convective regions, where these clouds are most abundant, more than 50% of the total heating is attributed to cirrus radiative heating (e.g., [Sohn, 1999]). These clouds often form mesoscale systems extending over thousands of kilometers, either as cirrus anvils from convective outflow or as isolated cirrus by *in situ* freezing. We want to gain a better understanding of how convection affects the UT clouds and vice versa. Critical to these interactions is cirrus radiative heating.

Slide 3 **Elements of UT cloud feedback:** Convective cloud systems (composed of a convective core, cirrus and thin cirrus anvil) over the warm surface near the Equator on one side. On the other side, low-level clouds and clear skies in cooler subsidence regions are coupled by the atmospheric large-scale circulation. Deep convective cloud systems are involved in a series of positive and negative feedback processes, mostly through their anvils. When tropical convection intensifies in a warming climate, it is hypothesized that this alters the properties of the cirrus anvils in such a way that the resulting atmospheric heating gradients between convective cloud systems and subsidence regions considerably influence the large-scale circulation, which itself then modifies patterns of intense precipitation. The height and amount of UT clouds are constrained by the clear sky mass and energy budget. Large-scale modeling is necessary to identify the most influential climate feedback mechanisms, in general, by confronting simulations of a present and a future climate. However, the outcome of such feedback studies relies upon the representation of convection and detrainment in these models. Therefore, simulations of our present climate should agree with observations. The UTCC PROES goals are hence to (1) understand and quantify the relation between convection, the outflowing cirrus anvil, and their radiative heating, and (2) provide observation-based metrics to evaluate detrainment processes in models.

Slide 4 **Hypotheses on UT cloud feedback to a warmer environment:** There exist several hypotheses concerning the UT cloud feedback to a warming climate. The Thermostat hypothesis (Ramanathan and Collins, 1991) assumes that increased boundary layer humidity leads to increased convective condensate, which produces larger, thicker cirrus anvils and an increased shortwave radiative forcing. This leads to negative cloud feedback and positive water vapor feedback. The Iris effect (Lindzen *et al.*, 2001; Lin *et al.*, 2002) assumes that increased convective condensate leads to a larger precipitation efficiency and thus reduced detrainment and a reduced anvil area. This corresponds to slightly positive cloud feedback and negative water vapor feedback (when radiative properties are accounted for). Detrainment (ice formation at upper levels) and precipitation (condensate removal at lower levels) offset each other in regulating the top of atmosphere (TOA) energy balance. An observational study, using data of the Tropical Rainfall Measuring Mission (TRMM), has revealed that both the ice water path and rainwater increase with sea surface temperature (SST) (Del Genio *et al.*, 2005). This suggests neutral cloud feedback, which is intermediate between the thermostat and IRIS extremes. A study by Mauritsen and Stevens (2015) indicates that enhanced convective aggregation may lead to an increased IRIS effect, and the inclusion of cloud radiative effects leads to a narrowing of the rainy areas. Bony *et al.* (2016) developed the hypothesis of the stability IRIS effect, using idealized radiative-convective equilibrium simulations: anvils rise in a warming environment, which leads to an increase in static stability and therefore to a reduction of convective outflow. On a shorter timescale, the Humidistat hypothesis (Stephens *et al.*, 2004; Lebsock *et al.*, 2010) describes the self-regulating radiative-convective feedback mechanism via cirrus anvil heating in three phases: (1) a warming destabilization phase with clear sky, calm winds, minimal evaporation and a radiative cooled UT, (2) a convective phase associated with heavy precipitation, and (3) a restoring phase characterized by UT clouds maintained by residual moisture. By absorbing infrared (IR) radiation, they warm the UT, thereby returning the atmosphere to a condition that will enable a new destabilization phase.

Slide 5 **UTCC PROES strategy:** The Global Energy and Water Exchanges (GEWEX) UTCC PROES working group links communities from observations, radiative transfer, transport modeling, cloud and climate modeling. The focus is first on tropical convective systems. A link to another World Climate Research Programme (WCRP) core project, SPARC (Stratosphere-troposphere Processes And their Role in Climate), arises by widening the focus on the role of isolated cirrus systems originating from *in situ* freezing and their contribution to radiative heating, and in studying the penetration of tropical convection into the stratosphere (Rossow and Pearl, 2007; Romps and Kuang, 2009; Takahashi and Luo, 2014) and its role for the stratospheric dynamics. As we are also interested in the thinner parts of cirrus, UTCC PROES introduces a new database of UT cloud systems (Protopapadaki *et al.*, 2017), based on cloud properties from IR sounders (AIRS [Atmospheric Infrared Sounder] and IASI [Infrared Atmospheric Interferometer]): It provides the horizontal emissivity structure within these systems and is being complemented by CloudSat-CALIPSO radar-lidar track vertical information, thermodynamic and dynamic information from meteorological reanalyses, and information on the life cycle stage from the TRMM radar and geostationary imagers. The Cloud System Concept, distinguishing convective cores, cirrus and thin cirrus anvil, permits to directly link the anvil properties to convective strength. Radiative heating rates, computed from active radar-lidar measurements along narrow nadir tracks will be expanded to the different parts of the UT cloud systems. The evolution of the systems can be studied using a Lagrangian transport model. This Cloud System Concept also allows one to investigate how cloud systems behave in Cloud Resolving Model studies and General Circulation Model (GCM) simulations, under different parameterizations of convection, detrainment and microphysics.

Slide 6 **Why use IR Sounders to derive cirrus properties?** Since the 1980s, IR Sounders have been providing reliable cloud properties. Their good spectral resolution makes them sensitive to cirrus, down to a visible optical depth of 0.2, also above low-level clouds, during daytime and nighttime. Compared to the International Satellite Cloud Climatology Project (ISCCP), geographical cloud structures are similar, but more UT clouds are identified. So far, the Cloud retrieval from IR Sounders (CIRS), developed at Laboratoire de Météorologie Dynamique (LMD), has been used to create cloud data records from Atmospheric Infrared Sounder (AIRS) (15 years) and IASI (Infrared Atmospheric Interferometer) (10 years), and production will be continued at the French data center AERIS. Recently, CIRS has been applied to 35 years of High Resolution Infrared Radiation Sounder (HIRS) data by the Climate Monitoring Satellite Application Facility of EUMETSAT. One way to study change patterns due to global warming is by relating global or tropical mean surface temperature anomalies to changes in other parameters (e.g., [Zhou *et al.*, 2015; Liu *et al.*, 2017]). By distinguishing between deep convective clouds and thin cirrus, we observe different geographical change patterns per degree Celsius of global warming: While the relative amount of deep convective clouds increases in a narrow band near the Equator, the one of thin cirrus increases around this band (Stubenrauch *et al.*, 2017). This may lead to changing heating gradients.

Slide 7 **From cloud retrieval to cloud systems:** The Cloud System Concept is based on two independent variables: After having filled data gaps between adjacent orbits, UT cloud systems are first built from adjacent elements of similar cloud height (given by cloud pressure, p_{cld} < 440 hPa); the horizontal emissivity structure then allows them to distinguish between convective cores, thick cirrus, and thin cirrus anvil. The map presents a snapshot of these UT cloud systems. Note that the yellow parts of these systems, with emissivity <0.6, can only be reliably identified by IR sounders. When using cells of 0.5° latitude x 0.5° longitude and demanding more than 70% UT clouds within these cells, UT cloud systems cover 25% of the latitude band 30°N to 30°S. Those without convective cores cover 5%, and according to Luo and Rossow (2004) and Riihimaki *et al.* (2012), about 50% of these originate from convection.

Slide 8 **Observation synergies:** We will first exploit the A-Train synergy: While the horizontal structure of the UT cloud systems is given by AIRS, the vertical structure is sampled along the radar-lidar nadir tracks of CALIPSO and CloudSat. The Advanced Microwave Scanning Radiometer–EOS (AMSR-E), providing the rain rate, has been used to establish the emissivity threshold for the definition of the convective cores.

Slide 9 **Life cycle of deep convective cloud systems:** As the maximum of convection over land is in the early evening and over the ocean in the early morning, most polar sun-synchronous observations do not catch these. Therefore geostationary imagers, having a good time resolution, have been used to track cold convective cores, defined by a threshold in IR brightness temperature (TB). Machado and Rossow (1993) analyzed sufficiently cold cloud systems (TB < 245 K and radius > 45 km) with convective cores defined by TB < 218 K and revealed that the coldest systems reach the longest lifetimes. More recent studies (e.g., [Yuan and Houze, 2010; Fiolleau and Roca, 2013]) include rain rate to define the convective cores, but the anvils are still defined via TB (<260 K and <233 K, respectively). However, TB depends on both cloud top temperature and cloud emissivity. This implies that only thick anvils can be analyzed by this method. Thinner, semi-transparent cirrus have a warmer TB, as they are partially transparent to the radiation of the underlying atmosphere and surface.

Slide 10 **Synergy with TRMM to analyze cloud system life evolution:** Recent observation analysis techniques of convection provide insight (e.g., [Masunaga, 2013; Masunaga and L'Ecuyer, 2014; Masunaga and Luo, 2016]) into processes here by building composite observations with respect to convective life stages. Therefore, TRMM, which observes the same location in the tropics at different day times on subsequent days, is used to determine time t_0 at which there is a deep convective rain event, and by combining this location with AIRS and CloudSat observations, the evolution of moisture and cloud structures can be followed in a statistical way. This revealed large humidity near the surface before the precipitation event, which then rises into the UT, and a well-defined convective cloud column at the time of precipitation, which then thins out. On the other hand, cirrus is ubiquitous in a moist environment.

Slide 11 **Cloud system properties in dependence of atmospheric environment:** Igel and van den Heever (2015) analyzed oceanic cloud systems along the narrow CALIPSO-CloudSat nadir tracks in combination with information from meteorological reanalyses and aerosol optical depth from the Moderate Resolution Imaging Spectroradiometer (MODIS) to establish correlations between atmospheric environment and cloud system properties. They found that the cloud top height and anvil width increase with increasing vertical updraft; SST seems to be an important factor for convective strength (as cloud top height increases). The anvil width seems to be smaller in a warmer environment, while it increases with increasing vertical wind shear and aerosol optical depth.

Slide 12 **How to achieve the goal to relate anvil properties to convective strength:** Various aspects of convection can be assessed using different measures of convective intensity, for instance, vertical updraft and radar echo top height (e.g., [Liu *et al.*, 2007; Takahashi and Luo, 2014]), level of neutral buoyancy (LNB) (Takahashi and Luo, 2012), area of heavy rainfall (Yuan and Houze, 2010), and mass flux (Tissier and Legras, 2016; Masunaga and Luo, 2016). To relate the cirrus anvil properties, obtained from AIRS, to the properties of the convective core, one also needs proxies to identify convective cores (here, it is cloud emissivity in comparison with the AMSR-E rain rate) and mature convective systems (so that one can decouple convective strength from the lifetime evolution of the system). The convective core fraction within the UT cloud system is used: mature systems reach maximum core size when the core fraction lies between 0.1 and 0.3.

Slide 13 **Cloud system properties in dependence of convective depth and strength:** For mature convective cloud systems, both cloud system size and maximum rain rate within the convective core increase with increasing convective depth, represented by decreasing minimum cloud top temperature within the convective core. However, the behavior is slightly different for continental and oceanic convective systems: at the same height, continental convective cloud systems have a stronger convective rain rate and smaller system size. As colder convective cores lead to a stronger convective rain rate, this indicates the minimum temperature as a good proxy for convective strength. These findings are in agreement with a TRMM study by Liu *et al.* (2007), revealing that continental convective cloud systems have, in general, a larger convective updraft and larger convective cores while building smaller systems, whereas oceanic convective systems have smaller convective updrafts and smaller convective cores while building larger systems. This can be understood in terms of entrainment: A CloudSat study by Takahashi *et al.* (2017) detected less entrainment for continental convective systems and stronger entrainment for oceanic convective systems, leading to more dilution.

Slide 14 **Anvil properties in dependence of convective strength:** According to the AIRS cloud data together with the applied Cloud System Concept, deeper convection leads to relatively more thin cirrus within larger anvils (Protopapadaki *et al.*, 2017). When one uses the height of the maximum anvil mass flux, determined from CloudSat, as a proxy for convective strength (Takahashi and Luo, 2012), the relationship still holds. This relationship is similar for continental and oceanic convective cloud systems. The question now is: Why does deeper convection lead to relatively thinner cirrus within larger anvils? Is it due to UT environmental predisposition, with larger relative humidity and temperature stratification at a higher altitude? Or is UT humidification from convection the more relevant factor? The Cloud System Concept, similarly applied to observations and cloud-resolving model (CRM) simulations, will enable insight into the processes that shape these clouds.

Slide 15 **Characteristics of deep convection from CRM simulations:** To unfold the underlying processes causing the observed relationships, model simulations of convective plumes and detrained anvils are necessary. These processes are well described by CRMs because they explicitly resolve convective updrafts (at a scale of 1 km). Moreover, they allow the testing of the sensitivity of anvil characteristics, including their atmospheric heating and cooling, to certain processes or parameterizations. At the GEWEX UTCC PROES meeting in 2017, Susan van den Heever presented characteristics of deep convection from CRM simulations, in particular, under different environmental conditions (surface temperature varying between 298 K and 302 K). To attain Radiative-Convective Equilibrium (RCE), one needs a large domain (e.g., 300 km times 200 km) and a relatively long time interval (about one month). It has been demonstrated with these idealized simulations that with increasing SST precipitable water, convective intensity (represented by vertical updraft), and UT cloud fraction increase, the resulting decrease in IR cooling is slowing down the radiatively driven atmospheric circulation (Posselt *et al.*,

2012; Igel *et al.*, 2014). Rachel Storer showed that the detrainment, presented by the mass rate of change, is located higher and is broader at larger SST.

Slide 16 **Convective and anvil heating**: Courtney Schumacher presented results from the TRMM: Latent heating is obtained from TRMM column precipitation and cloud profiles. One clearly distinguishes three different latent heating profiles: While tropical stratiform rain leads to heating in the upper troposphere and a cooling below, deep convective rain leads to broad atmospheric warming in the middle troposphere. Shallow convection only heats the lower troposphere (Schumacher *et al.*, 2004). Compared to the CloudSat radar, the TRMM radar, on average, misses a column of 5 km of cloud particles below cloud top and underestimates the horizontal extent of the convective systems by a factor of 4 (Li and Schumacher, 2011). In synergy with the ISCCP, to include the radiative heating, it was found that the total radiative heating enhances the gradient of latent heating at the upper levels (e.g., 250 hPa), especially over Africa, the Maritime Continent, and South America. The overall latent heating is enhanced by about 20% (Li *et al.*, 2013). A recent study (Stubenrauch *et al.*, 2021) estimated 22+3%, but with radiative heating throughout the troposphere from 250 hPa downward.

Slide 17 **Heating rates of UT cloud systems:** UT heating, which is induced by the anvils and cirrus, impacts the large-scale tropical atmospheric circulation. This heating is affected by: (1) areal coverage of UT clouds, (2) horizontal cloud emissivity structure within the UTC systems, and (3) vertical structure of the cirrus anvils (layering and microphysics). By categorizing the NASA CloudSat-lidar heating rates (L'Ecuyer *et al.*, 2008; Henderson *et al.*, 2013) with respect to cloud emissivity and cloud pressure, we demonstrated a clear distinction of the longwave (LW) heating associated with the convective core, cirrus, and thin cirrus anvil: Relatively, opaque clouds heat the atmospheric column below by trapping surface emissions but cool it above due to excess emission, in contrast to the thin cirrus heating the UT by intercepting the LW radiation coming from below. In addition, colder convective systems lead to much larger thin cirrus heating of the upper troposphere.

Slide 18 **UT Cloud System Concept to assess GCM parameterizations:** The UT Cloud System Concept can be used to evaluate GCM parameterizations of convection, detrainment and microphysics. Therefore, one should analyze the GCM clouds as seen from AIRS or IASI via a satellite simulator and construct the UT cloud systems at GCM resolution. Recently, we used this concept to assess parameterizations of coherent bulk ice-fall speed, v_m, and effective ice crystal size, D_e (Stubenrauch *et al.*, 2019). The first dictates the life time, while the latter strongly influences cloud radiative effects. As both depend on the ratio of mass over the area of the ice crystals in the cloud, they are strongly related. In the original LMDZ GCM (Hourdin *et al.*, 2013), v_m depends on ice water content (IWC) and D_e on temperature (T). In addition, v_m is scaled down by a factor of 0.3 to achieve radiation balance. Compared to AIRS data, this model version produces, in general, UT cloud systems with insufficient thin cirrus while using parameterizations of ice fall speed, which depend on both IWC and T to improve the agreement.

Slide 19 **Process-oriented UT cloud system behavior:** By using the convective core fraction within a cloud system as a proxy for the age of the system (decreasing with age) and minimum cloud top temperature within the convective core as a proxy for convective depth (decreasing with convective depth), the assessment becomes process-oriented. According to the data analysis, the anvil size increases until maturity is reached (when the convective core fraction lies between 0.2 and 0.3) and then decreases towards dissipation. The ratio of thin cirrus over total anvil size increases with increasing depth. All model versions produce this kind of behavior, but the nominal version has a much smaller anvil size and a smaller ratio of thin cirrus over total anvil size. However, the more coherent v_m – De parameterizations agree better with the data. To make a more realistic ice-fall speed possible (moving the scaling factor from 0.3 to 0.9), one had to decrease the threshold in UT subgrid variability in order to get enough UT clouds. For the next step, the constraint on this variable will be compared to observations (Kahn and Teixeria, 2009), and then the improved ice crystal single scattering property parameterization of (Baran *et al.*, 2016) will be tested.

Slide 20 **Summary and outlook:** First collaborations have been formed within GEWEX UTCC PROES, focusing on tropical convective cloud systems. AIRS and IASI cloud climatologies will be distributed by the French data center AERIS and be part of an updated GEWEX Cloud Assessment database (also to be distributed by AERIS). The AIRS-IASI simulator will be integrated into COSP (Bodas-Salcedo *et al.*, 2011). The UT cloud system concept, based on IR sounder data, is a powerful tool to study the relationship between convection and the resulting anvil properties and for process-oriented metrics to evaluate GCM parameterizations linked to microphysics, detrainment and convection. It has been demonstrated that the horizontal emissivity structure of mature convective systems changes with increasing convective depth towards relatively more thin cirrus within (or surrounding) the anvil. The categorization of heating rates with respect to cloud emissivity and pressure shows a clear distinction between those of convective core, cirrus and thin cirrus anvil. It also seems that the cirrus heating is larger for colder systems. For the next step, we will expand these heating rates, available only along the narrow radar-lidar nadir tracks, across the UT cloud systems and the surrounding environment in order to use them in feedback studies. To investigate the mechanisms leading to the horizontal emissivity structure, CRM RCE studies are foreseen.

References

Baran AJ, Hill P, Walters D, *et al.* (2016) The impact of two coupled cirrus microphysics–radiation parameterizations on the temperature and specific humidity biases in the tropical tropopause layer in a climate model. *Journal of Climate* **29**:5299–316. doi:10.1175/JCLI-D-15-0821.1

Bodas-Salcedo A, Webb MJ, Bony S, *et al.* (2011) COSP: Satellite simulation software for model assessment. *Bulletin of the American Meteorological Society* **92**:1023–43.

Bony S, Stevens B, Coppin D, *et al.* (2016) Thermodynamic control of anvil cloud amount. *Proceedings of the National Academy of Sciences* **113**:8927–32. doi:10.1073/pnas.1601472113

Del Genio AD, Kovari W, Yao M, Jonas J. (2005) Cumulus microphysics and climate sensitivity. *Journal of Climate* **18**:2376–87. doi:10.1175/JCLI3413.1

Deng M, Mace GG. (2008) Cirrus cloud microphysical properties and air motion statistics using cloud radar Doppler moments: Water content, particle size, and sedimentation relationships. *Geophysical Research Letters* **35**:L17808. doi:10.1029/2008GL035054

Field PR, Heymsfield AJ, Bansemer A. (2007) Snow size distribution parameterization for midlatitude and tropical ice clouds. *Journal of Atmospheric Sciences* **64**:4346–65. doi:10.1175/2007JAS2344.1

Fiolleau T, Roca R (2013) Composite life cycle of tropical mesoscale convective systems from geostationary and low Earth orbit satellite observations: Method and sampling considerations. *Quarterly Journal of the Royal Meteorological Society* **139**:941–53. doi:10.1002/qj.2174

Furtado K, Field PR, Cotton R, Baran AJ. (2015) The sensitivity of simulated high clouds to ice crystal fall speed, shape and size distribution. *Quarterly Journal of the Royal Meteorological Society* **140**:1546–59. doi:10.1002/qj.2457

Henderson DS, L'Ecuyer T, Stephens G, *et al.* (2013) A multisensor perspective on the radiative impacts of clouds and aerosols. *Journal of Applied Meteorology and Climatology* **52**:853–71. doi:10.1175/JAMC-D-12-025.1

Hourdin F, Grandpeix J-Y, Rio C, *et al.* (2013) LMDZ5B: The atmospheric component of the IPSL climate model with revisited parameterizations for clouds and convection. *Climate Dynamics.* **40**:2193–222.

Heymsfield AJ. (2003) Properties of tropical and midlatitude ice cloud particle ensembles. Part II: Applications for mesoscale and climate models. *Journal of Atmospheric Sciences* **60**:2573–91. doi:10.1175/1520-0469(2003)060<2592:POTAMI>2.0.CO;2

Igel MR, van den Heever SC. (2015) The relative influence of environmental characteristics on tropical deep convective storm morphology as observed by CloudSat. *Journal of Geophysical Research* **120**:4304–22.

Igel MR, van den Heever SC, Stephens GL, Posselt DJ. (2014) Convective-scale responses of a large-domain, modeled tropical environment to surface warming. *Quarterly Journal of the Royal Meteorological Society* **140**:1333–43.

Kahn BH, Teixeira J. (2009) A global climatology of temperature and water vapor variance scaling from the atmospheric infrared sounder. *Journal of Climate* **22**:5558–76. doi:10.1175/2009JCLI2934.1

Kahn BH, Teixeira J, Fetzer EJ, Gettelman A, Hristova-Veleva SM, Huang X, Kochanski AK, Köhler M, Krueger SK, Wood R, Zhao M. (2011) Temperature and water vapor variance scaling in global models: Comparisons to satellite and aircraft data, *Journal of Atmospheric Sciences* **68**:2156–68. doi:10.1175/2011JAS3737.1

Lebsock MD, Stephens GL, Kummerow C. (2010) An observed tropical oceanic radiative-convective cloud feedback. *Journal of Climate* **23**:2065–78. doi:10.1175/2009JCLI3091.1

L'Ecuyer TS, Wood NB, Haladay T. (2008) Impact of clouds on atmospheric heating based on the R04 CloudSat fluxes and heating rates data set. *Journal of Geophysical Research* **113**:D00A15. doi:10.1029/2008JD009951

Li W, Schumacher C. (2011) Thick anvils as viewed by the TRMM precipitation radar. *Journal of Climate* **24**:1718–35. doi:10.1175/2010JCLI3793.1

Li W, Schumacher C, McFarlane SA. (2013) Radiative heating of the ISCCP upper level cloud regimes and its impact on the large-scale tropical circulation. *Journal of Geophysical Research* **118**:592–604. doi:10.1002/jgrd.50114

Lin B, Wielicki BA, Chambers LH, *et al.* (2002) The iris hypothesis: A negative or positive cloud feedback? *Journal of Climate* **15**:3–7. doi:10.1175/1520-0442(2002)015<0003: TIHANO>2.0.CO;2

Lindzen RS, Chou M, Hou AY. (2001) Does the Earth Have an adaptive infrared iris? *Bulletin of the American Meteorological Society* **82**:417–32. doi:10.1175/1520-0477(2001)082<0417:DTEHAA>2.3.CO;2

Liu C, Zipser EJ, Nesbitt SW. (2007) Global distribution of tropical deep convection: Different perspectives from TRMM infrared and radar data. *Journal of Climate* **20**:489–503.

Liu R, Liou K-N, Su H, *et al.* (2017) High cloud variations with surface temperature from 2002 to 2015: Contributions to atmospheric radiative cooling rate and precipitation changes. *Journal of Geophysical Research: Atmospheres* **122**:5457–71. doi:10.1002/2016JD026303

Luo Z, Rossow WB. (2004) Characterizing tropical cirrus life cycle, evolution, and interaction with upper-tropospheric water vapor using Lagrangian trajectory analysis of satellite observations. *Journal of Climate* **17**:4541–63. doi:10.1175/3222.1

Machado LAT, Rossow WB. (1993) Structural characteristics and radiative properties of tropical cloud clusters. Mon. *Weather Rev.* **121**:3234–60. doi:10.1175/1520-0493(1993)121<3234:SCARPO>2.0.CO;2

Machado LAT, Rossow WB, Guedes RL, Walker AW. (1998) Life cycle variations of mesoscale convective systems over the Americas. *Monthly Weather Review* **126**:1630–54. doi:10.1175/1520-0493(1998)126<1630:LCVOMC>2.0.CO;2

Masunaga H. (2012) A Satellite Study of the Atmospheric Forcing and Response to Moist Convection over Tropical and Subtropical Oceans, *Journal of the Atmospheric Sciences* **69**:150–67. doi:10.1175/JAS-D-11-016.1

Masunaga H. (2013) A satellite study of tropical moist convection and environmental variability: A moisture and thermal budget analysis. *Journal of the Atmospheric Sciences* **70**:2443–66. doi:10.1175/JAS-D-12-0273.1

Masunaga H, L'Ecuyer TS. (2014) A mechanism of tropical convection inferred from observed variability in the moist static energy budget. *Journal of the Atmospheric Sciences* **71**:3747–66. doi:10.1175/JAS-D-14-0015.1

Masunaga H, Luo ZJ. (2016) Convective and large-scale mass flux profiles determined from synergistic analysis of a suite of satellite observations. *Journal of Geophysical Research* **121**:7958–74. doi:10.1002/2016JD024753

Mauritsen T, Stevens B. (2015). Missing iris effect as a possible cause of muted hydrological change and high climate sensitivity in models. *Nature Geoscience* **8**:346–51.

Mullendore GL, Homann AJ, Bevers K, Schumacher C. (2009) Radar reflectivity as a proxy for convective mass transport. *Journal of Geophysical Research* **114**(D16103). doi:10.1029/2008JD011431.

Posselt DJ, Van Den Heever S, Stephens GL, Igel MR. (2012) Changes in the interaction between tropical convection, radiation, and the large-scale circulation in a warming environment. *Journal of Climate* **25**:557–71. doi:10.1175/2011JCLI4167.1

Protopapadaki E-S, Stubenrauch CJ, Feofilov AG. (2017) Upper tropospheric cloud systems derived from IR sounders: Properties of cirrus anvils in the tropics. *Atmospheric Chemistry and Physics* **17**:3845–59. doi:10.5194/acp-17-3845-2017

Ramanathan V, Collins W. (1991) Thermodynamic regulation of ocean warming by cirrus clouds deduced from observations of the 1987 El Niño. *Nature* **351**:27–32. doi:10.1038/351027a0

Riihimaki LD, McFarlane SA, Liang C, *et al.* (2012) Comparison of methods to determine tropical tropopause layer cirrus formation mechanisms. *Journal of Geophysical Research* **117**:D06218. doi:10.1029/2011JD016832

Rossow WB, Pearl C. (2007) 22-yr survey of tropical convection penetrating into the lower stratosphere. *Geophysical Research Letters* **34**:L04803. doi:10/1029/2006GL028635

Schumacher C, Houze RA, Kraucunas I. (2004) The tropical dynamical response to latent heating estimates derived from the TRMM precipitation radar. *Journal of the Atmospheric Sciences* **61**:1341–58.

Sohn B-J. (1999) Cloud-induced infrared radiative heating and its implications for large-scale tropical circulations. *Journal of the Atmospheric Sciences* **56**:2657–72.

Stephens GL, Webster PJ, Johnson RH, *et al.* (2004) Observational evidence for the mutual regulation of the tropical hydrological cycle and tropical sea surface temperatures. *Journal of Climate* **17**:2213–24.

Stubenrauch, CJ, Chédin, A, Armante, R, Scott, NA. (1999) Clouds as seen by satellite Sounders (3I) and Imagers (ISCCP). Part II: A new approach for cloud parameter determination in the 3I algorithms, *Journal of Climate* **12**: 2214–23. doi:10.1175/1520-0442

Stubenrauch, CJ, Chédin, A, Rädel, G, Scott, NA, Serrar, S. (2006) Cloud properties and their seasonal and diurnal variability from TOVS Path-B. *Journal of Climate* **19**,5531–53. doi:10.1175/JCLI3929.1

Stubenrauch, CJ, Cros, S, Guignard, A, Lamquin, N. (2010) A 6-year global cloud climatology from the Atmospheric InfraRed Sounder AIRS and a statistical analysis in synergy with CALIPSO and CloudSat, *Atmospheric Chemistry and Physics* **10**, 7197–214. doi:10.5194/acp-10-7197-2010

Stubenrauch CJ, Rossow WB, Kinne S, *et al.* (2013) Assessment of global cloud datasets from satellites: Project and database initiated by the GEWEX Radiation Panel. *Bulletin of the American Meteorological Society* **94**:1031–49.

Stubenrauch CJ, Feofilov AG, Protopapadaki SE, Armante R. (2017) Cloud climatologies from the infrared sounders AIRS and IASI: Strengths and applications. *Atmospheric Chemistry and Physics* **17**:13625–44. doi:10.5194/acp-17-13625-2017

Stubenrauch CJ, Bonazzola M, Protopapadaki SE, Musat I. (2019) New cloud system metrics to assess bulk ice cloud schemes in a GCM. *Journal of Advances in Modeling Earth Systems* **11**:3212–34. doi:10.1029/2019MS001642

Stubenrauch CJ, Caria G, Protopapadaki SE, Hemmer F. (2021) 3D radiative heating of tropical upper tropospheric cloud systems derived from synergistic A-Train observations and machine learning. *Atmospheric Chemistry and Physics* **21**:1015–34. doi:10.5194/acp-21-1015-2021

Takahashi H, Luo ZJ. (2012) Where is the level of neutral buoyancy for deep convection? *Geophysical Research Letters* **39**. doi:10.1029/2012GL052638

Takahashi H, Luo ZJ. (2014) Characterizing tropical overshooting convection from joint analysis of CloudSat and geostationary satellite observations. *Journal of Geophysical Research* **119**:112–21. doi:10.1002/2013JD020972

Takahashi H, Luo ZJ, Stephens GL. (2017) Level of neutral buoyancy, deep convective outflow, and convective core: New perspectives based on 5 years of CloudSat data. *Journal of Geophysical Research: Atmospheres* **122**:2958–69. doi:10.1002/2016JD025969

Tissier AS, Legras B. (2016) Convective sources of trajectories traversing the tropical tropopause layer. *Atmospheric Chemistry and Physics* **16**:3383–98.

van den Heever SC. (2017) https://www7.obs-mip.fr/wp-content-aeris/uploads/sites/20/2017/12/03_anvils_van-den-Heever.pdf

Yuan J, Houze Jr RA. (2010) Global variability of mesoscale convective system anvil structure from A-train satellite data. *Journal of Climate* **23**:5864–88.

Zhou C, Zelinka MD, Dessler AE, Klein SA. (2015) The relationship between interannual and long-term cloud feedbacks. *Geophysical Research Letters* **42**:10463–9. doi:10.1002/2015GL066698

Lecture 15

Clouds and Precipitation in Extratropical Cyclones: A Global Satellite Climatology for GCM Evaluation

Catherine M. Naud

Columbia University/NASA-Goddard Institute for Space Studies, New York, NY, USA

Catherine Naud is a research scientist in the Applied Physics and Applied Mathematics Department of Columbia University and NASA Goddard Institute for Space Studies. She develops methods to combine observations of clouds and precipitation to help evaluate general circulation models. Her research focuses on both ground-based and satellite observations of clouds and precipitation in the mid-latitudes and more specifically in extratropical cyclones.

Introduction

The space race of the 1960s brought a new tool to meteorologists and climatologists: satellite-based observations of the Earth. With the advent of the Nimbus missions, the Earth could be observed globally and regularly. A new vantage point became available to observe clouds and other atmospheric components. Meteorologists could now follow weather systems, thanks to the strong contrast between bright clouds and the dark ocean surface in visible imagery and the strong link between the atmosphere dynamics and the condensation of water. Rapidly, methods emerged on how best to exploit the satellite imagery to obtain more detailed information on the clouds and continuously improve weather predictions. In parallel, climatological questions such as the impact of clouds on the global climate could be tackled finally, given the globally uniform view satellites offer. With this in mind, the International Satellite Cloud Climatology Project (ISCCP) was born; using multiple platforms that at any given time observe a significant portion of the planet, and by applying the same set of rules on similar measurements these platforms make, a global, 3-hourly, gridded dataset of cloud top pressure and optical depth was created by Robert Schiffer and William Rossow (Schiffer and Rossow, 1983). This was to be the first comprehensive, coherent (i.e., calibrated and normalized) and publicly accessible dataset specifically targeting cloud properties. It now spans four decades and has been used by thousands of researchers around the world. In its wake, more global climatologies have emerged, either based on different types of measurements or other atmospheric components, such as global datasets of precipitation, water vapor amounts, or aerosols.

The impact this influential dataset has had on current research is tremendous, but here we focus specifically on what this has implied for the interrelation between clouds and atmospheric dynamics. The choice of cloud top pressure and optical depth in the ISCCP was deliberate because these two parameters can uniquely inform on the atmospheric state. Inversely, specific weather phenomena have their own signature in the cloud fields, and such have extratropical cyclones in the mid-latitudes (Tselioudis *et al.*, 2000).

So Why Focus on Extratropical Cyclones?

Precipitation in the mid-latitudes depends strongly on the occurrence of extratropical cyclones. In the northern hemisphere, about 80% of the winter precipitation occurs in extratropical cyclones (Hawcroft *et al.*, 2012). Furthermore, in the United States, between 41% and 99% of the extreme precipitation events in the southeastern and northwestern states, respectively, are caused by either extratropical cyclones or their associated fronts (Kunkel *et al.*, 2012). Therefore, it is paramount to understand the dynamics of these systems and also the processes that lead to precipitation and its potentially devastating effects.

From a climatological standpoint, extratropical cyclones have a major regulating role, in terms of transferring energy and heat from the Equator to the pole and also through their radiative impact (Tselioudis *et al.*, 2000). Overall, they have a cooling effect (Polly and Rossow, 2016) of a magnitude that depends on their strength (Tselioudis and Rossow, 2006).

In a warming climate, the Equator-to-pole gradient is expected to decrease, lowering baroclinicity and the formation of cyclones. Therefore, a warming should entail a smaller number of cyclones. Coincidently, a warmer climate means a moister climate, which would imply greater precipitation in cyclones, resulting in greater latent heat release that would enhance the cyclones' intensity. However, past studies of the changes in extratropical cyclones give conflicting results — some studies find increases and other decreases in the number of cyclones, depending on the period considered (Feser *et al.*, 2015). Model predictions are also contradictory. Although some indicate a decrease in number and increase in intensity (Lambert and Fyfe, 2006), others find no changes at all (Bengtsson *et al.*, 2009), or different trends in different locations (Ulbrich *et al.*, 2008).

One aspect of the models that might in part explain the lack of consensus is their representation of moist processes, which have a large role in the development of the cyclones. Therefore, the representation of clouds and precipitation in extratropical cyclones has been the subject of extensive research. This is where satellite observations of clouds and precipitation prove extremely useful.

How Can Satellite Observations Help?

Conjunctly with the development of satellite-based datasets, new techniques have emerged to sort and exploit the wealth of information these observations provide. Specifically for extratropical cyclones, cyclone-centered compositing has been introduced. As far as we can tell, Lau and Crane (1995; 1997) were probably the first to arrange satellite observations of clouds in a frame of reference centered on a low pressure minimum. Thus, they were the first to introduce storm-centered composites of cloud observations. They used the ISCCP classification of clouds into the nine categories of high, mid, low top and thin, medium and thick clouds, and calculated the frequency of occurrence of each category by superimposing many events over a low and neighboring high pressure center. With this compositing framework, they could clearly show the large occurrence of high and thick clouds to the east of the low (location of strong ascent) and low clouds to the west (location of subsidence). Focusing only on the low pressure center, a whole series of studies used these composites to better understand cyclones, for example, to investigate the relationship between cyclone strength, environmental moisture, and precipitation in cyclones (Field and Wood, 2007), or the occurrence of supercooled droplets in the ascent region of the cyclones (Naud *et al.*, 2006). These techniques proved, in turn, very useful to evaluate general circulation models (e.g., [Klein and Jakob, 1999; Bauer and Del Genio, 2006; Field *et al.*, 2008; 2011; Govekar *et al.*, 2014]). This is because the composites allow the constraining of large-scale conditions before the model and observations are compared; they help alleviate the issue of non-uniform sampling in the observations and the lack of coincidence between observations and a free-running model.

How Can This Help Improve Climate Predictions?

Cyclones present a "great laboratory" for model evaluation: they include ascent and subsidence conditions, areas of moist warm advection and dry cold advection, high and low clouds, thick and thin clouds, and radiative cooling and warming. They can thus inform on the performance of the convection, boundary layer, or cloud microphysics schemes. By comparing composites of cloud properties or precipitation between satellite-observed cyclones and modeled cyclones, one can evaluate the performance of the convection, boundary layer, or cloud microphysics schemes (e.g., [Field *et al.*, 2011]). One can also assess the model's ability to correctly represent the sensitivity of these schemes to changes in the large-scale conditions, whether they are the cyclone dynamics itself or the environmental moisture or temperature (e.g., [Naud *et al.*, 2010; Booth *et al.*, 2013; Govekar *et al.*, 2014]). These types of tests help pinpoint possible deficiencies in specific aspects of the model or add confidence in a model's ability to correctly represent changes in the global climate, the frequency and intensity of extreme events, or in the large-scale dynamics of the model.

References

Bauer M, Del Genio AD. (2006) Composite analysis of winter cyclones in a GCM: Influence on climatological humidity. *Journal of Climate* **19**:1652–72.

Bengtsson L, Hodges KI, Keenlyside N. (2009) Will extratropical storms intensify in a warmer climate? *Journal of Climate* **22**:2276–301. doi:10.1175/2008JCLI2678.1

Booth JF, Naud CM, Del Genio AD. (2013) Diagnosing warm frontal cloud formation in a GCM: A novel approach using conditional subsetting. *Journal of Climate* **26**:5827–45.

Feser F, Barcikowska M, Krueger O, *et al.* (2015) Storminess over the North Atlantic and northwestern Europe — A review. *Quarterly Journal of the Royal Meteorological Society* **141**:350–82. doi:10.1002/qj.2364

Field PR, Wood R. (2007) Precipitation and cloud structure in midlatitude cyclones. *Journal of Climate* **20**:233–54. doi:10.1175/JCLI3998.1

Field PR, Bodas-Salcedo A, Brooks ME. (2011) Using model analysis and satellite data to assess cloud and precipitation in midlatitude cyclones. *Quarterly Journal of the Royal Meteorological Society* **137**:1501–15.

Field PR, Gettelman A, Neale R, *et al.* (2008) Midlatitude cyclone compositing to constrain climate model behavior using satellite observations. *Journal of Climate* **21**:5887–903.

Govekar PD, Jakob C, Catto J. (2014) The relationship between clouds and dynamics in Southern hemisphere extratropical cyclones in the real world and a climate model. *Journal of Geophysical Research* **119**:6609–28. doi:10.1002/2013JD020699

Hawcroft MK, Shaffrey LC, Hodges KI, Dacre HF. (2012) How much northern hemisphere precipitation is associated with extratropical cyclones? *Geophysical Research Letters* **39**:L24809. doi:10.1029/2012GL053866

Klein SA, Jakob C. (1999). Validation and sensitivities of frontal clouds simulated by the ECMWF model. *Monthly Weather Review* **127**:2514–31.

Kunkel KE, Easterling DR, Kristovich DAR, *et al.* (2012) Meteorological causes of the secular variations in observed extreme precipitation events for the conterminous United States. *Journal of Hydrometeorology* **13**:1131–41. doi:10.1175/JHM-D-11-0108.1

Lambert SJ, Fyfe JC. (2006) Changes in winter cyclone frequencies and strengths simulated in enhanced greenhouse warming experiments: Results from the models participating in the IPCC diagnostic exercise. *Climate Dynamics* **26**:713–28. doi:10.1007/s00382-006-0110-3

Lau N-C, Crane MW. (1995) A satellite view of the synoptic-scale organization of cloud properties in midlatitude and tropical circulation systems. *Monthly Weather Review* **123**:1984–2006.

Lau N-C, Crane MW. (1997) Comparing satellite and surface observations of cloud patterns in synoptic-scale circulation systems. *Monthly Weather Review* **125**:3172–89.

Naud CM, Del Genio AD, Bauer M. (2006) Observational constraints on the cloud thermodynamic phase in midlatitude storms. *Journal of Climate* **19**:5273–88.

Naud CM, Del Genio AD, Bauer M, *et al.* (2010) Cloud vertical distribution across warm and cold fronts in CloudSat-CALIPSO data and a general circulation model. *Journal of Climate* **23**:3397–415.

Polly JB, Rossow WB. (2016) Cloud radiative effects and precipitation in extratropical cyclones. *Journal of Climate* **29**:6483–507. doi:10.1175/JCLI-D-15-0857.1

Schiffer RA, Rossow WB. (1983) The International Satellite Cloud Climatology Project (ISCCP): The first project of the World Climate Research Programme. *Bulletin of the American Meteorological Society* **64**:779–84.

Tselioudis G, Rossow WB. (2006) Climate feedback implied by observed radiation and precipitation changes with midlatitude storm strength and frequency. *Geophysical Research Letters* **33**:L02704. doi:10.1029/2005GL024513

Tselioudis G, Zhang Y-C, Rossow WR. (2000) Cloud and radiation variations associated with northern midlatitude low and high sea level pressure regimes. *Journal of Climate* **13**:312–27. doi:10.1175/1520-0442(2000)013<0312:CARVAW>2.0.CO;2

Ulbrich U, Pinto JG, Kupfer H, *et al.* (2008) Changing northern hemisphere storm tracks in an ensemble of IPCC climate change simulations. *Journal of Climate* **21**:1669–79. doi:10.1175/2007JCLI1992.1

Slide 1

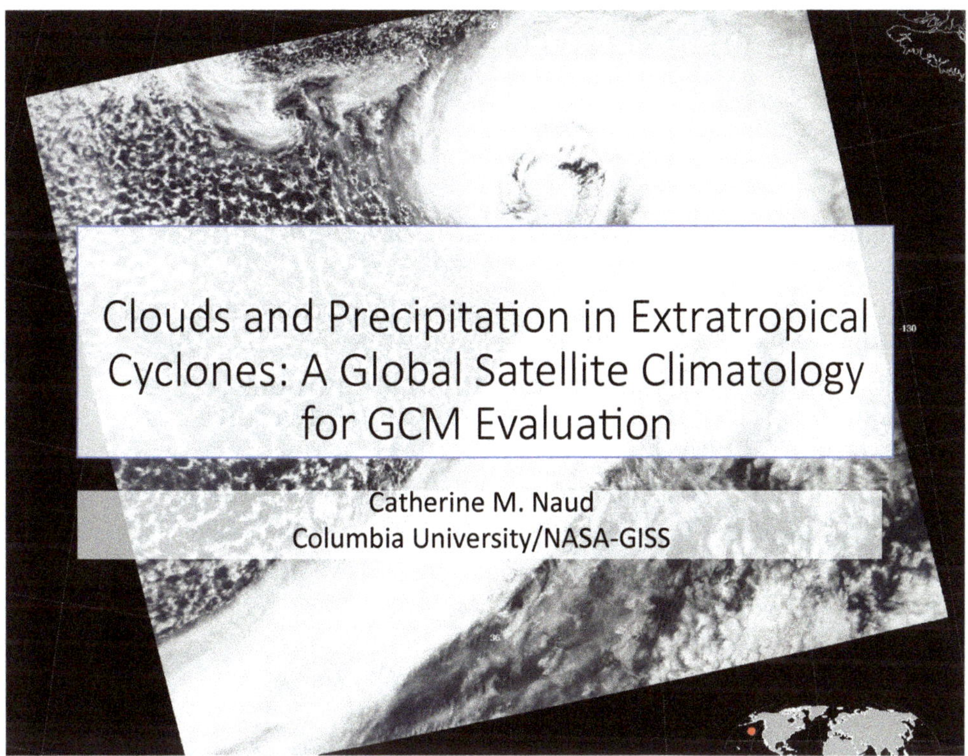

Slide 2

What is an Extratropical Cyclone?

- Baroclinic instability, occurring in the mid-latitudes, between approx. 30° and 60°N/S, first formalized by Bjerknes (1919) and colleagues from the Bergen School ([Bjerknes and Solberg, 1922], this figure).
- Characterized by a minimum in surface pressure ("depression") and often accompanied by cold and warm fronts.
- Clearly visible in satellite images, thanks to its cloud comma shape (c.f. Slide 1).

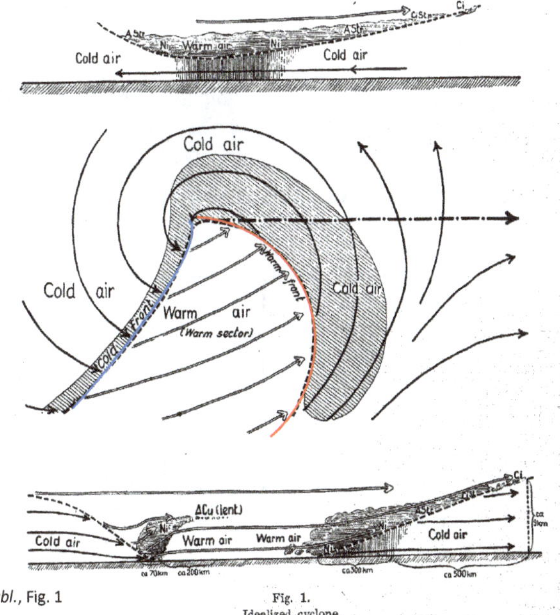

Bjerknes and Solberg (1922) *Geof. Publ.*, Fig. 1

Slide 3

The Role of Extratropical Cyclones in Climate

- Transport heat, moisture and energy poleward: regulatory role
- Regulate incoming solar and outgoing longwave radiation through associated cloud field (Tselioudis *et al.*, 2000)
- Major source of precipitation in the mid-latitudes (up to 80% in the northern hemisphere winter ([Hawcroft *et al.*, 2012]; figure below)
- Potentially catastrophic events: blizzard, floods, damaging winds (e.g., [Catto, 2016] and reference therein)

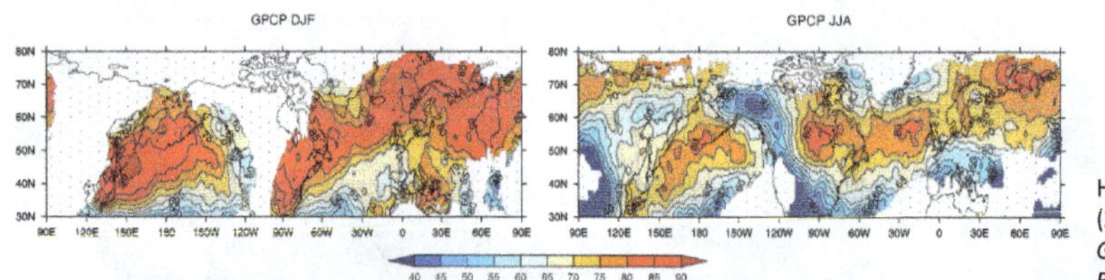

Contribution of storm-associated precipitation to the total precipitation (%) from the GPCP dataset for NH winter and summer. The masked and stippled areas are where the total climatological precipitation is less than 1 mm/day. The data is smoother over 3° to reduce noise.

Hawcroft *et al.* (2012) *Geophysical Research Letters*, adapted from Fig. 3

Slide 4

Satellite View of Cyclones

Global view:
Cloud cover highlights locations of extratropical cyclones (ETC)

MODIS, Terra, 19 Jan 2003

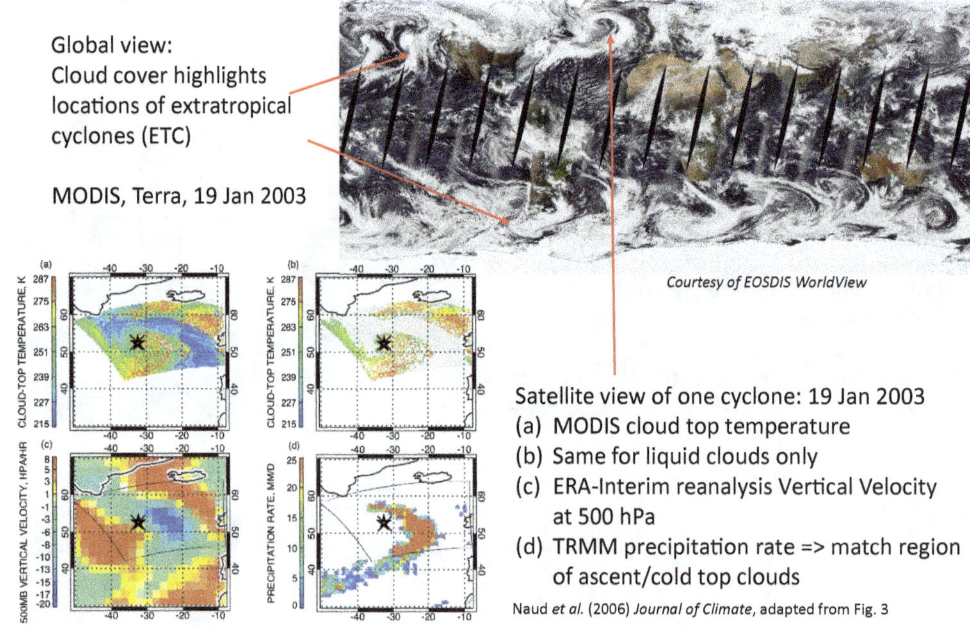

Courtesy of EOSDIS WorldView

Satellite view of one cyclone: 19 Jan 2003
(a) MODIS cloud top temperature
(b) Same for liquid clouds only
(c) ERA-Interim reanalysis Vertical Velocity at 500 hPa
(d) TRMM precipitation rate => match region of ascent/cold top clouds

Naud *et al.* (2006) *Journal of Climate*, adapted from Fig. 3

Slide 5

A Climatology of Extratropical Cyclone Properties: Compositing Methods

Pioneered by Lau and Crane (1995), compositing involves averaging together satellite observations in a frame of reference based on the cyclone morphology.

Figure: ISCCP cloud types with respect to locations of low pressure (L) and high pressure (H) zones.
([Schiffer and Rossow, 1983; Rossow and Schiffer, 1991])
Blue: high clouds, yellow: low clouds

Useful for:
1) Exploring the mean behavior of clouds and precipitation in cyclones
2) Evaluating GCMs

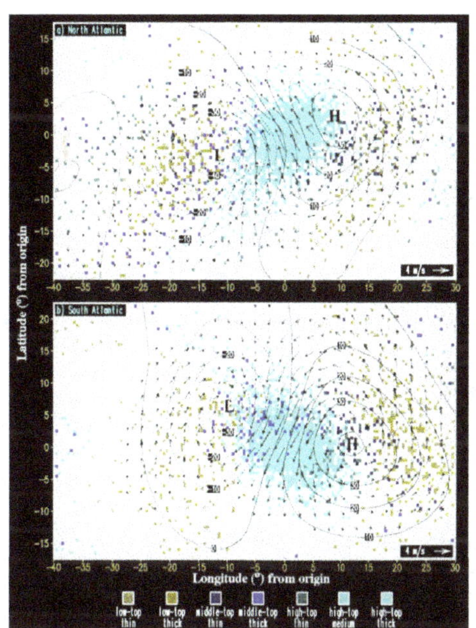

Lau and Crane (1995) *Monthly Weather Review*, adapted from Fig. 6

Slide 6

Step by Step Compositing

Step 1: Track and locate cyclones (Bauer and Del Genio, 2006; Neu *et al.*, 2013; Bauer *et al.* 2016) + for each cyclone locate cold and warm fronts (e.g., Naud *et al.*, 2010; 2016)

Step 2: Match cyclones to near coincident satellite observations

Step 3: Regrid data so all observations on same cyclone centered grid and average multiple cases together using (a) center or (b) surface fronts as an anchor.

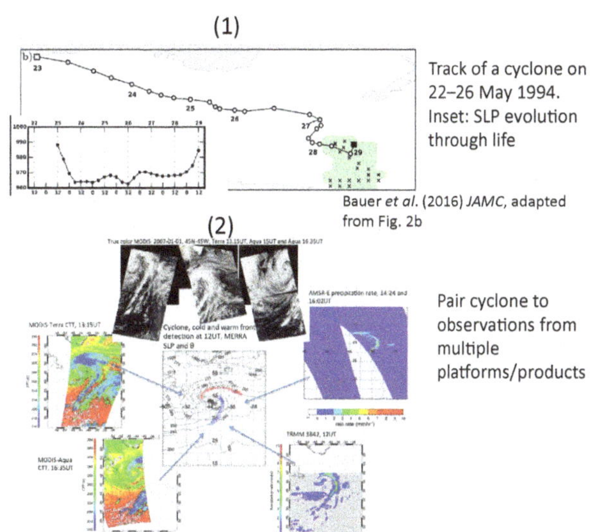

Track of a cyclone on 22–26 May 1994. Inset: SLP evolution through life

Bauer *et al.* (2016) *JAMC*, adapted from Fig. 2b

Pair cyclone to observations from multiple platforms/products

Slide 7

Composites of cloud properties in extratropical cyclones: plan view

Instruments: MODIS, MISR, CloudSat, CALIPSO

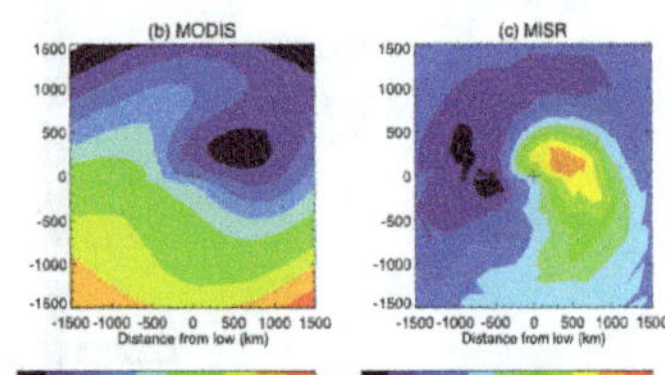

Naud *et al.* (2014) *Journal of Climate*, adapted from Fig. 5

Cloud top temperature (MODIS; left) and height (MISR; right)
=> Highest clouds just east of the storm center in the warm frontal zone, also the coldest
⇒ High clouds in the warm sector, relatively warm
⇒ Low level clouds in the cold sector

Comma-shaped and well-defined, when observing cloud top height (or pressure), not so much for temperature or fraction.
=> Ascent drives high level clouds

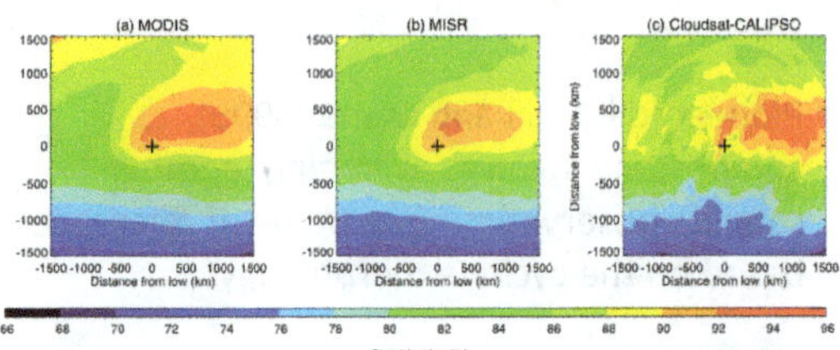

Naud *et al.* (2013) *Journal of Geophysical Research*, Fig. 2

Cloud cover from different data sources: no clear signature of comma shape for this property.
Comparing different products gives some measure of observational uncertainty (below), e.g., MISR more successful at detecting low level clouds in a cold sector.

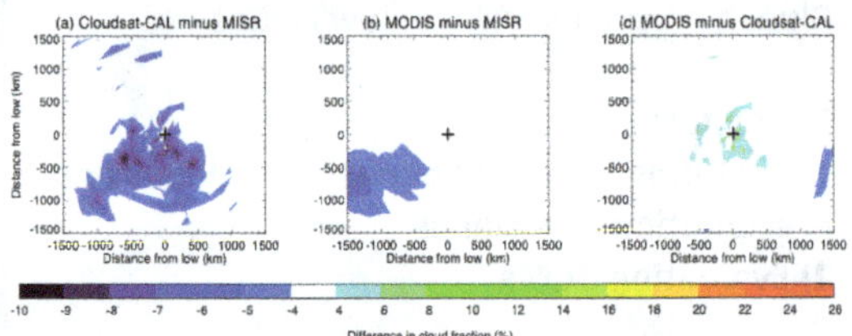

Naud *et al.* (2013) *Journal of Geophysical Research*, Fig. 8d to 8f

Slide 8

Composites Across Cold and Warm Fronts: Vertical Transects

Cold front transect

Warm front transect

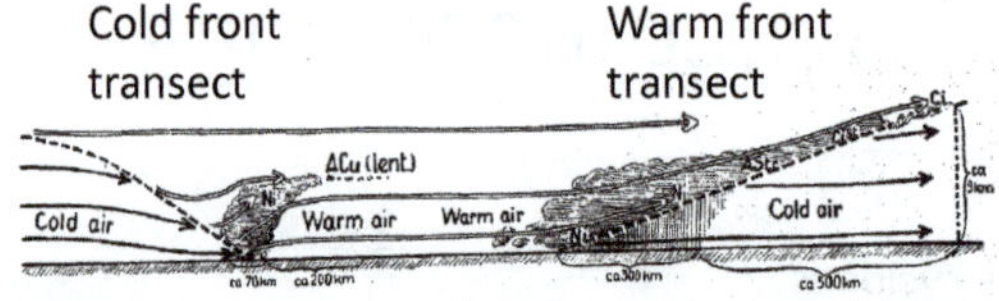

Fig. 1.
Idealized cyclone.

Bjerknes and Solberg (1992) *Geof. Publ.*, Fig. 1

Above: Bjerknes and Solberg model
Below: CloudSat-CALIPSO transects of cloud frequency of occurrence for north and south hemisphere fronts
Differences could be the result of composites smoothing out a lot of the details => the composites are not expected to look like an individual cyclone

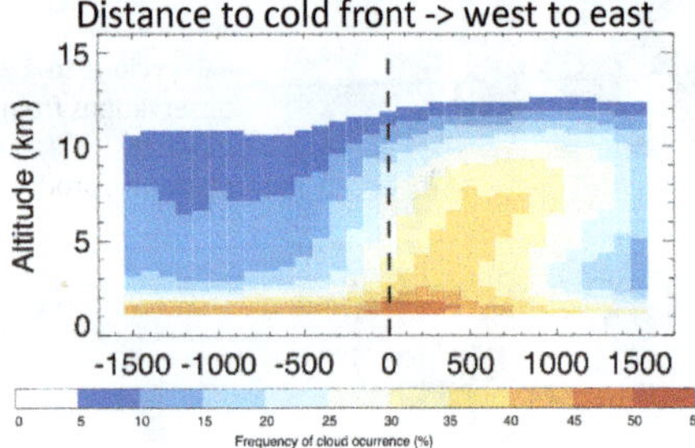

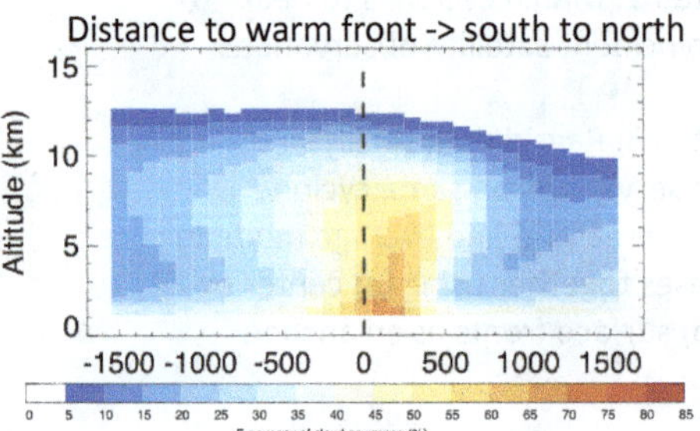

Slide 9

Relations between Dynamics, Cloud Types and Precipitation

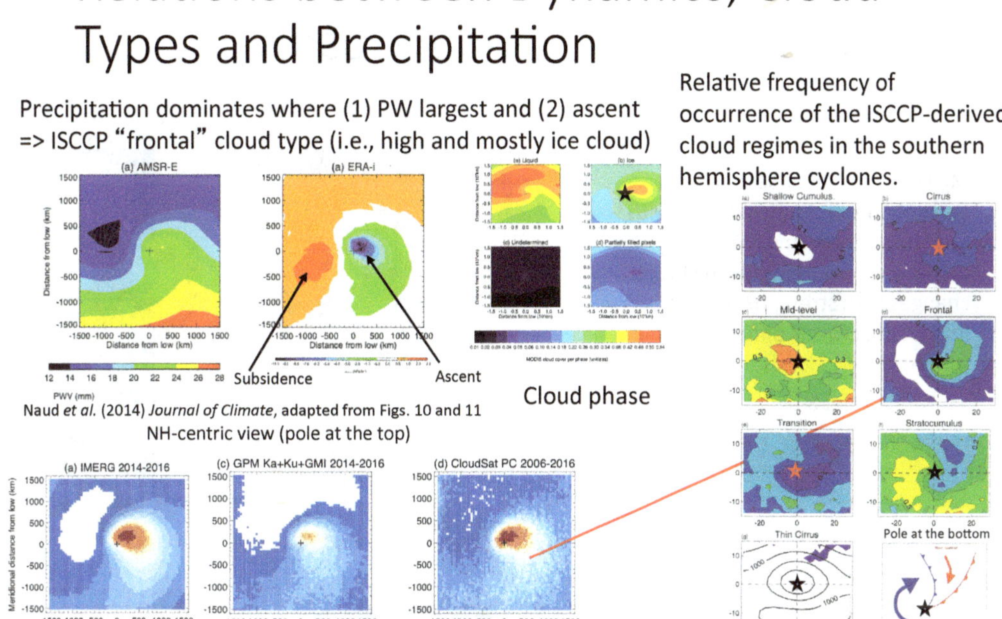
Precipitation dominates where (1) PW largest and (2) ascent => ISCCP "frontal" cloud type (i.e., high and mostly ice cloud)
(a) AMSR-E
(a) ERA-i
Subsidence
Ascent
Cloud phase
Naud *et al.* (2014) *Journal of Climate*, adapted from Figs. 10 and 11
NH-centric view (pole at the top)
(a) IMERG 2014-2016
(c) GPM Ka+Ku+GMI 2014-2016
(d) CloudSat PC 2006-2016
Zonal distance from low (km)
Precipitation rate (mm hr⁻¹)
Relative frequency of occurrence of the ISCCP-derived cloud regimes in the southern hemisphere cyclones.
Shallow Cumulus
Cirrus
Mid-level
Frontal
Transition
Stratocumulus
Thin Cirrus
Pole at the bottom

Naud *et al.* (2018) *JAMC*, adapted from Figs. 2 and 3

Bodas-Salcedo *et al.* (2012) *Journal of Climate*, adapted from Fig. 6

Slide 10

Sensitivity of Clouds and Precipitation to: Environmental Precipitable Water (PW) and Cyclone Strength

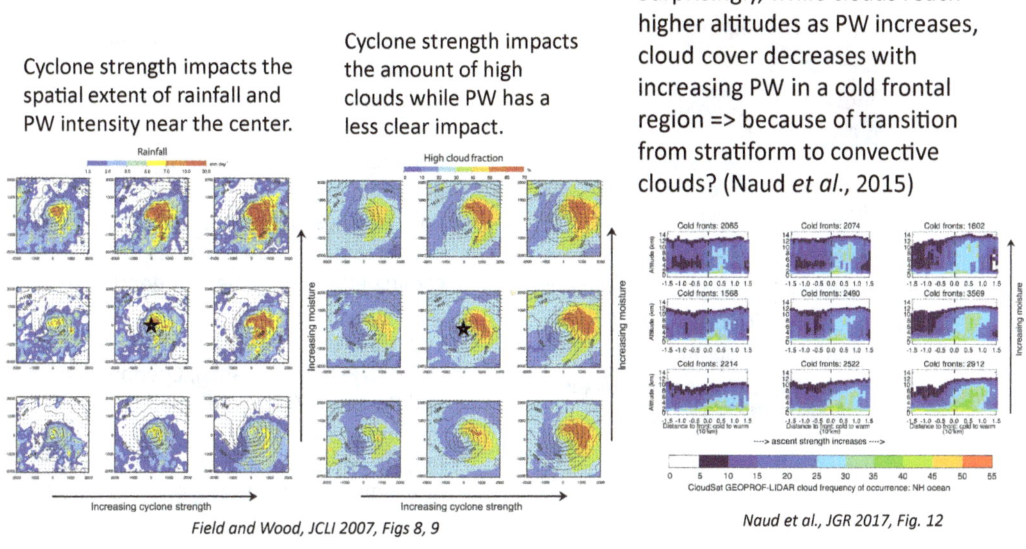
Cyclone strength impacts the spatial extent of rainfall and PW intensity near the center.
Cyclone strength impacts the amount of high clouds while PW has a less clear impact.
Surprisingly, while clouds reach higher altitudes as PW increases, cloud cover decreases with increasing PW in a cold frontal region => because of transition from stratiform to convective clouds? (Naud *et al.*, 2015)
Rainfall
High cloud fraction
Increasing moisture
Increasing cyclone strength
Cold fronts: 2065
Cold fronts: 2074
Cold fronts: 1602
Cold fronts: 1568
Cold fronts: 2490
Cold fronts: 3569
Cold fronts: 2214
Cold fronts: 2522
Cold fronts: 2912
----> ascent strength increases ---->
CloudSat GEOPROF-LIDAR cloud frequency of occurrence: NH ocean

Field and Wood, JCLI 2007, Figs 8, 9

Naud et al., JGR 2017, Fig. 12

Slide 11

Sensitivity of Cloud Cover to PW, Cyclone Strength and Aerosols

MODIS cloud cover sensitivity to aerosol optical depth changes with cyclone strength and PW
=> Because amount of high clouds changes with strength/PW and impact of aerosols different on high (more) vs. low clouds (less)

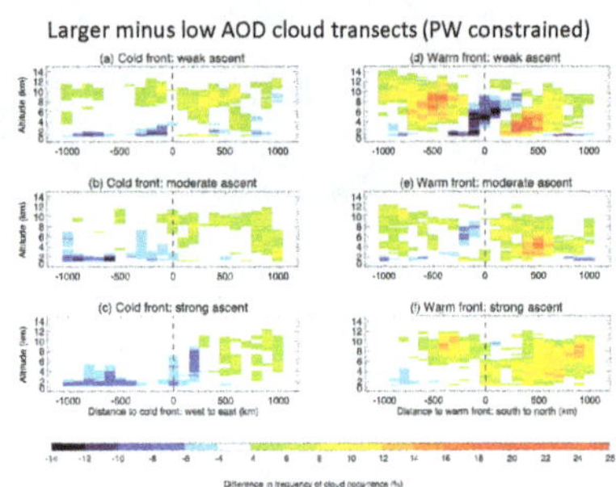

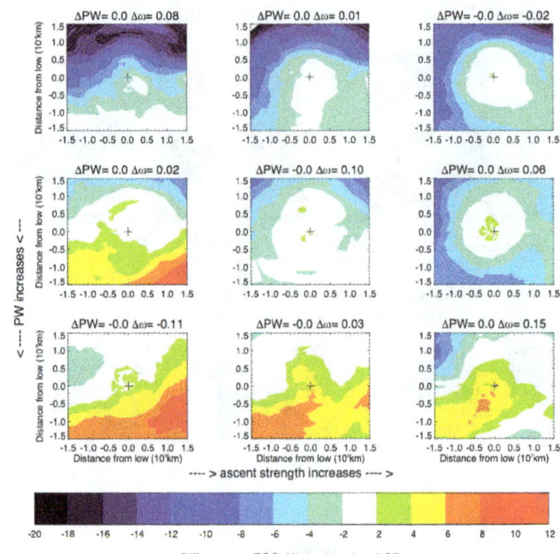

Naud *et al.* (2017) *Journal of Geophysical Research*, Figs. 10 and 13

Slide 12

Impact on Climate: Radiative Effect of Cyclones

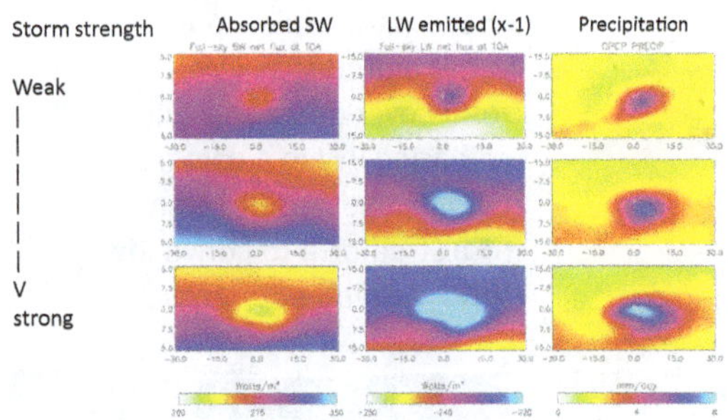

Radiative impact of cyclones (ISCCP-FD) + precipitation (GPCP) as function of strength for NH summer

Tselioudis and Rossow (2006) *Geophysical Research Letters*, adapted from Fig. 1

⇒ Tselioudis *et al.* (2000) used ISCCP cloud and radiation to estimate the radiative impact of cyclones and anticyclones.
⇒ Tselioudis and Rossow (2006) found that if cyclones were to intensify but reduce in number in a warming climate, the overall impact of intensification dominates for both radiation and precipitation.
⇒ Polly and Rossow (2016) found that the overall impact of the cyclones is a cooling effect, regardless of depth (or strength).
⇒ Using weather states (based on ISCCP cloud CTP-TAU diagrams), Oreopoulos and Rossow (2011) confirmed the importance of frontal systems for mid-latitude radiation budget, while Haynes *et al.* (2011) highlight the importance of low clouds (cold sector).

Slide 13

Application: Evaluation of General Circulation Models using Cyclones

- No consensus on the evolution of cyclones in a warming climate; probably fewer cyclones, but intensity change is unclear + depends on location and time period (Feser *et al.*, 2015).
- Although climate models can resolve the scales of these systems, it is not necessarily fine enough for mesoscale structures and frontal boundaries and parameterizations of moist processes required for the correct representation of clouds and precipitation.
- In themselves, an ideal mixture of conditions for model testing: subsidence and ascent, cold and warm advection, convective and stratiform clouds and precipitation, high/ice clouds, and low/liquid clouds.
- Two issues related to the representation of clouds and precipitation in extratropical cyclones:
  - Precipitation in the midlatitudes (Stephens *et al.*, 2010).
  - Excess SW absorption in the southern oceans (Trenberth and Fasullo, 2010).

Slide 14

Evaluation of Precipitation in Extratropical Cyclones

Example of the GISS model: evolution of skill over the past decade with ETC precipitation

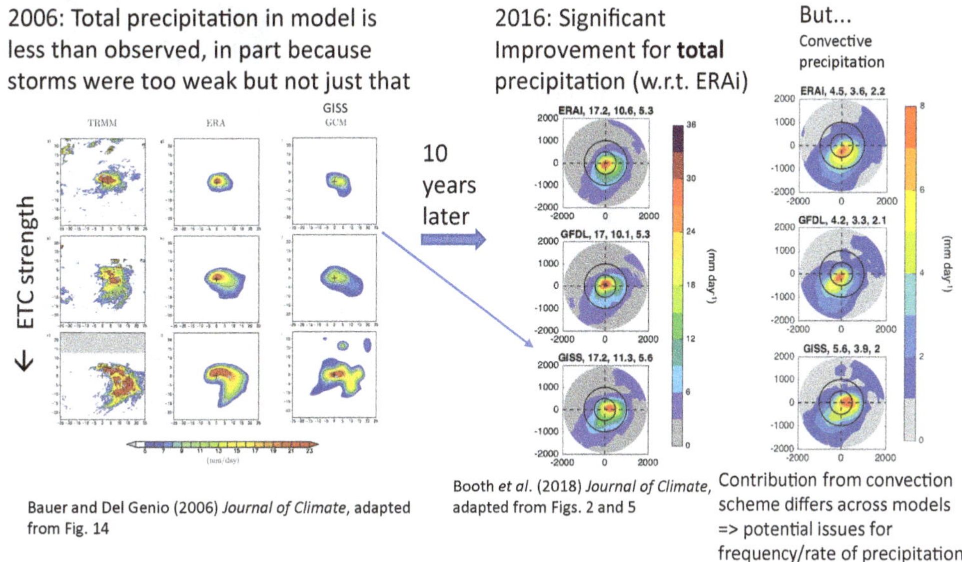
2006: Total precipitation in model is less than observed, in part because storms were too weak but not just that
TRMM
ERA
GISS GCM
ETC strength
10 years later
2016: Significant Improvement for **total** precipitation (w.r.t. ERAi)
ERAI, 17.2, 10.6, 5.3
GFDL, 17, 10.1, 5.3
GISS, 17.2, 11.3, 5.6
(mm day⁻¹)
But...
Convective precipitation
ERAI, 4.5, 3.6, 2.2
GFDL, 4.2, 3.3, 2.1
GISS, 5.6, 3.9, 2
(mm day⁻¹)

Bauer and Del Genio (2006) *Journal of Climate*, adapted from Fig. 14

Booth *et al.* (2018) *Journal of Climate*, adapted from Figs. 2 and 5

Contribution from convection scheme differs across models => potential issues for frequency/rate of precipitation

Slide 15

Precipitation in Models: Total Close to Observations, But Too Frequent and Too Weak

Example of MERRA-2 reanalysis: forced by extensive amount of observations but clouds and precipitation from model parameterizations

MERRA-2 very close to observations (here IMERG) for total precipitation in cyclones

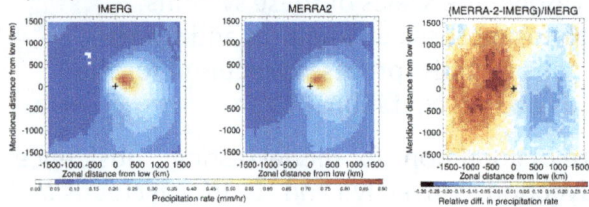

But: overestimate in area where precipitation is rare and light (cold sector) and underestimate where precipitation frequent and strong

=> Despite setting a threshold on precipitation rates to match IMERG sensitivity, frequency of precipitation in MERRA-2 exceeds 75% (less than 70% in IMERG) (lower left)
c.f. (Sun *et al.*, 2006; Catto *et al.*, 2015; Stephens *et al.*, 2010) for other models

Naud *et al.* (2020) *Journal of Climate*

Frequency of occurrence of precipitation

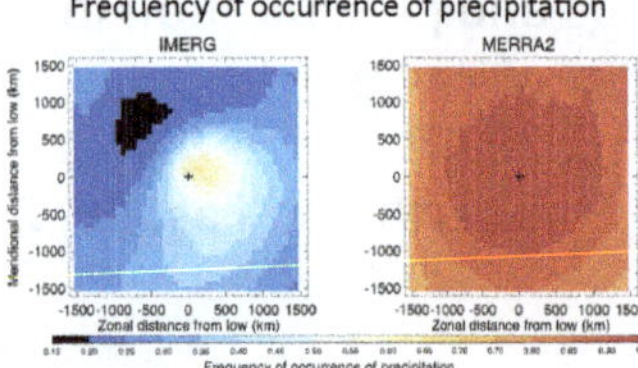

Rain rate when raining

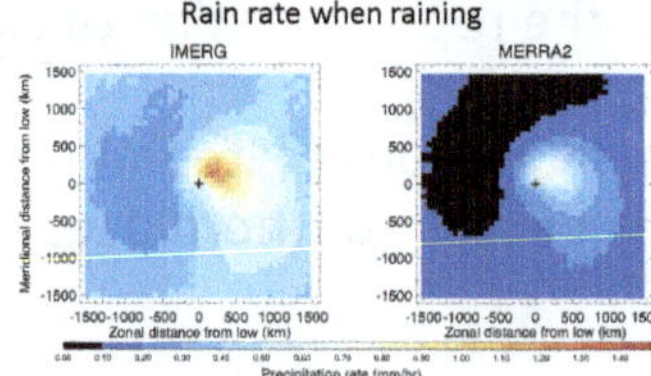

Issue with **convection** parameterization for correct representation of precipitation in mid-latitudes?

Slide 16

Excess in Modeled Shortwave Absorption at Southern Oceans Surface

Vexing issue, still not fully resolved:
Too much shortwave radiation absorbed in southern hemisphere oceans because, in part, GCMs underestimate cloud cover

Bodas-Salcedo *et al.* (2014) identified that cloud cover bias in cyclones cold sector (right).
Naud *et al.* (2014): same issue in reanalyses (below)

Difference in cloud cover in SH cyclones: ERA-interim/MERRA minus MODIS/MISR

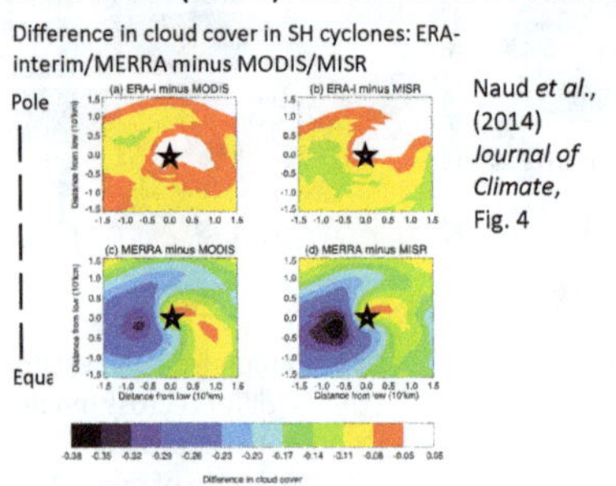

Naud *et al.*, (2014) *Journal of Climate*, Fig. 4

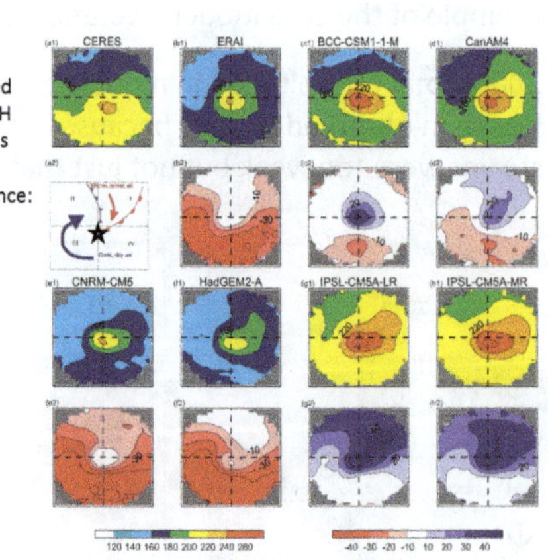

TOA reflected SW flux from CERES vs. ERA-interim and six GCMs, + diff = model minus CERES (red means excess absorption)

Bodas-Salcedo *et al.* (2014) *Journal of Climate*, adapted from Fig. 3

Slide 17

SW Absorption Issue: Low-level cloud Issue in Subsidence Regions

Recent model versions from GFDL: same model, 2 different parameterizations for convection (“2PM” vs. “MPM”)

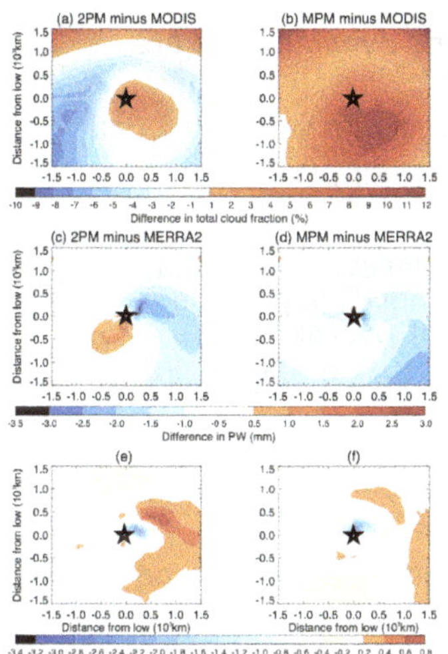

Cyclone centered bias in cloud cover: in “2PM” model, still present in cold sector.
Unrelated to differences in PW or vertical velocities.

Bias found in post-cold frontal region where observations indicate low-level cloud cover > 40% but models < 35%.

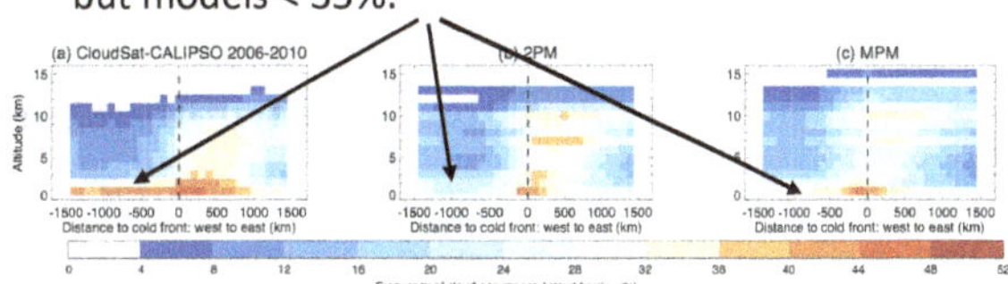

Analysis reveals impact of convection parameterization but also tuning parameters (cloud erosion, entrainment, precipitation efficiency).

Naud *et al.* (2019) *Journal of Climate*

Slide 18

Low-level Cloud Issue: Impact of Boundary Layer and Cloud Microphysics

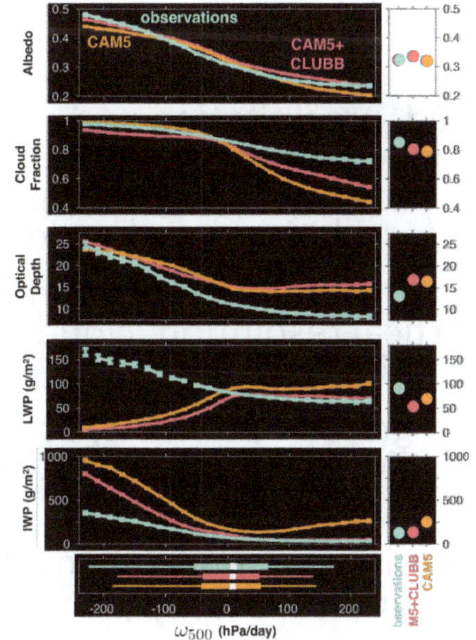

Wall *et al.* (2017) *Journal of Climate*, adapted from Fig. 9

Using the NCAR CAM5 model, the standard model is modified to include the “CLUBB” scheme that handles boundary layer and cloud microphysics in a different way.
For SH oceans, Wall *et al.* (2017) partitioned grid cells according to vertical velocity and compared albedo, cloud fraction, optical depth, LWP and IWP to observations:
⇒ CLUBB significantly improves the representation of LWP and IWP in subsiding regions (i.e., microphysics scheme more skillful).
⇒ However, CLUBB only slightly improves cloud fraction and even less so optical depth in subsiding regions.
⇒ No big improvement in ascending regions for LWP, IWP or optical depth.

Improving microphysics is not enough to improve the representation of low-level cloud cover in models; convection and PBL play an important role, especially in cold sectors of extratropical cyclones.

Slide 19

Conclusions

Over the past decade, a large number of studies have utilized satellite observations of cloud properties, precipitation, precipitable water, wind, and many others to build a global climatology of extratropical cyclones. This proved especially important to explore remote regions, e.g., Southern Oceans. Composites in extratropical cyclones or across fronts have revealed:

• Cloud types and precipitation spatially align with cyclone dynamics and are sensitive to storm strength and environmental moisture.

• Some evidence of the impact of aerosols on clouds in cyclones was also found.

• Radiatively, cyclones have an overall cooling effect of magnitude strongly dependent on the cyclone intensity.

With these composites, process-oriented metrics for model evaluation and development were designed to exploit observations in a novel way. These evaluations highlighted and help diagnose some recurrent issues:

• Models tend to produce too much high-level but not enough low-level clouds.

• Precipitation in modeled cyclones is too frequent and too light.

• These issues are to some extent related to cloud microphysics representation, but convection and boundary layer parameterizations also play a role.

• These issues in cyclones have an impact on radiation budget estimates and extreme event predictions.

Slide 20

Extratropical cyclones in a changing climate: How can satellite observations help improve predictions?

- Observational period of cloud, precipitation and meteorological variables (e.g., water vapor, winds) getting longer and allowing better sampling of large scale interannual variability (oscillations) and cyclones.
- Profiles becoming available, and need to be continued and improved (e.g., finer vertical resolution, cloud clearing for meteorological variables).
- Development of merged and gridded products, as pioneered by the ISCCP (e.g., IMERG for precipitation, MAC-LWP, NVAP for water vapor profiles, ...).
- Constantly improving reanalysis by assimilating more satellite information (e.g., cloudy vs. clear sky radiances).

Slide Captions

Slide 1 The background image was acquired with MODIS-Aqua on 1 April 2012 at 22.45 UT in the east Pacific. It reveals the complexity of the cloud field associated with an extratropical cyclone while clearly delineating the various dynamical conditions it encompasses. The overall comma shape is clearly defined, with the center of the storms at the head and the tail following the cold front. The area of thick and high clouds reveals the cyclonic circulation, with a spiraling shape around the center indicating the overall area of air ascent. Clouds on both sides of the cold front are clearly stratocumulus fields, open cell to the west, closed cell to the east. Cirriform clouds can also be observed in advance of the cold front.

Slide 2 Extratropical cyclones form in regions of baroclinic instability, triggered by the temperature gradient between the Equator and the poles. They form mostly between 30 and 60 degrees north and south and were first described by scientists from the Norwegian Bergen School (Bjerknes, 1919; Bjerknes and Solberg, 1922). The model they proposed a century ago is still being used today, albeit with some refinements, by weather forecasters. These cyclones are characterized by an area of low pressure, thus the term "depression", and anti-clockwise air circulation. They are often accompanied by surface fronts, cold and warm, which are areas of steep gradients in surface temperature forming a boundary between the poleward flow of warm moist air and the equatorward flow of cold dry air. These systems can easily be spotted in satellite images because of the classic comma shape of the clouds they help form (c.f. Slide 1). The Bjerknes and Solberg model includes a description of the locations in the cyclones where clouds and precipitation occur: the warm and cold fronts. The warm fronts usually form along an east-west line to the east of the surface low-pressure minimum (the storm center) and are characterized by an isentropic slope into the cold sector. Clouds increase in altitude as they form along the isentrope. Precipitation occurs mostly at the location of the surface front. A strong flow of warm moist air usually runs parallel to the cold front, which tends to be more often oriented meridionally, with the most intense precipitation occurring at or slightly in advance of the surface front, thus forming the tail of the comma. Around the storm center, the circulation produces clouds and precipitation into the west sector, forming the head of the comma.

Slide 3 Extratropical cyclones have a regulatory role in the climate system. They transport heat, moisture and energy poleward. Because of the clouds they form, they also have an important role in the radiative budget for the mid-latitude regions (Tselioudis *et al.*, 2000). They are also the main purveyor of precipitation in the mid-latitudes, with contributions in excess of 80% of total precipitation in the northern hemisphere winter (Hawcroft *et al.*, 2012). Therefore, they are also the most frequent cause of catastrophic meteorological events in the mid-latitudes, such as floods, blizzards or damaging winds (Catto, 2016, and references therein).

Figure from

Slide 4 As mentioned earlier, extratropical cyclones can easily be spotted in satellite imagery (here is a MODIS-Terra true-color mosaic image acquired on 1 April 2012). At any given time, of the order of 10, cyclones can be seen in these images, more clearly so over the ocean than land, in part because the former are larger (Polly and Rossow, 2016), but mostly thanks to the large contrast in brightness between the clouds and dark ocean waters. Weather forecasters have made use of these images since the first satellites of the 1970s started providing such capabilities. Beyond a picture of the cloud field, satellite imagery has been processed to provide information on the nature of the clouds. Such algorithms were introduced for the ISCCP project and have been implemented for each instrument that could be exploited for this kind of application. Here, we show an example of one of the data fields retrieved with the MODIS satellite: cloud top temperature, and using the information on the cloud type also derived with the MODIS instrument, the cloud top temperature for liquid clouds only. We also show for the same cyclone the ERA-Interim (Dee *et al.*, 2011) vertical velocity. This field reveals the dynamics of these systems, with a region of strong ascent, to the east and south of the center that wraps around to the north and a region of relatively strong subsidence to the west. By comparing the vertical velocity field to the cloud top temperature retrievals, one can see how cold-top clouds are collocated with the region of strong ascent, while relatively warm and liquid clouds are found in the region of subsidence. Finally, the NASA-TRMM precipitation rates are also shown, as retrieved using a combination of radar and microwave radiometer data. The precipitation field clearly shows the comma shape as well, indicating the area of the cyclones where moisture convergence and lift form clouds and precipitation.

Slide 5 To fully exploit the vast amount of satellite observations and compensate for their partial view of a cyclone and irregular sampling in time, methods of compositing have been proposed. The first individuals to combine satellite observations of cloud properties in a cyclone-centered frame of reference were most probably Lau and Crane (1995; 1997). In their work, they exploit the ISCCP cloud top pressure and optical depth information to derive a cloud type classification (e.g., Fig. 4 in [Rossow and Schiffer, 1991]). Then using the cloud classification, they composite each cloud type in a low versus high-pressure system and describe which type dominates with respect to the location of the low and high centers. Their composites clearly show the predominance of high clouds in advance of the low, where the ascent is occurring, and the predominance of low clouds to the west of the low, where subsidence prevails. Following their lead a whole range of satellite-derived cloud and precipitation products, along with meteorological variables, for some produced with reanalyses, have been combined in such cyclone-centered composites to understand extratropical cyclones better and to help in GCM evaluation.

Slide 6 One method for building composites of satellite observations is as follows: first, the cyclones need to be located and tracked. This is done using various quantities (e.g., vorticity, sea level pressure) as provided with a reanalysis. A comprehensive survey of the techniques currently utilized and a comparison of their respective merits are discussed in (Neu *et al.*, 2013). Here, we chose to use an algorithm developed by Bauer and Del Genio (2006) and further refined and assessed in (Bauer *et al.*, 2016). It is based on sea level pressures from ERA-Interim and provides 6-hourly cyclone center locations. It also tracks the cyclones, therefore indicating past and future locations of the 6-hourly snapshots (which we call cyclones here, as opposed to "tracks", which refers to the entire life cycle), information that is useful for applications where the age of the cyclone during its development, peak and dissipative stages is needed. Such applications might be the evolution of precipitation through the life cycle (e.g., [Booth *et al.*, 2018]) or the study of extreme events where the cyclone is only studied at peak intensity (e.g., [Bengtsson *et al.*, 2009]). More recently, automated detection algorithms of cold and warm fronts locations have also been proposed (e.g., [Schemm *et al.*, 2018, and reference therein]). Therefore, for each cyclone in the database, cold and warm front locations might also be known. The second step is to match the observations of interest to the cyclones in space and time, and this can be done using instantaneous (i.e., Level 2) observations (as in the examples given in Fig. 2 of the slide) or using gridded products (i.e., Level 3). The matching can be tuned in terms of the time difference between cyclone detection and observation acquisition. But whether Level 2 or Level 3 products are used, the matching is rather simply described as cookie cutting. Step 3 consists of transposing the observations into a common grid, centered on the cyclone center or the frontal boundaries, and then the averaging of multiple cyclones.

Slide 7 Cloud observations from satellites are available for multiple platforms and instruments that might use different imaging techniques and retrieval algorithms for the same cloud properties. This redundancy presents the great advantage of permitting the estimation of observational uncertainties. These observations can be: cloud top pressure, temperature or height, cloud cover, cloud phase, liquid and ice water path, and optical depth. Here, we only show a few of these variables composited in a cyclone-centered grid. Using MODIS cloud top temperature, one can clearly see the very cold core of high-level clouds that populate the warm frontal area and center and a rather large area of cloud top temperatures that are relatively warmer on the equator side of the cyclones, as well as in the western quadrants. These clouds have warm tops either because they are forming in relatively warm air or because they are close to the surface. By comparing this composite to the MISR cloud top height composite, one can delineate these clouds further, where it is now clear that in the warm sector (equator-east quadrant), clouds extend to high altitudes, but being in relatively warm air, have relatively warm tops, while clouds to the west of the storm center are at a low level and their top temperature follows more closely the surface temperature gradient.

The dynamical structure of the cyclones is clearly seen in the cloud top height composite, while the thermodynamic-dynamic coupling is better observed when considering cloud top temperatures. In contrast, the cyclone properties are less well defined when examining the cloud fraction (or cover in GCM terminology), somewhat akin to what cloud top temperatures are indicating. Clouds show maximum coverage in the warm frontal and central areas but minimum coverage on the equator side and more uniform fractions anywhere else. Here, we show different composites using different types of instruments. One, MISR, tends to be more successful at detecting low-level clouds than the others and gives cloud fractions greater than the others in the western quadrant on the equator side where low-level clouds dominate. Overall, the intercomparison indicates differences of the order of up to 5%, a quantity that can be used when using these composites for model evaluation. That is to say, if a model gives cloud fractions within 5% of any of these observations, then it can be considered quite skillful in its representation of cloud cover in extratropical cyclones.

Slide 8 Here, we recall the Bjerknes and Solberg (1922) model of cloud distribution and compare it to the observed composites of cloud frequency of occurrence transects across cold and warm fronts, as obtained with the radar onboard CloudSat and lidar onboard CALIPSO. This slide discusses two aspects of these composites. One aspect is how the idealized cyclone model, which was based on ground-based observations of multiple cyclones in the north of Europe, compares to observations obtained from space globally. For cold fronts, the CloudSat-CALIPSO composites show an area of relatively large cloud fraction just east of the surface front that extends to large altitudes, as suggested in the idealized model of Bjerknes and Solberg (1922). The difference here is in the amount of low-level clouds throughout the cold frontal region, which is not indicated in the idealized model. Across the warm front, the sloping is visible in the composite, but the low clouds are also more frequent in the composite than the idealized model. These differences lead us to the second point of this slide: The idealized model is representing what could happen in any given cyclone, while the composites are an average over many disparate cyclones, causing the details of each to be smoothed out while only retaining the most salient features. This means that, for example, as the slope of warm fronts varies from one storm to the next, the maximum in the cloud cover above is somewhat washed out. The relatively large occurrence of low-level clouds in the composites does not indicate whether these are stratocumulus decks happening some of the time, or cumuli happening a lot of the time. Composites do not look like an individual cyclone.

Slide 9 As evoked already a few times, the cyclone-centered composites are a great tool to explore the relationship between dynamics, clouds and precipitation. Here, we illustrate these relations. We first show composites in southern hemisphere cyclones from [Bodas-Salcedo *et al.*, 2012] of ISCCP-based weather regimes. Note that their composites respect the SH orientation of the cyclones, with the pole side at the bottom. These regimes are defined with a clustering technique applied to ISCCP cloud top pressure/cloud optical depth histograms. By calculating the frequency of occurrence of seven regimes in the cyclone-centered region, one can clearly see how these regions relate to the circulation. In particular, we point out the area where the "frontal" type dominates, which includes the warm front and cold frontal regions, and delineate the classic comma shape. When compositing the precipitation information, the same comma shape emerges, regardless of the dataset being used (here, an NH-centric view with the pole at the top). The correspondence between the frontal regime and precipitation maximum is strong and matches the region of ascent indicated in the ERA-interim 850 hPa vertical velocity composite shown on the slide. It also coincides with the region of relatively larger precipitable water (PW) in the AMSR-E based composite. Using the cloud phase information from MODIS, one can also see that liquid clouds dominate in the areas where the "stratocumulus" cloud regime dominates (i.e., the cold sector), while the ice cloud frequency is at a maximum where the "frontal" and "mid-level" regimes are most frequent. With MODIS observations, one can derive a quantity that measures the level of the brokenness of the clouds, and this quantity is at maximum in an area of the cyclones where the dominant regimes are "mid-level" and "transition".

Slide 10 Sensitivity analyses are important to understand the main drivers of clouds and precipitation in extratropical cyclones and for model evaluation. Also, because of the smoothing issue with compositing, one can separate by sorting cyclones based on some large-scale property, very disparate systems, and, eventually, refine the average picture. Field and Wood (2007) explored the impact of cyclone strength (using surface wind speed from WindScat) and PW using AMSR-E to explore the sensitivity of precipitation in the warm conveyor belt (the flow of moist air that runs parallel to the cold front into the warm frontal region). They verified that both PW and storm strength impact precipitation and the spatial extent of high clouds in the comma region. Surprisingly, when we investigated how these two factors influence cloud cover in cold frontal regions (i.e., across the tail of the comma), we found the following: while cyclone strength showed a similar impact on clouds in the cold frontal area with a maximum frequency in cloud occurrence, the PW had the opposite effect — more moisture meant fewer clouds (Naud *et al.*, 2017). This is possibly caused by a transition from stratiform to the convective cloud as moisture increases (Naud *et al.*, 2015), thus causing more localized cloud distributions, which might end up giving the overall impression of a decrease in the cloud in the composites.

Slide 11 Using a similar approach, we examined the impact of environmental aerosol optical depth (AOD) on the cloud fraction in the cyclones (Naud *et al.*, 2017). For this, the moisture distribution across the cyclones had to be constrained as aerosols can swell in humid conditions and give larger optical depth for a given aerosol concentration. The work revealed a dependency of cloud sensitivity to aerosol amounts to both cyclone strength and environmental moisture. In relatively dry environments, cloud cover was lower in more polluted environments, while in wet environments, the cloud cover was larger. By inspecting cloud cover changes across the cold and warm front in large versus low AOD environments, we found that the different cloud response to changes in AOD was, in fact, caused by opposite changes at different altitudes: in dry environments, clouds are relatively low, and for these clouds, cloud cover was less for high than low AOD environments. However, moist environments tend to exhibit significantly higher clouds, and for these clouds, cloud cover is larger for high than low AOD environments.

Slide 12 Satellite observations in cyclones can also inform on their radiative impact, which has important implications in climate systems. Recently, Polly and Rossow (2016) showed that the overall impact of the cyclones is a cooling effect, regardless of cyclone strength. Tselioudis and Rossow (2006) explored the impact on absorbed shortwave and emitted longwave TOA radiation of the cyclones and estimated how these might change in a changing climate. They found that if cyclones intensify but reduce in number, then the overall impact would still be dominated by intensification. Using the ISCCP weather states mentioned earlier, Oreopoulos and Rossow (2011) confirmed the importance of frontal systems for mid-latitude radiation budget; however, Haynes *et al.* (2011) found a rather large impact of low-level clouds (in subsidence) as well.

Slide 13 As just explained, the radiative impact of an extratropical cyclone is significant and must be considered in future climate predictions. This leads to the question of whether GCMs can represent extratropical cyclones realistically, in particular, their associated cloud field and precipitation. At present, there is no consensus on how cyclones will evolve in a changing climate. Most recent studies indicate a decrease in the number of cyclones. This is because the polar amplification implies a decrease in equator to pole temperature gradient, i.e., baroclinicity. However, it is at present unclear whether the intensity of the cyclones will change, and this seems to depend on regions too (e.g., [Feser *et al.*, 2015]). In effect, GCMs can resolve the scale of the cyclones but they cannot resolve the scale of the frontal boundaries and associated mesoscale precipitation (and cloud) organization. As parameterizations have to handle these scales, the issue of the correct representation of moist processes in GCMs makes predictions of intensity changes less certain.

This motivates detailed studies of modeled cloud and precipitation in extratropical cyclones. In addition, because these systems encompass a wide variety of conditions (e.g., ascent versus subsidence, low/high clouds, ice/liquid/mixed clouds, stratiform/convective clouds, cold/dry versus warm/humid air masses), evaluating the modeled representation informs on more than just the cyclones themselves. Here we will focus on two issues that have been recognized in a large number of models but are still yet unsolved: (1) the representation of precipitation in the mid-latitudes (Stephens *et al.*, 2010) and (2) the excess in shortwave radiation absorption in the southern oceans (Trenberth and Fasullo, 2010).

Slide 14 Here we focus on one model, the GISS GCM. In the model's version available in 2006, composites of precipitation in extratropical cyclones were constructed for three different storm strengths (weak, medium and strong), and comparisons were done between the GISS model and the TRMM observations, and two reanalyses

(Bauer and Del Genio, 2006). With that version, the GCM had difficulty producing as much precipitation as observed, and while part of this problem was attributed to cyclones that were too weak, as the figure demonstrates, the issue was present even in strong cyclones. For reference, the reanalysis (which models precipitation but assimilates observations that constrain the dynamics and thermodynamic of the systems) was producing precipitation fairly close to observations. A decade later, with a different set of parameterizations, i.e., a different model altogether, Booth *et al.* (2018) found that the GISS-GCM represents precipitation on par with other models and similarly to ERA-Interim.

However, when comparing the convective precipitation contribution, the GISS-GCM composites differ from ERA-interim or another GCM. In other words, different models might successfully represent the total amount of precipitation in extratropical cyclones, but the reasons for this success might be very different from one model to another. In this case, the parameterization of convection might help for total amounts but will also impact the frequency of occurrence and precipitation rates (when precipitating). This needs to be further explored because the impact of precipitation is very different between situations of light steady rain and situations of hard, localized rain, even if, in the end, both situations can give the same total accumulation. Therefore, the details of the precipitation in cyclones must be examined.

Slide 15 As an example of current model performance for the representation of precipitation in extratropical cyclones, we examine here the NASA reanalysis MERRA-2. Compared to a free-running GCM, we can evaluate the performance of the model (as mentioned earlier, precipitation is parameterized in reanalysis, not assimilated) by matching in time and space the cyclones between MERRA-2 and observations. Here we used the IMERG product. When we compare the overall composite of precipitation, we find great similarities between MERRA-2 and IMERG, i.e., same spatial distribution and fairly small differences (close to observational uncertainties, Naud *et al.*, 2018). However, the relative difference (normalized by the IMERG precipitation rate), we can clearly see that for areas where precipitation is rare/light, MERRA-2 tends to overestimate the precipitation. For areas where precipitation is frequent/strong, MERRA-2 underestimates the precipitation. When we separately compare the frequency of occurrence of precipitation and the precipitation rate when precipitating, the cause is quite clear: MERRA-2 predicts precipitation everywhere and all the time in the cyclones. Note that for this comparison we set MERRA-2 precipitation rates to zero when they are less than IMERG sensitivity (IMERG does not report rain rates of less than $\sim 10^{-2}$ mm/hr). And yet, the frequency of precipitation exceeds 75% when IMERG indicates frequencies no larger than 70% (found at the center). The comparison of precipitation rates indicates much lower rates in MERRA-2. Therefore, the problem highlighted by Stephens *et al.* (2010) of too much drizzle in models appears to also affect MERRA-2 — it rains too often and too lightly. The same series of tests was applied to ERA-Interim, as well as two recent free-running GCMs and they all showed similar biases. Of course, a much more detailed analysis would be needed to pinpoint in the models the root cause of the issue, but one might speculate that the representation of convection could participate.

Slide 16 The second problem that plagues a number of GCMs is the excessive shortwave absorption at the surface in the southern hemisphere. Trenberth and Fasullo (2010) used ISCCP data to attribute this issue to a lack of clouds in these regions. Bodas-Salcedo *et al.* (2014) pointed out that a significant portion of this region is affected by extratropical cyclones, especially south of 50°S. Indeed, they found that the shortwave absorption issue was clearly visible when inspecting cyclone-centered composites for a wide range of models. They found that the issue occurred in the cold sector of the cyclones, an area where low and mid-level clouds dominate (as we showed earlier). This issue is also present in recent reanalyses, which, with various degrees of bias, underestimate the cloud cover in cyclone cold sector, and more specifically in the wake of the cold fronts (Naud *et al.*, 2014).

Slide 17 A recent version of the GFDL model was tested for its ability to represent the low-level cloud cover at the back of the cold fronts. With this model, two different versions of the cumulus parameterization were used so that the impact of the representation of convection could be evaluated for these regions. Indeed, by changing the convection scheme, the cloud cover in the post-cold frontal region changes, suggesting that this aspect of the model is contributing to the bias discussed here. In addition, the various cloud-tuning parameters were also found to contribute to differences between the two models in the post-cold frontal region. However, the two versions of the model both underestimate the cloud cover, indicating that convection is not the only aspect of the model that affects these clouds, and/or neither convection schemes is well suited for the representation of these clouds.

Slide 18 Other studies have explored the impact of other aspects of the model. In particular, a version of the NCAR model CAM5 was tested with the implementation of a new scheme that handles both the boundary layer and cloud microphysics scheme differently than what was initially implemented. Focusing on the southern oceans and separating conditions of ascent from conditions of subsidence, Wall *et al.* (2017) found that while the new parameterization helped the model be more realistic for the representation of liquid and ice water path in clouds in subsidence conditions, more modest improvements were found for the cloud fraction or the optical depth of the clouds. The issue of clouds in cyclone cold sectors ties in with the representation of low-level clouds in models overall, whether they are found in the subtropics or the mid-latitudes, and the difficulty is exacerbated by the complex interactions between various aspects of the model, i.e., convection, boundary layer, and microphysics. As new schemes are proposed, metrics focused on extratropical cyclones should prove very useful to measure progress.

Slide 19 Over the past two decades or so, a significant number of studies have made use of satellite observations of cloud, precipitation, atmospheric state, and other fields to build a comprehensive picture of extratropical cyclones. The unique vantage point of satellites has allowed us to observe these systems from new angles, in particular, in inaccessible places or in systems that would be too dangerous to directly observe. This is especially true of the southern oceans, where cyclones are ubiquitous, rendering conditions on the ground challenging for observations and too expansive to fully document. This new perspective has provided a wealth of new information on these systems and allowed to generalize or supplement models that were based on specific locations. Here, we have shown examples of recent advances in our understanding of the climatological characteristics of these systems: (1) how clouds, in particular cloud type, and precipitation are strongly linked to the dynamical characteristics of cyclones, and conjunctly to the environmental moisture of these systems; (2) how aerosols might impact clouds in these systems, and how this impact changes with altitude and across cold and warm fronts; and (3) what the radiative effect of these systems is like on a global basis, with an overall cooling effect that is strongly dependent on the storms' intensity rather than the number. This has been valuable in climatology and provided a better understanding of the sensitivities of moist variables to the large-scale environment. This, in turn, has proved fruitful for model evaluation and enriched the recent development of process-oriented metrics. Here, we showed how compositing in extratropical cyclones helped demonstrate that current models are still having issues at representing low-level clouds while overestimating high-level clouds. This technique helped establish that this is not uniquely caused by the representation of cloud microphysics in the models but also by issues with convection and boundary layer parameterizations. Compositing in cyclones also illustrated that models tend to overestimate the frequency of occurrence of precipitation while underestimating rain intensity.

Slide 20 Looking forward, there are many ways satellite observations can help improve predictions of climate in the future, and extratropical cyclones in particular. The observational period of clouds, precipitation and atmospheric variables, in general, has lengthened to a point where interannual variability can be better characterized and different phases of the major climate oscillations can now be contrasted (e.g., North Atlantic Oscillation, ENSO). In addition, latitude range and temporal frequency of observations allow us to sample more cyclones through their life cycles in more disparate environments. Profiles are now available, not just of atmospheric variables, but with active instruments, also for clouds and precipitation. More importantly, following the lead of the ISCCP for clouds, more gridded, global, high-resolution (temporal and spatial) products are becoming available, e.g., IMERG for precipitation, MAC-LWP for cloud liquid water path, or NVAP for water vapor. Finally, their inclusion in data assimilation has helped improve NWP models and reanalyses, and more complex products are being tested, including cloudy radiances. All of these developments will not only be beneficial for improving our understanding of the drivers of mid-latitude weather but also, by helping improve models, for a more reliable prediction of the mid-latitude climate evolution.

References

Bauer M, Del Genio AD. (2006) Composite analysis of winter cyclones in a GCM: Influence on climatological humidity. *Journal of Climate* **19**:1652–72.

Bauer M, Tselioudis G, Rossow WB. (2016) A new climatology for investigating storm influences in and on the extratropics. *Journal of Applied Meteorology and Climatology* **55**:1287–303.

Bengtsson L, Hodges KI, Keenlyside N. (2009) Will extratropical storms intensify in a warmer climate? *Journal of Climate* **22**:2276–301. doi:10.1175/2008JCLI2678.1

Bjerknes J. (1919) On the structure of moving cyclones, *Monthly Weather Review* **47**:95–99.

Bjerknes J, Solberg, H. (1922) Life cycle of cyclones and the polar front theory of atmospheric circulation. *Geophysisks Publikationer* **3**:3–18.

Bodas-Salcedo A., Williams KD, Field PR, Lock AP. (2012) Downwelling solar radiation surplus over the southern ocean in the Met Office model: The role of mid-latitude cyclone clouds. *Journal of Climate* **25**:7467–86. doi:10.1175/JCLI-D-11-00702.1

Bodas-Salcedo A, Williams KD, Ringer MA, *et al.* (2014) Origins of the solar radiation biases over the Southern Ocean in CFMIP2 Models. *Journal of Climate* **27**:41–56. doi:10.1175/JCLI-D-13-00169.1

Booth JF, Naud CM, Jeyaratnam J. (2018) Extratropical cyclone precipitation life cycles: A satellite-based analysis. *Geophysical Research Letters* **45**:8647–54. doi:10.1029/2018GL078977

Booth JF, Naud CM, Willison J. (2018) Evaluation of extratropical cyclone precipitation in the North Atlantic Basin: An analysis of ERA-Interim, WRF, and two CMIP5 models. *Journal of Climate* **31**(6):2345–60.

Catto JL. (2016) Extratropical cyclone classification and its use in climate studies. *Reviews of Geophysics* **54**:486–520. doi:10.1002/2016RG000519

Catto JL, Jakob C, Nicholls N. (2015) Can the CMIP5 models represent winter frontal precipitation? *Geophysical Research Letters* **42**:8596–604. doi:10.1002/GL2015GL066015

Dee DP, Uppala SM, Simmons AJ, *et al.* (2011) The ERA-Interim reanalysis: Configuration and performance of the data assimilation systems. *Quarterly Journal of the Royal Meteorological Society* **137**:553–97.

Feser F, Barcikowska M, Krueger O, *et al.* (2015) Storminess over the North Atlantic and northwestern Europe — A review. *Quarterly Journal of the Royal Meteorological Society* **141**:350–82. doi:10.1002/qj.2364

Field PR, Wood R. (2007) Precipitation and cloud structure in midlatitude cyclones. *Journal of Climate* **20**:233–54. doi:10.1175/JCLI3998.1

Hawcroft MK, Shaffrey LC, Hodges KI, Dacre HF. (2012) How much northern hemisphere precipitation is associated with extratropical cyclones? *Geophysical Research Letters* **39**:L24809. doi:10.1029/2012GL053866

Haynes JM, L'Ecuyer TS, Stephens GL, *et al.* (2009) Rainfall retrieval over the ocean with spaceborne W-band radar. *Journal of Geophysical Research* **114**:D00A22. doi:10.1029/2008JD009973

Haynes JM, Jakob C, Rossow WB, Tselioudis G, Brown J, (2011) Major characteristics of southern ocean cloud regimes and their effects on the energy budget. *Journal of Climate* **24**:5061–80. doi: 10.1175/2011JCLI4052.1

Lau N-C. Crane MW (1995) A satellite view of the synoptic-scale organization of cloud properties in midlatitude and tropical circulation systems. *Monthly Weather Review* **123**:1984–2006.

Lau N-C, Crane MW. (1997) Comparing satellite and surface observations of cloud patterns in synoptic-scale circulation systems. *Monthly Weather Review* **125**:3172–89.

Naud CM, Booth JF, Del Genio AD. (2014) Evaluation of ERA-interim and MERRA cloudiness in the Southern Ocean. *Journal of Climate* **27**(5):2109–24. doi:10.1175/JCLI-D-13-00432.1

Naud CM, Del Genio AD, Bauer M, et al. (2010) Cloud vertical distribution across warm and cold fronts in CloudSat-CALIPSO data and a general circulation model. *Journal of Climate* **23**:3397–415.

Naud CM, Booth JF, Jeyaratnam J, Donner LJ, Seman CJ, Zhao M, Guo H, Ming Y. (2019) Extratropical cyclone clouds in the GFDL climate model: Diagnosing biases and the associated causes. *Journal of Climate* **32**:6685–701., doi:10.1175/JCLI-D-19-0421.1

Naud CM, Booth JF, Lebsock M, Grecu M. (2018) Observational constraint for precipitation in extra-tropical cyclones: Sensitivity to data sources. *Journal of Applied Meteorology and Climatology* **57**:991–1009. https://doi.org/10.1175/JAMC-D-17-0289.1

Naud CM, Booth JF, Posselt DJ, van den Heever SC. (2013) Multiple satellite observations of cloud cover in extratropical cyclones. *Journal of Geophysical Research: Atmospheres* **118**:9982–96. doi:10.1002/jgrd.50718

Naud CM, Del Genio AD, Bauer N. (2006) Observational constraints on the cloud thermodynamic phase in midlatitude storms. *Journal of Climate* **19:**5273–88.

Naud CM, Jeyaratnam J, Booth JF, Zhao M, Gettelman A. (2020) Evaluation of modeled precipitation in oceanic extratropical cyclones using IMERG. *Journal of Climate* **33**(1):95–113. doi:10.1175/JCLI-D-19-0369.1

Naud CM, Posselt DJ, van den Heever SC. (2015) A CloudSat-CALIPSO view of cloud and precipitation properties across cold fronts over the global oceans. *Journal of Climate* **28**:6743–62.

Naud CM, Posselt DJ, van den Heever SC. (2017) Observed covariations of aerosol optical depth and cloud cover in extratropical cyclones. *Journal of Geophysical Research: Atmospheres* **122**:10338–56. https://doi.org/10.1002/2017JD027240

Neu U, *et al.* (2013) IMILAST, a community effort to Intercompare extratropical cyclone detection and tracking algorithms, *Bulletin of the American Meteorological Society* **94**:529–47. doi:10.1175/BAMS-D-11-001541

Oreopoulos L, Rossow WB. (2011) The cloud radiative effects of International Satellite Cloud Climatology Project weather states. *Journal of Geophysical Research* **116**:D12201. doi:10.1029/2010JD015472

Polly JB, Rossow WB. (2016) Cloud radiative effects and precipitation in extratropical cyclones. *Journal of Climate* **29**:6483–507. doi:10.1175/JCLI-D-15-0857.1

Rossow WB, Schiffer RA. (1991) ISCCP cloud data products. *Bulletin of the American Meteorological Society* **71**:2–20.

Schemm S, Sprenger M, Wernli H. (2018) When during their life cycle are extratropical cyclones attended by fronts? *Bulletin of the American Meteorological Society* **99**:149–65. doi:10.1175/BAMS-D-16-0261.1

Schiffer RA, Rossow WB. (1983) The International Satellite Cloud Climatology Project (ISCCP): The first project of the World Climate Research Programme. *Bulletin of the American Meteorological Society* **64**:779–84.

Stephens GL, L'Ecuyer T, Forbes R, *et al.* (2010) Dreary state of precipitation in global models. *Journal of Geophysical Research* **115**:D24211. doi:10.1029/2010JD014531

Sun Y, Solomon S, Dai A, Portmann RW. (2006) How often does it rain? *Journal of Climate* **19**:916–34.

Trenberth KE, Fasullo J. (2010) Simulation of present day and 21st century energy budgets of the southern oceans, *Journal of Climate* **23**:440–54.

Tselioudis G, Rossow WB. (2006) Climate feedback implied by observed radiation and precipitation changes with midlatitude storm strength and frequency. *Geophysical Research Letters* **33**:L02704. doi:10.1029/2005GL024513

Tselioudis G, Zhang Y-C, Rossow WR. (2000) Cloud and radiation variations associated with northern midlatitude low and high sea level pressure regimes. *Journal of Climate* **13**:312–27. doi:10.1175/1520-0442(2000)013<0312:CARVAW>2.0.CO;2

Wall C, Hartmann DL, Ma P-L. (2017) Instantaneous linkages between clouds and large-scale meteorology over the Southern ocean in observations and a climate model, *Journal of Climate* **30**:9455–74. doi:10.1175/JCLI-D-17-0156.1

About the Editors

Zhengzhao Johnny Luo is a Professor at the Department of Earth and Atmospheric Sciences of City College of the City University of New York (CUNY), USA. He received his Ph.D. in Atmospheric Sciences from Columbia University, USA in 2003. His postdoctoral research was conducted at Colorado State University, USA. Prof. Luo's research focuses on satellite remote sensing of clouds and convection, and use of satellite observations for studying convective dynamics. He also participates in airborne field campaigns flying in and around convective storms to study convective transport of trace gases. Prof. Luo served as a Science Leader of the NASA field mission SEAC^{4}RS, for which he received the NASA Group Achievement Award in 2015. He is currently a science team member of the NASA Earth Venture mission INCUS. He has been an Editor of *Journal of the Meteorological Society of Japan* (JMSJ) since 2014.

George Tselioudis is a Research Physical Scientist at the NASA Goddard Institute for Space Studies, and an Adjunct Professor at the Department of Applied Physics and Applied Mathematics of Columbia University, USA. He received his Ph.D. in Earth Sciences from Columbia University in 1992 and went on to do climate change research at NASA and Columbia. His research focuses primarily on the use of satellite observations and general circulation models to understand physical processes and their interactions in the Earth's atmosphere and to quantify the resulting feedbacks on climate change, with emphasis on cloud, radiation, and precipitation processes and feedbacks. He has authored numerous scientific papers and book chapters and was honored in 2004 by the American Geophysical Union with the Charles Faklenberg Award, for 'his contribution to the quality of life, economic opportunities, and stewardship of the planet through the use of Earth science information'. He served as both author and reviewer of the Intergovernmental Panel for Climate Change (IPCC) Assessment Reports, and currently serves as co-chair of the Cloud Feedback Model Intercomparison Project (CFMIP) of the World Climate Research Program.

William B. Rossow was Distinguished Professor of Engineering at The City College of New York in the CUNY Remote Sensing Earth System Institute (2006–2017). He was a Senior Research Scientist at NASA Goddard Institute for Space Studies (1978–2006) as Head of Earth Observations and led the Global Processing Center for the World Climate Research Program's International Satellite Cloud Climatology Project. He received the NASA Exceptional Scientific Achievement Medal (1988) and the AMS Verner E. Suomi Award (2005). He is a founding member of the Planetary Society and Fellow of the American Meteorological Society and the American Geophysical Union.

Appendix

Most of the chapters in this book derived from a Symposium sponsored by the NOAA Cooperative Remote Sensing and Technology Institute at the City College of New York and NASA Goddard Institute for Space Studies. The symposium was held in New York City On 6–8 June 2017 to celebrate the career and retirement of Prof. William B. Rossow. The objective was to discuss progress that has been made over the past decades in understanding cloud properties, processes, and their weather and climate feedbacks, and to explore strategies to tackle the issues that remain unresolved. Below are some photos from the event.

The more than 60 international participants of the Symposium are pictured.

Before the Symposium, some senior colleagues had a good Mexican dinner. Pictured clockwise from left: Claudia Stubenrauch (LMD, France), William B. Rossow, George Tselioudis (NASA GISS, USA), Zhengzhao Johnny Luo (CCNY, USA), Christian Jakob (Monash Univ., Australia), Ademe Mekonnen (North Carolina A&T State Univ, USA), Graeme Stephens (NASA JPL, USA) and Lynne M. Kemen (William B. Rossow's wife).

During the Symposium, William B. Rossow had another dinner with former students and post-docs. Pictured clockwise from left: Luiz Machado (INPE, Brazil), Zhengzhao Johnny Luo (CCNY, USA), Joy Romanski (Columbia Univ., USA), Lynne M. Kemen (William B. Rossow's wife), Ademe Mekonnen (North Carolina A&T State Univ, USA), George Tselioudis (NASA GISS, USA), Narges Shahroudi (CCNY, USA), William B. Rossow, Junhong Wang (SUNY Albany, USA) and Rong Fu (UCLA, USA).

At a reception after the first day of the Symposium, William B. Rossow told stories of international adventures in ISCCP.

www.ingramcontent.com/pod-product-compliance
Lightning Source LLC
Chambersburg PA
CBHW081051220326
41598CB00038B/7055

* 9 7 8 9 8 1 1 2 5 7 9 4 0 *